FREMONT UNIFIED SCHOOL DISTRICT
Fremont, California

Scott, Foresman

Physical Science

Authors

Jay M. Pasachoff
Director, Hopkins Observatory
Williams College
Williamstown, Massachusetts

Naomi Pasachoff
Research Associate
Williams College
Williamstown, Massachusetts

Timothy M. Cooney
Science Chairperson
Malcolm Price Laboratory School
University of Northern Iowa
Cedar Falls, Iowa

Scott, Foresman and Company
Editorial Offices: Glenview, Illinois

Regional Offices: Palo Alto, California
Tucker, Georgia • Glenview, Illinois
Oakland, New Jersey • Dallas, Texas

Cover: Lightning over
the Arizona desert
Credit: Thomas Ives © 1982

Series Consultant

Irwin L. Slesnick
Professor of Biology
Western Washington University
Bellingham, Washington

Program Consultant

John Hockett
Professor of Science Education
Governors State University
Park Forest, Illinois

Reading Consultant

Robert A. Pavlik
Reading Department Chairperson
Cardinal Stritch College
Milwaukee, Wisconsin

Editorial Advisors

Abraham S. Flexer
Educational Consultant
Boulder, Colorado

Karin L. Rhines
Educational Consultant
Bedford Hills, New York

Feature Writer

David Newton
Professor of Chemistry and Physics
Salem State College
Salem, Massachusetts

86-151

ISBN: 0-673-14118-7
Copyright © 1986, Scott, Foresman and Company, Glenview, Illinois
All Rights Reserved. Printed in the United States of America.
This publication is protected by Copyright and permission should be obtained from the publisher prior to any prohibited reproduction, storage in a retrieval system, or transmission in any form or by any means, electronic, mechanical, photocopying, recording, or otherwise. For information regarding permission, write to: Scott, Foresman and Company, 1900 East Lake Avenue, Glenview, Illinois 60025.

10—RMI—9493929190898887 86

Reviewers and Contributors

LeVon Balzer
Dean of Arts and Sciences
Seattle Pacific University
Seattle, Washington

Marsha Barber
Earth Science Teacher
Lockport Township High School
Lockport, Illinois

Herb Bassow
Chemistry Teacher
Germantown Friends School
Philadelphia, Pennsylvania

Rose Mary Castro
Science Teacher
L. W. Fox Academic and
Technical High School
San Antonio, Texas

Emily Fast Christensen
Science Teacher
Kirby Junior High School
Hazelwood School District
St. Louis County, Missouri

York Clamann
Science Consultant
Abilene Independent School District
Abilene, Texas

Obe Hofer
Science Department Representative
Whittier Junior High School
Sioux Falls, South Dakota

Michelle M. Kovac
Physical Science Teacher
Maple School
Northbrook, Illinois

Sol Krasner
Associate Professor of Physics
University of Chicago
Chicago, Illinois

Kara May
Biology Teacher
James B. Conant High School
Hoffman Estates, Illinois

Frederick Rasmussen
Science Education Consultant
Boulder, Colorado

Susanne Rego
Science Teacher
Belleville Senior High School
Belleville, New Jersey

Victor Showalter
Director of FUSE Center
Capital University
Columbus, Ohio

R. A. Slotter
Assistant Chairperson
Department of Chemistry
Northwestern University
Evanston, Illinois

Lucy Smith
Coordinator of Science Education
Atlanta Public Schools
Atlanta, Georgia

Katherine Taft
Science Education Services
Highland Park, Illinois
Teacher
American International School,
Austria

William D. Thomas
Science Supervisor
Escambia County Schools
Pensacola, Florida

Dorothy Wallinga
Jenison, Michigan

Les Wallinga
Physical Science Teacher
Calvin Christian Junior High School
Wyoming, Michigan

Scott A. Welty
Laboratory Director
Museum of Science and Industry
Chicago, Illinois

Lawrence Zambrowski
Science Department Chairperson
Churchill Junior High School
East Brunswick, New Jersey

UNIT ONE

Motion *1*

Chapter 1
Studying Physical Science *3*

1-1 Investigations in Science *4*
1-2 Physical Science Research Today *6*
1-3 Solving Problems Scientifically *8*
ACTIVITY: Doing Physical Science *11*
1-4 Making Measurements *12*
ACTIVITY: Graphing in SI Units *16*
DID YOU KNOW? Lucky Accidents in Science *17*
1-5 Safety in the Physical Science Laboratory *18*
Chapter Summary *20*
Interesting Reading *20*
Questions/Problems *20*
Extra Research *20*
Chapter Test *21*

Chapter 2
Moving Objects *23*

2-1 Describing Motion *24*
2-2 Contrasting Distance and Displacement *26*
2-3 Comparing Speed and Velocity *28*
ACTIVITY: Measuring Speed *30*
DID YOU KNOW? Speed Comparisons *31*
2-4 Describing Acceleration *32*
ACTIVITY: Graphing Motion *35*
Chapter Summary *36*
Interesting Reading *36*
Questions/Problems *36*
Extra Research *36*
Chapter Test *37*

Chapter 3
The Laws of Motion 39

3-1 Newton's First Law of Motion 40
ACTIVITY: Newton's First Law 42
BREAKTHROUGH: Special Relativity 43
3-2 Newton's Second Law and Momentum 44
ACTIVITY: Newton's Second Law 47
3-3 Newton's Third Law of Motion 48
ACTIVITY: Newton's Third Law 51
3-4 Circular Motion 52
Chapter Summary 54
Interesting Reading 54
Questions/Problems 54
Extra Research 54
Chapter Test 55

Chapter 4
Balanced Forces, Friction, and Rotation 57

4-1 Balanced Forces 58
DID YOU KNOW? Center of Gravity 61
4-2 Friction as a Force 62
ACTIVITY: Studying Friction 65
4-3 Forces on Rotating Objects 66
ACTIVITY: Studying Center of Gravity 69
Chapter Summary 70
Interesting Reading 70
Questions/Problems 70
Extra Research 70
Chapter Test 71

Chapter 5
Gravitation 73

5-1 The Motion of Falling Objects 74
ACTIVITY: The Effects of Gravity and Air Resistance 77
5-2 Gravity, Mass, and Weight 78
ACTIVITY: Studying Microgravity 80
DID YOU KNOW? Life Under Different Gravitational Forces 81
5-3 How Gravity Affects the Planets 82
5-4 The Law of Gravity 84
Chapter Summary 86
Interesting Reading 86
Questions/Problems 86
Extra Research 86
Chapter Test 87
CAREERS 88

UNIT TWO

Energy *91*

Chapter 6
Work and Machines *93*

- **6-1** Doing Work *94*
- **6-2** Some Simple Machines *96*
- **ACTIVITY:** Using Levers and Pulleys *100*
- **DID YOU KNOW?** The Bicycle—A Combination of Simple Machines *101*
- **6-3** Forms of Inclined Planes *102*
- **ACTIVITY:** Experimenting with Inclined Planes *105*
- **6-4** Compound Machines and Efficiency *106*
- **Chapter Summary** *108*
- **Interesting Reading** *108*
- **Questions/Problems** *108*
- **Extra Research** *108*
- **Chapter Test** *109*

Chapter 7
Energy and Power *111*

- **7-1** Defining Energy *112*
- **7-2** Kinetic Energy *114*
- **7-3** Energy Is Conserved *116*
- **ACTIVITY:** Investigating Kinetic and Potential Energy *119*
- **7-4** Forms of Energy *120*
- **7-5** Measuring Power *122*
- **ACTIVITY:** Measuring Human Power *124*
- **DID YOU KNOW?** Amusement Park Science *125*
- **Chapter Summary** *126*
- **Interesting Reading** *126*
- **Questions/Problems** *126*
- **Extra Research** *126*
- **Chapter Test** *127*

 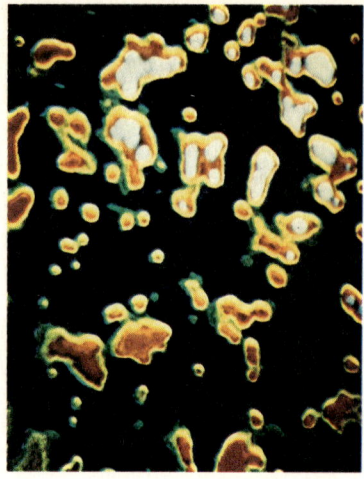 

Chapter 8
Heat and Temperature 129

- **8-1** Defining Heat and Temperature 130
- **8-2** Detecting Heat and Temperature 132
- **ACTIVITY:** Investigating Temperature Changes 135
- **8-3** Producing Heat 136
- **8-4** Energy Transfer 138
- **ACTIVITY:** A Material's Effect on Absorbing Radiation 144
- **BREAKTHROUGH:** Cryogenics 145
- **8-5** Expansion and Contraction 144

Chapter Summary 146
Interesting Reading 146
Questions/Problems 146
Extra Research 146
Chapter Test 147
CAREERS 148

UNIT THREE

The Structure of Matter 151

Chapter 9
Matter and Its States 153

- **9-1** The States of Matter 154
- **DID YOU KNOW?** Plasma 157
- **9-2** Characteristics of Gases 158
- **9-3** Characteristics of Solids and Liquids 160
- **ACTIVITY:** Liquids and Solids 163
- **9-4** Changes in the State of Matter 164
- **ACTIVITY:** Changing Physical State 167

Chapter Summary 168
Interesting Reading 168
Questions/Problems 168
Extra Research 168
Chapter Test 169

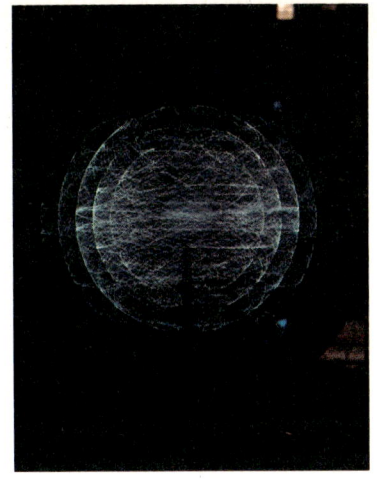

Chapter 10
Properties of Matter
171

- **10-1** The Elements *172*
- **10-2** Physical Properties and Changes *176*
- **ACTIVITY:** Physical Properties *179*
- **10-3** Chemical Properties and Changes *180*
- **ACTIVITY:** A Chemical Change *182*
- **DID YOU KNOW?** Naming the Elements *183*
- **Chapter Summary** *184*
- **Interesting Reading** *184*
- **Questions/Problems** *184*
- **Extra Research** *184*
- **Chapter Test** *185*

Chapter 11
The Atom *187*

- **11-1** Theories About the Atom *188*
- **11-2** How Atoms Differ *192*
- **ACTIVITY:** Atomic Models *195*
- **11-3** Classifying Elements *196*
- **ACTIVITY:** Using the Periodic Table *202*
- **DID YOU KNOW?** Alchemy *203*
- **Chapter Summary** *204*
- **Interesting Reading** *204*
- **Questions/Problems** *204*
- **Extra Research** *204*
- **Chapter Test** *205*

Chapter 12
The Atomic Nucleus
207

- **12-1** The Structure of the Nucleus *208*
- **BREAKTHROUGH:** Particle Accelerators *211*
- **12-2** Radioactivity *212*
- **ACTIVITY:** Radioactive Decay *215*
- **12-3** Nuclear Fission *216*
- **ACTIVITY:** Chain Reactions *219*
- **12-4** Nuclear Fusion *220*
- **12-5** Mass, Energy, and the Speed of Light *222*
- **Chapter Summary** *224*
- **Interesting Reading** *224*
- **Questions/Problems** *224*
- **Extra Research** *224*
- **Chapter Test** *225*
- **CAREERS** *226*

UNIT FOUR

Changes in Matter 229

Chapter 13
Compounds and Mixtures 231

- **13-1** Identifying Compounds 232
- **ACTIVITY:** Forming a Compound 235
- **13-2** Identifying Mixtures 236
- **13-3** Solutions—One Kind of Mixture 238
- **13-4** Suspensions—Another Kind of Mixture 240
- **ACTIVITY:** Investigating Suspensions, Solutions, and Emulsions 242
- **BREAKTHROUGH:** Superblanket 243
- **Chapter Summary** 244
- **Interesting Reading** 244
- **Questions/Problems** 244
- **Extra Research** 244
- **Chapter Test** 245

Chapter 14
Holding Atoms Together 247

14-1 Bonding Atoms *248*

ACTIVITY: Bonds Make the Difference *251*

14-2 Ionic Bonding *252*

14-3 Covalent Bonding *254*

14-4 Chemical Equations *256*

ACTIVITY: A Chemical Reaction and Its Equation *258*

DID YOU KNOW? Rocket Propellants *259*

Chapter Summary *260*

Interesting Reading *260*

Questions/Problems *260*

Extra Research *260*

Chapter Test *261*

Chapter 15
Chemical Reactions 263

15-1 Recognizing Chemical Reactions *264*

ACTIVITY: Controlling the Speed of Chemical Reactions *268*

DID YOU KNOW? Fireworks *269*

15-2 Synthesis and Decomposition Reactions *270*

15-3 Replacement Reactions *272*

ACTIVITY: Determining Chemical Activity *275*

15-4 Carbon Compounds *276*

Chapter Summary *280*

Interesting Reading *280*

Questions/Problems *280*

Extra Research *280*

Chapter Test *281*

Chapter 16
Acids, Bases, and Salts 283

16-1 Properties of Acids and Bases *284*

16-2 Explaining the Properties of Acids and Bases *286*

ACTIVITY: Recognizing Acids and Bases *289*

16-3 Indicator Colors and the pH Scale *290*

ISSUES IN PHYSICAL SCIENCE: Water Table Pollution *293*

16-4 Neutralization and Salts *294*

ACTIVITY: A Neutralization Reaction *297*

Chapter Summary *298*

Interesting Reading *298*

Questions/Problems *298*

Extra Research *298*

Chapter Test *299*

CAREERS *300*

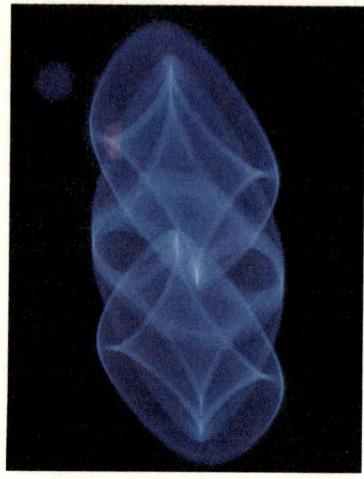

UNIT FIVE

Wave Motion *303*

Chapter 17

Waves *305*

17-1 Properties of Waves *306*

ACTIVITY: Properties of Waves *309*

17-2 Wave Motions *310*

17-3 The Behavior of Waves *312*

ACTIVITY: Reflection *315*

17-4 Electromagnetic Waves *316*

ISSUES IN PHYSICAL SCIENCE: Microwave Pollution *319*

Chapter Summary *320*

Interesting Reading *320*

Questions/Problems *320*

Extra Research *320*

Chapter Test *321*

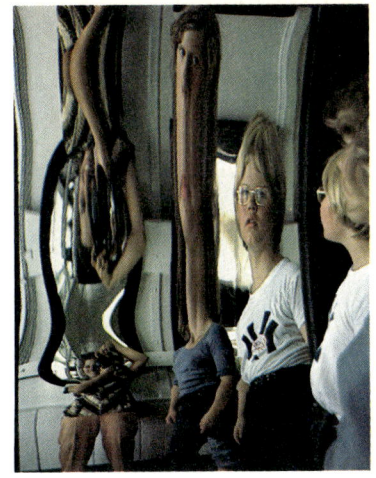

Chapter 18
Light *323*

- **18-1** The Nature of Light *324*
- **ACTIVITY:** The Path of Light Rays *327*
- **18-2** The Visible Spectrum *328*
- **ACTIVITY:** Colors in White Light *330*
- **DID YOU KNOW?** The Colors in the Sky *331*
- **18-3** Lasers *332*
- **18-4** Using Lasers *334*
- **Chapter Summary** *338*
- **Interesting Reading** *338*
- **Questions/Problems** *338*
- **Extra Research** *338*
- **Chapter Test** *339*

Chapter 19
Light and Its Uses *341*

- **19-1** Plane Mirrors *342*
- **19-2** Curved Mirrors *344*
- **19-3** Lenses *348*
- **ACTIVITY:** Lenses *351*
- **19-4** Eyes and Lenses *352*
- **DID YOU KNOW?** Mirages *355*
- **19-5** Using Lenses and Mirrors *356*
- **ACTIVITY:** Making a Pinhole Camera *359*
- **Chapter Summary** *360*
- **Interesting Reading** *360*
- **Questions/Problems** *360*
- **Extra Research** *360*
- **Chapter Test** *361*

Chapter 20
Sound *363*

- **20-1** Sound as a Wave *364*
- **ACTIVITY:** Properties of Sound *368*
- **DID YOU KNOW?** Designs for Better Listening *369*
- **20-2** Characteristics of Sound *370*
- **20-3** The Sound of Music *372*
- **ACTIVITY:** Making Music *375*
- **Chapter Summary** *376*
- **Interesting Reading** *376*
- **Questions/Problems** *376*
- **Extra Research** *376*
- **Chapter Test** *377*
- **CAREERS** *378*

UNIT SIX

Electricity and Magnetism 381

Chapter 21
Electricity 383

21-1 Electric Charge 384
ACTIVITY: Electric Charge 387
21-2 Electric Current— Charges on the Move 388
21-3 Two Kinds of Circuits 392
21-4 Using Electricity 394
ACTIVITY: Switches and Fuses 396
BREAKTHROUGH: Superconductivity 397
Chapter Summary 398
Interesting Reading 398
Questions/Problems 398
Extra Research 398
Chapter Test 399

Chapter 22
Magnetism 401

22-1 Magnetic Properties 402
ACTIVITY: Magnetic Poles 405
22-2 Magnets from Magnetic Substances 406
22-3 Changing Electricity into Magnetism 408
ACTIVITY: Electromagnets 411
22-4 Changing Magnetism into Electricity 412
BREAKTHROUGH: Supertrain 415
Chapter Summary 416
Interesting Reading 416
Questions/Problems 416
Extra Research 416
Chapter Test 417

Chapter 23
The Electronic Revolution *419*

23-1 Electronic Devices *420*

ACTIVITY: Sending Messages *423*

23-2 Making Electronic Components Smaller *424*

23-3 Using Chips *426*

ACTIVITY: The Computing Process *430*

BREAKTHROUGH: The Bionic Revolution *431*

Chapter Summary *432*

Interesting Reading *432*

Questions/Problems *432*

Extra Research *432*

Chapter Test *433*

CAREERS *434*

UNIT SEVEN

Frontiers *437*

Chapter 24
Energy Resources *439*

24-1 Energy in Your Life *440*

ACTIVITY: Investigating Energy Use *443*

24-2 Fossil Fuels *444*

24-3 Looking for More Energy Sources *446*

ACTIVITY: Light Energy Collector *450*

BREAKTHROUGH: Alternative Energy Sources for Airplanes *451*

24-4 Making Decisions About Energy Sources *452*

Chapter Summary *456*

Interesting Reading *456*

Questions/Problems *456*

Extra Research *456*

Chapter Test *457*

Chapter 25
Exploring the
Universe *459*

25-1 The Development of Stars *460*

ACTIVITY: Using a Spectroscope *463*

25-2 Aging Stars *464*

25-3 A Universe of Galaxies *468*

ACTIVITY: The Expanding Universe *471*

25-4 Theories About the Universe *472*

ISSUES IN PHYSICAL SCIENCE: Messages from Outer Space *475*

Chapter Summary *476*

Interesting Reading *476*

Questions/Problems *476*

Extra Research *476*

Chapter Test *477*

CAREERS *478*

ACKNOWLEDGEMENTS
480

GLOSSARY *482*

INDEX *490*

UNIT ONE
MOTION

What do you think caused the colorful shapes in the photograph? Do you think that whatever caused them was moving or standing still?

The pattern formed when a delta-shaped wing was tested in a fluid tunnel near Paris, France. Dyes in the water show how the fluid moves around the wing. As you read Unit One, you will encounter many ideas about how we view and study motion.

Chapter 1 Studying Physical Science

Physical science affects your life in many important ways. Certain skills will help you study physical science.

Chapter 2 Moving Objects

We can observe and describe the motion of objects. These motions range from simple to complex.

Chapter 3 The Laws of Motion

Forces cause changes in motion. Explaining an object's motion involves applying just a few basic rules.

Chapter 4 Balanced Forces, Friction, and Rotation

Analyzing an object's motion becomes more interesting when more forces act on the object. A simple motion might be the result of many interacting forces.

Chapter 5 Gravitation

The force of gravity affects everyday events on earth as well as the motions of the planets.

Chapter 1
Studying Physical Science

Physical scientists try to understand the basic ways all things work on the earth and in the rest of the universe. Physical scientists developed the intense beam of light the worker shown is using. They are investigating using such light beams to produce a new energy source.

This chapter begins with an explanation of what physical science is. The chapter describes areas of current research in physical science and how physical scientists do their jobs. The chapter continues with some tools physical scientists use. The chapter ends by explaining what you should do as you perform your own experiments.

Chapter Objectives

1. Describe what physical science is.
2. Discuss current research projects in physical science.
3. Describe the methods scientists use to investigate an idea.
4. Explain the International System of measurement.
5. Explain how to work safely in a laboratory.

1–1 Investigations in Science

Science is the study of the universe and its laws. Physical science answers questions such as "Why is the sky blue?" and "How does salt form?" Science begins with simple curiosity and observations about the many fascinating events in the universe. These events might lead you to think about the following questions:

a. What is physical science?
b. Why should I study physical science?

The Study of Matter and Energy

Different arrangements of matter and energy make up the entire universe. Physical science is the study of changes in matter and energy. Physics, chemistry, and astronomy are areas of study in the physical sciences. Since everything is made of matter and energy, physical science can answer questions about almost every topic you can think of. People doing research in physical science study both living and nonliving things. The study of physical science even helps art experts analyze the painting shown. Then they can determine whether the painting is an original or a fake.

Why You Should Study Physical Science

Long ago, striking stones together to make a fire was a daily chore. Over the years, people improved the way they lit a fire. But they did not always know that finding these improvements was an experiment in physical science. Today, results of physical science research are all around us.

Advances in physics research are responsible for the safety of seatbelts. Understanding the movement of objects or the stresses on materials makes possible launching a spacecraft or building the bridge shown.

Chemical researchers invented artificial turf for baseball and football fields and nylon and polyester for our clothes. These materials did not exist fifty years ago.

The curiosity of today's physical science students will lead to more exciting discoveries and new areas of study. Your introduction to physical science will help you understand the importance and effect of these discoveries on our future.

Review It

1. What are three areas of study in physical science?
2. What are some reasons for studying physical science?

1–2
Physical Science Research Today

Research is a continuing process. Scientists are always revising old ideas and developing new understandings based on new information. Today's research is the bridge between past ideas and future possibilities. Keep these questions in mind as you read about studies in physical science:

a. What are some research topics in physical science now?
b. How can I benefit from physical science research?

Doing Physical Science Research

About forty years ago, a computer was the size of a classroom. Through research, scientists developed smaller and more capable computers. Today, the computer is both a subject of study and a tool for research.

Computers are used to launch a spacecraft and to direct its safe return to earth. Computers even help athletes. Through computer analysis, Mac Wilkins, a discus thrower, increased his throwing ability by 3.6 meters. In the picture below, a researcher uses a computer to work with mentally handicapped children.

Current research in physical science has provided many new materials. Researchers recently developed a new lightweight plastic that can be used for automobile parts. Because the plastic is made from grain instead of oil, it saves fuel use.

Benefits of Physical Science Research

In the past, pictures like this one did not exist. Physical science researchers discovered how to make this "picture" of heat. Because of escaping heat, windows and many other spots in the city appear brightly colored. By using such pictures, building owners can prevent heat loss. So they save energy and reduce their heating costs.

Other advances in physical science research have produced fireproof fabrics, shatterproof glass, and a way to perform surgery using intense beams of light.

Space technology has provided us with such different discoveries as longer-lasting light bulbs, inexpensive smoke and fire alarms, better fans, improved pavement materials, and advanced car designs.

Review It

1. How do computers help physical science researchers?
2. List some benefits of physical science research.

1–3 Solving Problems Scientifically

You may picture a scientist as someone wearing a long, white coat and working in a laboratory. Or perhaps you imagine a scientist as someone daydreaming about living in outer space. Another image of a scientist is shown on this page. As you read about how scientists do their jobs, consider these questions:

a. How are problems solved scientifically?
b. How is scientific knowledge revised?

The Scientific Method

Without thinking about it, you are involved in problem solving every day. You decide how to comb your hair and what to wear. In thinking about a problem, you probably consider many possible solutions and their effects. For more important decisions, you probably seek advice from a friend, teacher, or parent. After considering all the information, you decide on a solution to your problem.

Scientists solve problems in much the same way. Often scientists find answers to questions about one problem while investigating other problems. They stumble upon still other discoveries. No perfect rule guides scientists in their research. However, they always try to use a logical method. This **scientific method** limits the possible answers and decreases the chances for making errors.

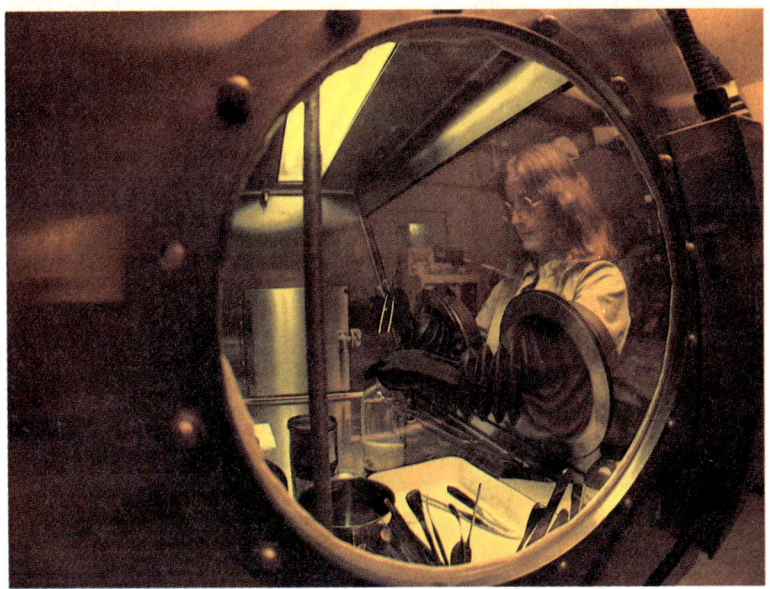

Problem solving with a scientific method starts with observations. Detailed observations depend on the five senses. Because just one observation may not always be reliable, scientists must be able to repeat an observation many times.

Observations can describe the amount, size, shape, color, and odor of a substance. These observations become scientific **data**. Scientists, such as the one shown above, collect data to analyze a problem. Then they make a reasonable guess, called a **hypothesis** (hī poth′ə sis), about how or why an event happens. Hypotheses (hī poth′ə sēs) form the basis for further investigations.

Testing distinguishes a scientific method from our everyday way of solving problems. Scientists test a hypothesis with experiments. To reduce the chance for error, scientists keep some of the quantities unchanged while experimenting with another quantity.

Sometimes scientists test several hypotheses at the same time. If observations and experiments support one hypothesis, the other hypotheses may be rejected. Even hypotheses that have been accepted for years may prove to be incorrect. So researchers experiment continually to collect more data to improve or reject a hypothesis.

Revising Scientific Information

Suppose a hypothesis has been tested many times and found to be true again and again. It then becomes a more general explanation of observations known as a **theory** (thē′ər ē). The difference between a hypothesis and a theory is not always clear. Usually, however, a theory not only explains what has been observed but also predicts the results of new experiments.

For centuries, Aristotle's statement that a heavy object falls faster than a light object was accepted by most people. From everyday observations, that statement seemed to be true. In the 1500s, however, Galileo observed that shape and surface area affect how fast an object falls. He began to wonder if Aristotle was correct. Galileo was one of the first scientists to do more than just think about an idea. He actually tested his hypothesis. He rolled balls of different weights down a ramp. This experiment led him to believe that objects would fall at the same rate, regardless of their weights, if the air did not slow them down. He disproved Aristotle's statement. His own hypothesis became a theory. Over the centuries, Galileo's theory has not been proved false. Therefore, it has become a **scientific law**—an accepted explanation that should apply over and over again throughout the universe.

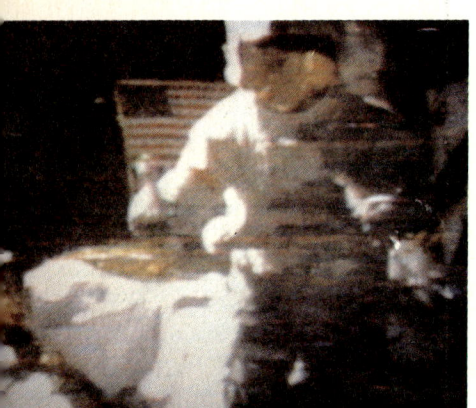

As scientists discover new information, they change their ideas to explain the facts. While on our airless moon, one of the *Apollo 12* astronauts tested Galileo's idea. He is pictured holding a hammer in his right hand and a feather in his left hand. Both objects fell to and hit the ground at the same time.

Review It

1. How do scientists try to answer scientific questions?
2. How do recent experiments affect old theories?

Activity

Doing Physical Science

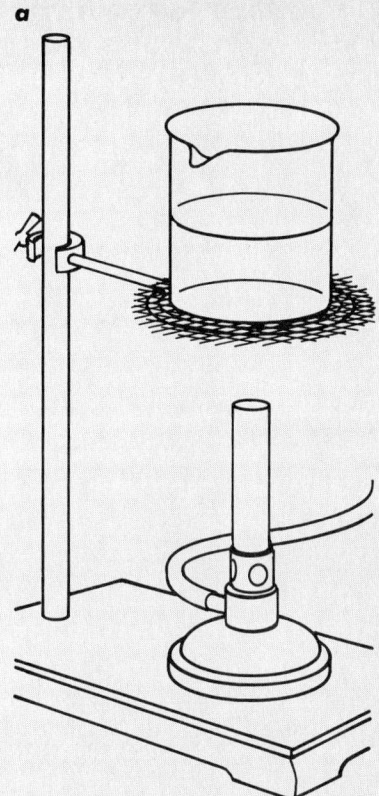

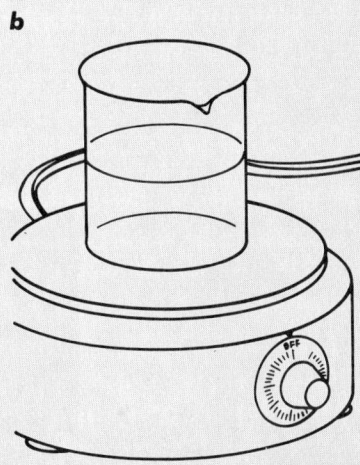

Purpose
To practice using the scientific method.

Materials
- 100-mL beaker
- hot plate, or Bunsen burner, matches, ring stand, ring, and wire gauze
- distilled water
- salt
- spoon
- clock or watch
- safety goggles
- grease pencil
- tongs or hot pad

Procedure
1. Put on your safety goggles.
2. Set up your equipment to boil water, as shown. Your teacher will tell you whether to use a burner or a hot plate.
3. Fill the beaker about half full. Mark the height of the water with your grease pencil.
4. Heat the water until it is at a full, bubbling boil. Time how long the water takes to reach this state. Record this observation.
5. Carefully remove the hot beaker with the tongs or hot pad. Set the beaker aside to cool.
6. Repeat steps 3-5, but add a spoonful of salt to the water before you begin heating it. Be sure to make the water level even with your grease pencil mark from the first trial.

Analysis
1. What hypothesis did you test as you performed this activity?
2. Which quantities or characteristics did you control (maintain) in both trials?
3. Which quantities or characteristics did you change?
4. What conclusion can you draw from this experiment? Explain why you think it is correct.
5. Explain how you could make the experiment more accurate.

11

1-4
Making Measurements

In the past, people used familiar objects for measuring. The width of a thumb was an inch. The length of the king's foot was called a foot. Since these amounts changed from person to person (or king to king), measurements changed too. A better way of measuring solved the problems of changing units of measure. As you learn how we measure, think about the following questions:

a. What is the metric system?
b. How do you measure with SI units?

The Metric System

Measuring is a tool of observation and research. Measurements tell how far, how large, how many, and how much. When you measure, you compare a known amount, called a standard, to an unknown amount. If everyone uses the same standard, people everywhere will get the same answer when they measure the same unknown amount.

All scientists and most people in the world have agreed to use one system of measurement. The **metric system** consists of standard units and prefixes that are multiples of ten. The version of the metric system now agreed on is called the International System or **SI** from its French initials. The metric measures of several objects are listed below.

The metric unit of length is the **meter** (symbol: *m*). A football field is just over 90 meters long, and an Olympic swimming pool is 50 meters long. In your classroom, you probably have a meter stick for measuring lengths.

Measuring with the metric system

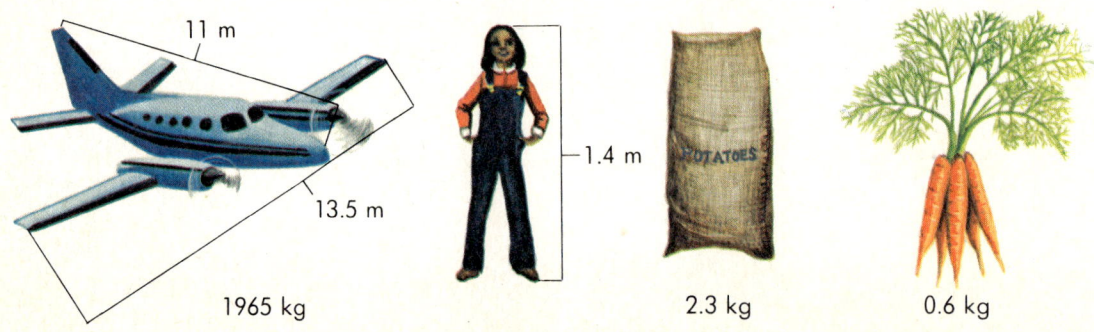

11 m
13.5 m
1965 kg

1.4 m
2.3 kg

0.6 kg

Volume is the amount of space an object takes up. The unit of liquid volume is the **liter** (symbol: *L*). Some beakers and soda pop bottles have a volume of one liter. In many places, gasoline is sold by the liter.

Mass is the amount of matter an object has. The **gram** (symbol: *g*) is the metric unit to measure mass. A paper clip is about a gram. A pharmacist measures some medicines in grams. Vitamins and other medicines are measured in smaller units called milligrams.

The table shown lists some prefixes that can be attached to a unit of measurement. Notice that metric symbols do not end with a period.

Time and temperature are two other units in the metric system. Time is measured in seconds (symbol: *s*), and temperature is measured in **degrees Celsius** (symbol: °*C*).

Your family car may have the unit "km/hr" on the speedometer. This unit measures how fast the car travels. It is derived from two units. We can combine units to measure many other quantities.

Some Metric Prefixes

Prefix	Symbol	Amount	Example and Symbol
milli-	m	0.001 x	millimeter (mm)
centi-	c	0.01 x	centimeter (cm)
deci-	d	0.1 x	decimeter (dm)
kilo-	k	1,000 x	kilometer (km)
mega-	M	1,000,000 x	megameter (Mm)

Measuring with the metric system

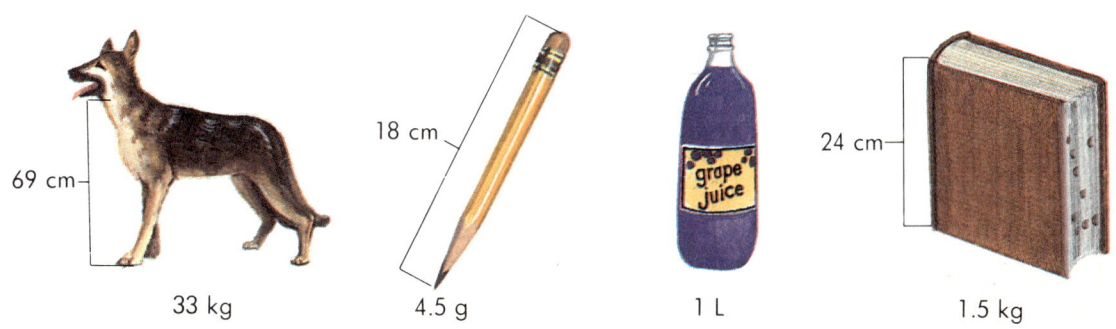

69 cm — 33 kg
18 cm — 4.5 g
1 L
24 cm — 1.5 kg

Measuring with SI Units

Scientists use instruments to measure quantities. A **balance** is an instrument that measures the mass of an object.

Centuries ago, people determined the mass of an amount of grain with a stone and a balance. The stone was put in one pan of the balance. A buyer poured an unknown mass of grain in the second pan. When the two pans of the balance were level, the mass of the grain equaled the mass of the stone. People bought a "stone of grain." Today a balance measures the mass of any object in some multiple of the gram, usually the kilogram.

To measure the volume of liquids, scientists use a cylinder with one open end and markings along its height. This instrument is a **graduated cylinder.** The graduated cylinders shown have equal spaces marked in milliliters. Notice that the surface of each liquid curves. This curve is the meniscus (mə nis′kəs). Always read the height of a liquid at the meniscus, as marked in the photograph.

Scientists measure the volume of some solid objects in another way. Multiplying an object's length by width by height tells you its volume.

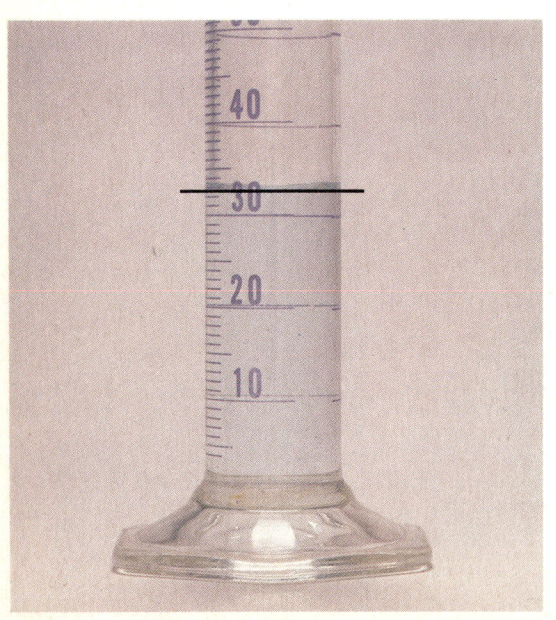
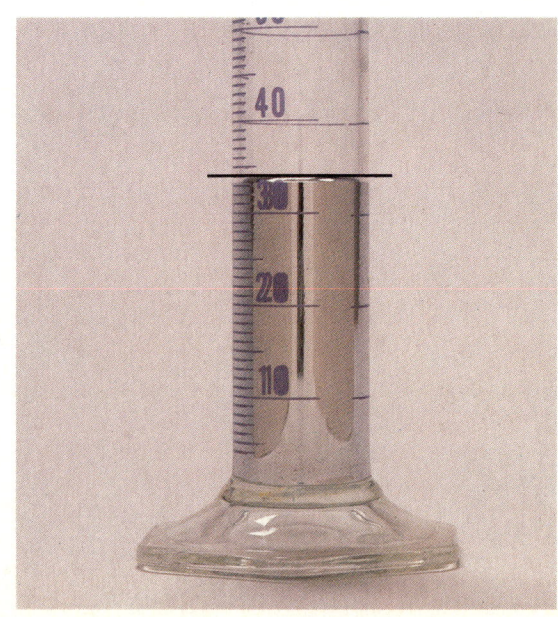

For example, a block of wood measures 2 meters long, 1 meter wide, and 0.5 meter high. Its volume is:

Volume = length × width × height
= 2 meters × 1 meter × 0.5 meter
= 1 meter^3.

Notice that multiplying the units with the numbers gives another metric unit for volume, the meter^3.

Measuring how closely mass is packed gives another important quantity called **density**. It is the amount of mass in a certain volume. The picture shows three objects of equal mass. The density of the wrench is greatest because its mass takes up the least volume.

You calculate density using

$$\text{density} = \frac{\text{mass}}{\text{volume}}.$$

If 100 milliliters of water have a mass of 100 grams,

$$\text{density} = \frac{100 \text{ grams}}{100 \text{ milliliters}}$$
$$= 1 \text{ gram/milliliter}.$$

One milliliter equals one centimeter^3, so the unit for density is either the gram per milliliter (g/mL) or the gram per centimeter^3 (g/cm^3).

For Practice

Use the proper equation to calculate the following:
- the volume of a box whose length = 3 m, width = 1 m, and height = 0.3 m.
- the density of 144 g of steel that take up 20 cm^3.
- the density of 1,360 g of mercury that take up 100 mL.

Review It

1. What measurement system do most people in the world use?
2. Name the instrument used to measure mass.

Activity

Graphing in SI Units

Purpose
To practice measuring and graphing quantities.

Materials
- 2 meter sticks
- masking tape
- kilogram bathroom scale
- ruler
- graph paper
- balance
- small stone
- graduated cylinder
- 3 blocks of different sizes and materials, marked 1, 2, and 3
- water

Procedure

Part A
1. Tape the meter sticks on the wall with masking tape, as shown in *a*.
2. Stand with your back against the wall and ask your partner to read your height. Record your height to the nearest centimeter.
3. Measure your mass on the scale. Record your mass.
4. Write your height and mass on the blackboard with those of your classmates. Record these data.
5. Copy and label the grid, as shown in *b*.
6. Plot your data as follows. Put your pencil on the horizontal line on the mark for the first height. Move your pencil up this line until you reach the horizontal line for the first mass. Put a dot there.

a

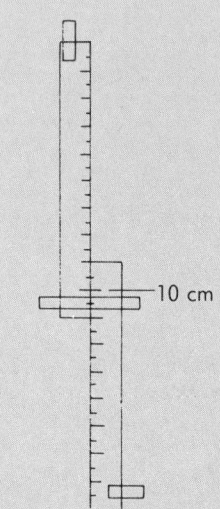

b. Comparing mass and height

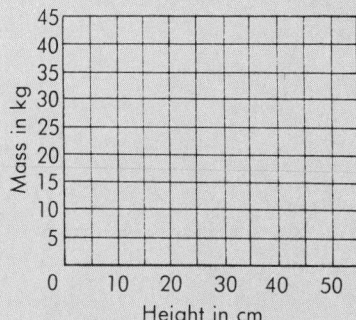

7. Repeat step 6 for each pair of weight and height values recorded.
8. Connect the dots, starting with the lowest value of height.

Part B
1. Measure the mass of each block on the balance. Record your data.
2. Measure the length, width, and height of each block with the ruler. Record your data with the correct mass.
3. Calculate each block's volume, using volume = length × width × height. Record each volume.
4. Measure the stone's volume using water displacement as follows. Note the height of water in a graduated cylinder before and after you drop the stone into it. Subtract the readings to find the stone's volume. Change the unit from mL to cm³.
5. Calculate each block's density using density = mass ÷ volume. Record each density.

Analysis
1. How accurately can you measure distances and mass with the equipment you are using?
2. Some people think of a graph as a picture of the data. Explain how looking at a graph can be more helpful than looking at the actual data.
3. When is measuring volume by water displacement necessary?

Did You Know?

Lucky Accidents in Science

Hurray for accidents! This comment may sound strange in a science book. Usually scientific research is carefully and accurately planned and carried out. However, accidents can help the development of science. For example, the American inventor Charles Goodyear made a great accidental discovery. The picture shows Goodyear experimenting on his kitchen stove.

Goodyear was trying to find a way to improve natural rubber. At normal temperatures, rubber is a tough, pliable solid. However, heated rubber gets as soft and sticky as tar. On a very hot day, car tires made from natural rubber would melt and stick to the highway.

Goodyear had heard that adding sulfur made the rubber less sticky. But he found that the sulfur did not completely solve the problem.

One day in 1839, while heating a batch of rubber, Goodyear spilled some of the liquid on the stove. The liquid burned, smoked, and turned into a charred mess. The result of his accident surprised him. When the rubber cooled, it was even tougher than natural rubber. The spilled rubber did not get soft or sticky at high temperatures. In addition, cold did not stiffen the rubber. Goodyear's mistake improved natural rubber.

Goodyear named his process vulcanization after Vulcan, the Roman god of fire. Now, vulcanized rubber is used to make tires, hoses, and other rubber products. Being able to find valuable things (such as vulcanization) accidentally is called *serendipity*.

Scientists, of course, should be careful in their research. However, sometimes accidents happen, and occasionally experimental results are a little different from what you expect. The important thing is to take advantage of unusual results of accidents. The great scientist Louis Pasteur once said, "Chance favors the prepared mind." He meant that accidents alone do not lead to great discoveries. You must also be prepared to understand and use what you learn from an accident.

For Discussion
1. What is serendipity?
2. Explain why scientists should try to understand mistakes and accidents that happen during their experiments.

1-5
Safety in the Physical Science Laboratory

A physical science class is different from other classes. Learning physical science often means performing experiments, as the students pictured are doing. You must learn how to conduct experiments carefully and safely. As you read about laboratory safety, remember these questions:

a. What safety rules should you practice in a physical science laboratory?
b. What should you do in case of an accident in a physical science laboratory?

Practicing Laboratory Safety

Safety and caution are the keys to successful experiments. In physical science, most laboratory accidents result from carelessness and impatience. Your teacher is responsible for your safety. But he or she cannot do this job alone. You must act in a thoughtful and responsible way. To help you in your laboratory work, study the list of LABORATORY PROCEDURES that follows:

1. **REPORT ACCIDENTS OR HAZARDS TO YOUR TEACHER AT ONCE.**
2. Read and understand all directions before you begin an activity. If you have any questions, ask your teacher for help.
3. Look for and follow all CAUTIONS in the activity.

4. Know the location of all the safety equipment (first-aid kit, fire extinguisher, and fire blanket). Know how to use the safety equipment, and learn your classroom's emergency procedures.
5. Wear safety goggles and a safety apron when heating, pouring, or mixing chemicals and water.
6. Handle glassware with care. Broken glass has sharp edges. Hot glass does not show its heat until it burns your fingers.
7. Handle Bunsen burners with care. Tie long hair back. If a fire starts, DO NOT RUN and DO NOT PANIC. Cover the fire with a safety blanket and wash the affected area with water.
8. Laboratory chemicals can be dangerous. Read all labels. Do not taste or inhale any chemicals. Notify your teacher if you spill any chemicals. Always point a test tube you are heating away from yourself and others, as shown.
9. Do not touch metal on electrical equipment in use.
10. Clean and return all materials to their proper places after you finish every activity. Clean your laboratory area, as the students shown are doing. Do not throw harmful chemicals into the sink.

Point test tubes away from yourself and others

Always clean up after you finish laboratory activities

In Case of an Accident

Physical science activities can be fun. But the fun stops when someone gets hurt. Always tell your teacher about any accidents. Above all, remain calm. Getting upset will not help the person who is hurt.

With some thought and care, your physical science laboratory can be an interesting and safe place to learn.

Review It

1. Restate the LABORATORY PROCEDURES in your own words.
2. What should you do if an accident happens in your physical science laboratory?

Chapter Summary

- Physical science is the study of changes in matter and energy. (1–1)
- Physical science research revises old ideas, develops new understandings, and increases our scientific and technical knowledge about changes in matter and energy. (1–2)
- Researchers usually follow a scientific method as they conduct experiments and draw conclusions. (1–3)
- A hypothesis is a reasonable guess about how or why an event happens. (1–3)
- A theory is an explanation of events that has been proved true many times and that also predicts new events. (1–3)
- A scientific law is an accepted explanation of events that is thought to apply throughout the universe. (1–3)
- Throughout the world, scientists use the International System of measurement. (1–4)
- The metric unit for length is the meter, for volume is the liter or meter^3, for mass is the gram, for time is the second, and for temperature is the degree Celsius. (1–4)
- Density is how closely mass is packed. (1–4)
- Density can be calculated using density = mass ÷ volume. (1–4)
- To carry out safe laboratory experiments, you must follow safety guidelines. (1–5)

Interesting Reading

Knowledge and Wonder: The Natural World as Man Knows It. MIT, 1979. Overall view of the place of science today, our current understanding of natural phenomena, and their relationship to human life.

Vergara, William C. *Science in Everyday Life.* Harper, 1980. Question-and-answer format covering a wide range of topics, both practical and theoretical.

Questions/Problems

1. What topics are physical scientists interested in? Why do they investigate them?
2. Suppose you wanted to know if salting water makes the water take longer to freeze. Describe the experiment(s) you would perform to find an answer. Be sure to explain how to control the experiment to reduce errors. What is your hypothesis?
3. What important contributions did Galileo make to the study of science?
4. List four ways to prevent accidents in the physical science laboratory.
5. Calculate the volume of a shoe box that measures 19 cm in length, 16 cm in width, and 9 cm in height.
6. Calculate the density of 270 g of aluminum that take up 100 cm^3 of space.
7. What is the importance of the computer in physical science research?

Extra Research

1. Use reference books in your library to learn about some other topics of research that are of current interest to physical scientists.
2. Find out what the National Bureau of Standards does.
3. Use the newspaper to find out the maximum and minimum temperatures every day for two weeks. Graph this information on the same piece of graph paper, using different colors for the maximum and the minimum temperatures.
4. Use metric units to measure the length, volume, or both of your kitchen floor, your bed, and a friend.

Chapter Test

A. Vocabulary Write the numbers 1–10 on a piece of paper. Match the definition in Column I with the term it defines in Column II.

Column I

1. an explanation of fact thought to apply over and over throughout the universe
2. a logical way of solving scientific problems
3. a reasonable or scientific guess
4. an explanation of fact that predicts new events
5. the metric unit for temperature
6. the amount of mass in a given volume
7. the metric unit for length
8. how closely packed the particles of matter are
9. a metric unit for volume
10. the metric unit that measures mass

Column II

a. degree Celsius
b. density
c. gram
d. hypothesis
e. law
f. liter
g. mass
h. meter
i. scientific method
j. theory

B. Multiple Choice Write the numbers 1–10 on your paper. Choose the letter that best completes the statement or answers the question.

1. Physical science includes a) chemistry. b) biology. c) astrology. d) none of the above.

2. Observations taken during an experiment are called a) a hypothesis. b) data. c) a theory. d) meters^3.

3. Scientific experiments a) test hypotheses. b) provide data. c) use measurements. d) all of the above.

4. If 106 g of silver take up 10 cm^3, the density of silver is a) 106 g/cm^3. b) 0.10 g/cm^3. c) 10.6 g/cm^3. d) 1060 g/cm^3.

5. You measure your height in a) meters. b) centimeters^3. c) grams. d) liters.

6. A milliliter is a) 0.1 L. b) 10 L. c) 0.01 L d) 0.001 L.

7. You measure liquid volume with a a) meter stick. b) balance. c) graduated cylinder. d) meniscus.

8. A balance measures a) mass. b) volume. c) degrees. d) weight.

9. The statement "Mercury has a greater density than water" is a(n) a) observation. b) theory. c) law. d) hypothesis.

10. To avoid accidents in the physical science laboratory you should a) talk with your friends as you work. b) follow safety procedures. c) ignore the scientific method. d) drop glass in the sink.

Chapter 2
Moving Objects

The students in the picture are competing in a bicycle race. As the cyclists speed toward the finish line, they think of the distance they must cover and the course they must follow. Each rider tries to cover the course in less time than the others. The cyclists must understand many basic ideas about movement.

This chapter describes the connections among movement, distance, direction, and speed. The chapter explains how we measure and compare these factors. It also discusses changes in speed and direction.

Chapter Objectives
1. Define motion and explain its dependence on frames of reference.
2. Distinguish between distance and displacement.
3. Distinguish between speed and velocity and explain their relationship to distance and time.
4. Contrast constant speed and accelerated motion.

2–1
Describing Motion

Before you learned to crawl, everything that moved seemed to move around you. About the time you first learned to walk, your ideas of movement began to change. Suddenly, you knew that you moved, not the objects around you. This section challenges you to think about movement in a new way. Consider these questions as you read:

a. How is motion defined?
b. How does motion depend on frames of reference?

How Motion Is Defined

When you ride a bicycle, you know you are moving because you pass trees, houses, and other fixed objects. But suppose you were in a well-cushioned vehicle with no windows. You might have a hard time deciding if you were moving because you could neither see objects outside your vehicle nor feel the vehicle moving.

Relative motion, which we simply call motion, is defined as a change in position in relation to some fixed object. In other words, you can determine if an object is moving by comparing its position to that of another object. In the two pictures, for example, the mailbox is a fixed object. You know the car in the pictures has moved because its position has changed in relation to the mailbox. In everyday life, you use fixed objects, such as bushes, fence posts, and buildings, to determine motion. These fixed objects are called **reference** (ref′ər əns) **points.**

24

No movement relative to the bus or to the road

Movement relative to the bus and to the road

Movement relative to the road only

Motion Depends on Frames of Reference

How you view an object's motion depends on how you move. A **frame of reference** is the reference points that are not moving in relation to an observer and things moving with the observer. The diagrams show two different frames of reference: the bus and the road. In the first diagram, the girl is not moving in relation to the bus and the road. In the second diagram, the girl is moving in relation to both the bus and to the road. In the third diagram, the girl is moving in relation to the road, but not in relation to the bus. She is moving in the road's frame of reference, but not in the bus's.

Sitting at your desk, you say you are not moving. But the earth carries you along as it speeds around the sun. Whether or not you say you are moving depends on your frame of reference. If you choose the earth as your frame of reference, you say you are sitting still. If you choose the solar system as your frame of reference, you say you are moving—fast—around the sun.

Review It

1. How do you determine if an object is moving?
2. What affects how you view motion?

2–2 Contrasting Distance and Displacement

Imagine that you are taking a dog for a walk. As you stroll slowly down the sidewalk, your dog runs back and forth between you and the corner. Sometimes the dog even stands still. Your movements and those of the dog represent two measurable quantities described in this section. Think about these questions as you read:

a. How do distance and displacement differ?
b. How can you represent displacement?

How Distance and Displacement Differ

The diagram at the left shows the imaginary paths traveled by a little boy and a dog. Even though the boy and the dog start out and end up at the same place, the dog has had much more exercise than the boy has had. The dog has traveled a greater **distance,** which is the length along a path between two points. The difference is that the dog has traveled in many directions, while the boy has traveled in a straight line.

In studies of motion, a slightly different measure from distance is sometimes used. This measure gives both the length and direction of an object's path from its starting point straight to its ending point. It is known as **displacement.** The boy and the dog end up at the same place, in the same direction away from their starting point. Therefore, even though the dog traveled a greater distance than the boy did, his displacement is the same as the dog's. The horizontal black line marks the place where the displacements of both the boy and the dog are equal.

The boy could round the corner and continue walking on the blocks that lead back to his house. Then he would travel a distance of four blocks. But he would end up where he started out, and his displacement would be zero.

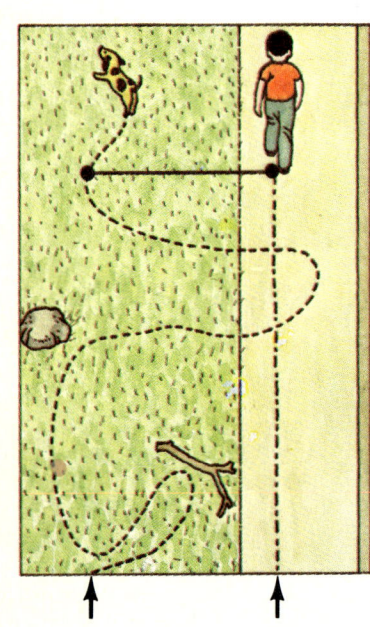

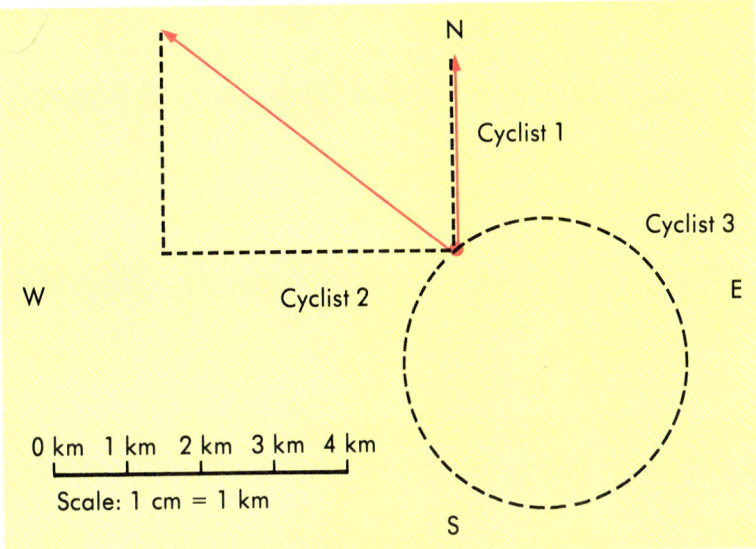

Distance and displacement for the cyclists

Representing Displacement

Displacement involves both the amount of distance traveled and the direction traveled. The diagram represents three cyclists who have traveled different distances in different directions. The black dashes represent the paths traveled by the cyclists. The red arrows represent the displacements of the cyclists. They are drawn to the scale of one centimeter equals one kilometer.

Cyclist One rode north in a straight line for 2.5 kilometers. A 2.5-centimeter arrow pointing north shows his displacement. Cyclist Two rode 4 kilometers west and then 3 kilometers north. A 5-centimeter arrow pointing approximately northwest shows her displacement. Cyclist Three rode in a circle, ending up at the starting point. The diagram shows just a red point for Cyclist Three because his displacement is zero.

Have You Heard?

The U.S. has about 6.5 million kilometers of roads—more than any other country in the world.

Review It

1. How can the amount of your displacement be less than the distance you travel?
2. How can you draw an object's displacement?

2–3 Comparing Speed and Velocity

In most races, the first person to cross the finish line wins the race. Officials time the racers to determine how fast they moved and if a new record is set. As you continue reading, keep these questions in mind:

a. How is speed determined?
b. How do speed and velocity differ?

Speed Depends on Distance and Time

Speed describes how quickly a person or an object changes position. The greater an object's speed, the greater the distance it travels during each second. Therefore, fast-moving objects change position more quickly than slow-moving objects.

To determine speed, you divide the distance traveled by the time spent covering that distance:

$$\text{speed} = \frac{\text{distance}}{\text{time}}.$$

If the girl in the picture swims 50 meters in 25 seconds, her speed is:

$$\text{speed} = \frac{50 \text{ meters}}{25 \text{ seconds}}$$
$$= 2 \text{ meters/second}.$$

Challenge!
Using an encyclopedia or books about cars and airplanes, find out how speedometers and air speed indicators work.

Does this speed mean the girl swam 2 meters during each second of the race? She may have traveled faster at the beginning or end of the race than she did in the middle. The total distance traveled divided by the total time is the **average speed.** In this case, the swimmer's average speed was 2 meters per second. But she may never have traveled 2 meters in any one second.

You can describe an object's motion with a graph. At the right is a graph of the distance a car traveled during a trip. After three hours, the car had covered a distance of 90 kilometers. From the equation, the car's speed was:

$$\text{speed} = \frac{\text{distance}}{\text{time}}$$

$$= \frac{90 \text{ kilometers}}{3 \text{ hours}}$$

$$= 30 \text{ kilometers/hour}.$$

If you calculate the speed from each point on this graph, you will find that it does not change. This type of graph results when an object moves at a **constant** (unchanging) speed.

Speed and Velocity Differ

Speed is how fast an object moves. **Velocity** (və los′ə tē) is both how fast and in what direction an object moves. Therefore, velocity tells us both speed and direction. For example, the speed of a car might be 45 kilometers per hour. Its velocity, however, would be 45 kilometers per hour due east. Another car with the same speed would have a different velocity if it were traveling due west.

Review It

1. How do distance and time affect speed?
2. How does velocity differ from speed?

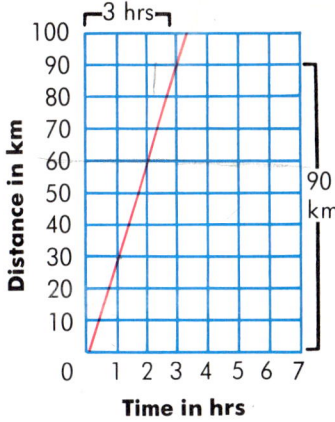

Graphing speed

For Practice

Use the speed equation to solve these problems.
• What is the average speed of a car that travels 440 km in 5 hours?
• How fast do you walk if you take 3 hours to cover 15 km?

29

Activity

Measuring Speed

Purpose
To measure the speed of a moving object.

Materials
- small ball
- 2 meter sticks
- piece of rubber band
- watch or clock to measure seconds.
- masking tape

Procedure

Part A
1. Tape the meter sticks to the table or the floor, as in *a*. They should form a narrow track for the ball.
2. Hold the rubber band and the ball as in *b*.
3. Release the ball.
4. Observe the ball's motion and record your observations about it.
5. Copy the table in *c*.
6. Repeat step 3 while your partner records the time for the ball to roll along the track.
7. Repeat step 6 two more times. Be sure to stretch the rubber band about the same amount each time.
8. Calculate the speed for each trial by dividing the distance by the time.
9. Record the three speeds.
10. Average your speeds by adding them and dividing by 3.
11. Record your averaged speed.

Part B
1. Repeat Part A, steps 5–11, but make the ball go faster by stretching the rubber band more.
2. Repeat Part A, steps 5–11, but make the ball go slower by stretching the rubber band less.

Analysis
1. Was the ball's speed constant as it moved along the track? If not, explain.
2. Why is it better to average the speeds, rather than choose one?

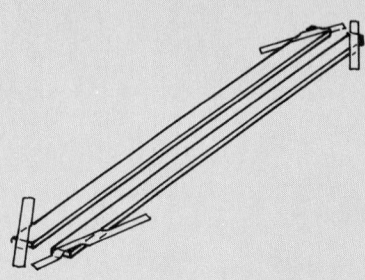

a

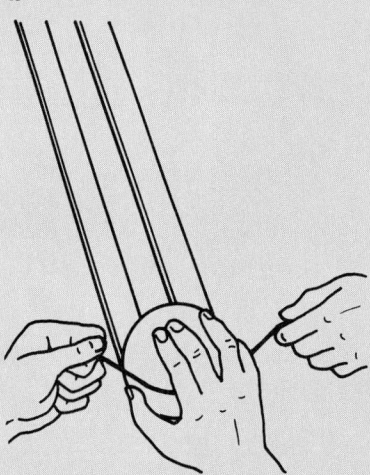

b

c. Speed records

Trial	Time	Speed
1		
2		
3		

Distance

Averaged Speed

Did You Know?

Speed Comparisons

How fast is fast? The answer depends on who or what you are. Some animals do not move very fast. A "mad dash" for a caterpillar is about 0.005 kilometer per hour. At this rate, a speedy caterpillar could make it across this page in about 3 minutes. The caterpillar will do much better when it becomes a butterfly. Then it will fly about 19 kilometers per hour.

Some larger animals are slowpokes too. The sloth, a furry tree dweller, spends most of its life hanging upside down. When it decides to move, 0.8 kilometer per hour is about the best it can do. At this rate, the sloth would take nearly seven minutes to "run" down a football field. A good football player can go the same distance in about 10 seconds!

Size is not what determines how fast an animal can move. For example, some small birds fly very quickly. The tiny spine-tailed swift is only 15 centimeters from beak to tail. But its flight has been timed at 171 kilometers per hour! Vultures, eagles, and falcons all can fly nearly 160 kilometers per hour.

The shape of an animal helps determine its speed. Notice the shape of the bird pictured. The bird's body shape is well adapted for flight. The body shape of a fish helps it move swiftly through water. Swordfish and sailfish have been timed at speeds of more than 100 kilometers per hour. Only a fast boat can keep up with these speedsters of the deep!

Humans are not very quick compared to some other animals. During a 100-meter dash, the fastest human speed was recorded to be 39 kilometers per hour. Humans do better at making other things move quickly. Human speed records include the following:
- riding a race horse at 70 kilometers per hour.
- throwing a baseball at 160 kilometers per hour.
- pedaling a bicycle at 225 kilometers per hour.
- hitting a golf ball at 275 kilometers per hour.
- driving an automobile at 1185 kilometers per hour.

For Discussion
1. How important is size in determining how fast an animal can move?
2. What does the shape of an animal's body have to do with its speed?

2–4 Describing Acceleration

If you have ever ridden a roller coaster, you may remember the thrill of being pulled up the first hill. Suddenly, you plunged downward, moving faster and faster. Consider these questions as you read about such changes in velocity:

a. What is acceleration?
b. In what ways can an object accelerate?

Changing Velocity

The changing speed and direction of the roller coaster involve changes in velocity. The rate of change in velocity is **acceleration** (ak sel′ə rā′shən). Because velocity consists of speed and direction, acceleration can involve a change in speed, direction, or both.

An accelerating object can travel a different distance each second than it did in the past second. Examine the pictures of the moving balls. On the top, the ball rolls at a constant velocity. It moves the same distance during each second it rolls. On the bottom, the ball rolls down a slope. The ball gains velocity as it rolls. As the ball accelerates, it travels farther during each second. It moves with an ever changing velocity.

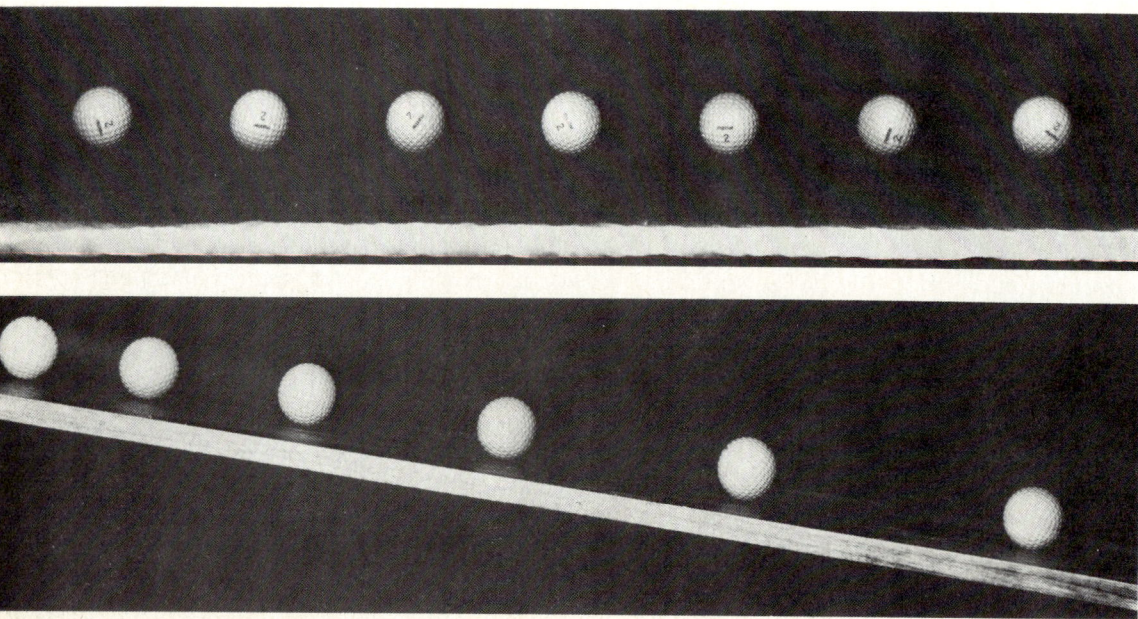

32

An object with a changing velocity can have a constant acceleration. The object can change speed, direction, or both, but it does so by the same amount each second. The roller coaster shown above and the rolling ball have a constant acceleration when they go downhill. Their velocities increase at a constant rate.

You can calculate acceleration resulting from a change in speed by first subtracting the object's original speed from its final speed. Then you divide this number by the time for the change in speed:

$$\text{acceleration} = \frac{\text{final speed} - \text{original speed}}{\text{time for the change}}.$$

Suppose two cyclists are traveling in the same direction at 10 kilometers per hour. Then they increase their velocity to 20 kilometers per hour. Cyclist One reaches the new speed in 10 seconds. His acceleration is:

$$\text{acceleration} = \frac{20 \text{ kilometers/hour} - 10 \text{ kilometers/hour}}{10 \text{ seconds}}$$
$$= 1 \text{ kilometer/hour/second}.$$

Cyclist One increases his velocity by 1 kilometer per hour each second.

Cyclist Two reaches 20 kilometers per hour in 5 seconds. Her acceleration is:

$$\text{acceleration} = \frac{20 \text{ kilometers/hour} - 10 \text{ kilometers/hour}}{5 \text{ seconds}}$$
$$= 2 \text{ kilometers/hour/second}.$$

Cyclist Two has twice the acceleration of Cyclist One.

For Practice

Calculate the acceleration in each problem below.
• A car accelerates from 66 km/hr to 88 km/hr in 11 s. What is the acceleration?
• What is the acceleration of a car that changes speed from 20 km/hr to 50 km/hr in 4 s?

The Meaning of Acceleration

As with speed, an acceleration of 2 kilometers per hour per second does not always mean that the object accelerated 2 kilometers per hour during each second. The acceleration may have been more or less during some seconds. So the equation gives you the **average acceleration**—the total change in the object's speed during a certain period of time.

Acceleration does not have to mean an increase in velocity. Slowing down is also an acceleration. This kind of acceleration is a **deceleration** (dē sel′ə rā′shən).

You can find an object's deceleration just as you find its acceleration. For example, a cyclist slows from 10 kilometers per hour to a stop in 5 seconds. The graph represents the cyclist's motion. The horizontal line at 10 kilometers per hour shows that the cyclist travels at that constant speed. Then the acceleration is:

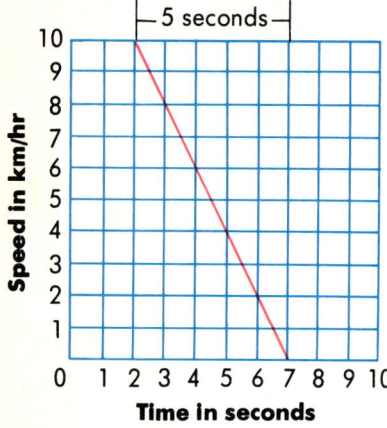

Graphing acceleration

$$\text{acceleration} = \frac{\text{final speed} - \text{original speed}}{\text{time for the change}}$$

$$= \frac{0 \text{ kilometers/hour} - 10 \text{ kilometers/hour}}{5 \text{ seconds}}$$

$$= -2 \text{ kilometers/hour/second}.$$

Here the negative sign indicates a deceleration. The cyclist decelerates 2 kilometers per hour each second.

Acceleration may also mean a change in direction. A car making a turn is accelerating because it is changing the other part of its velocity—direction. Even though the car moves at a constant speed during the turn, it is accelerating because it is turning.

Review It

1. How do constant speed and constant acceleration differ?
2. How can an object accelerate without changing speed?

Activity

Graphing Motion

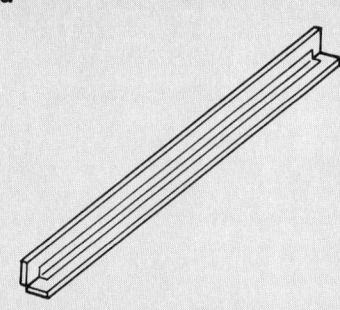

a

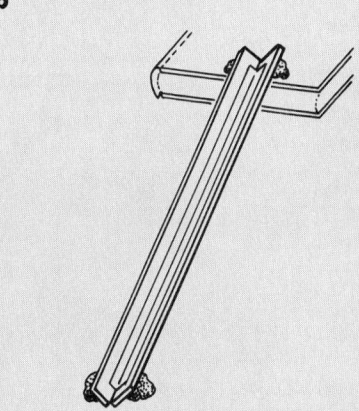

b

Purpose
To graph the motion of an accelerating object.

Materials
- small ball
- 2 meter sticks
- 3 hardcover books
- watch or clock to measure seconds
- masking tape
- clay

Procedure

Part A
1. Tape the meter sticks together at right angles, as in a.
2. Place a book under one end of the sticks, and secure the sticks with clay and tape, as in b.
3. Calculate the distance in centimeters from the top of the sticks to the bottom. Record the distance.
4. Hold the ball at the top of the elevated end of the sticks.
5. Release the ball and time how many seconds it takes to reach the bottom of the sticks. Record the time.
6. Repeat steps 4–5 twice and average the three results. Record the average time.
7. Repeat steps 4–6 using two books to elevate the metersticks.
8. Repeat steps 4–6 using three books to elevate the sticks.

Part B
1. Copy the grid in c.
2. On the grid, put a small dot at the point that represents the distance the ball rolled and the average time it took when the sticks were elevated by one book. Draw a line from the 0 point on the grid to the dot you plotted. Label the line "1 book."
3. Plot the results from Part A, steps 7–8, labeling the lines appropriately.

Analysis
1. Compare the steepness of the three lines on your graph, and explain their significance.
2. Calculate the average speed of the ball at each of the three elevations, and compare the average speeds with the steepness of the lines on the graph.

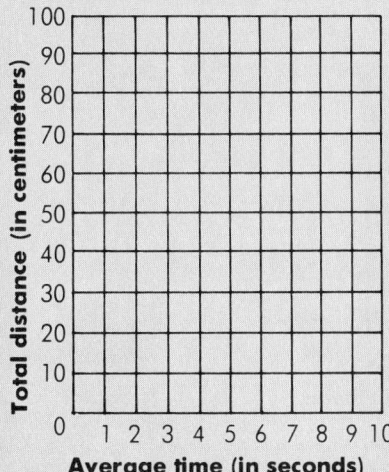

c. Changes in distance over time

35

Chapter Summary

- The relative motion of an object is determined in relation to reference points. (2–1)
- All statements about motion depend on the observer's frame of reference. (2–1)
- The distance and direction of an object's motion from the starting point straight to the end point is the object's displacement. (2–2)
- The speed of an object tells how quickly the object changes position. (2–3)
- Speed can be determined using speed = distance ÷ time. (2–3)
- Velocity is the speed and the direction of a motion. (2–3)
- The motion of an object can be represented by a graph. (2–3)
- Any change in velocity is an acceleration. (2–4)
- Acceleration can be calculated using acceleration = (final speed − original speed) ÷ time. (2–4)
- Deceleration is a decrease in speed. (2–4)

Interesting Reading

Apfel, Necia H. *It's All Relative.* Lothrop, 1981. Find out how Einstein's ideas affected our understanding of motion, from frames of reference to the shape of the universe.

Gardner, Robert, and Webster, David. *Moving Right Along. A Book of Science Experiments and Puzzles About Motion.* Doubleday, 1978. Explains motion experiments you can do at home. (Be sure to ask your parent's permission before performing any experiments at home.)

Questions/Problems

1. You wave to a friend who is riding past you in a car. Explain how the scene appears to you from these frames of reference: a) you are standing still on the sidewalk; b) you are riding your bicycle alongside the car at the same speed.
2. Draw a diagram showing some of the possible paths of a runner who has a displacement of 3 km to the southeast. Be sure to mark the distance of each path.
3. A cyclist travels 2 km due west, then 2 km due north, and then 2 km due east. What is the cyclist's displacement?
4. An airplane travels from San Francisco to San Diego, a distance of about 740 km, in 2 hours. Determine its average speed.
5. If the speedometer on a car indicates a constant speed, can you be certain the car is not accelerating? Explain.
6. Calculate the accelerations of the following: a) a car increasing from 0 km/hr to 100 km/hr in 20 s; b) a bicycle going from 50 km/hr to 0 km/hr in 25 s; c) a runner going from 0 m/s to 10 m/s in 2 s.

Extra Research

1. Check an astronomy text to find out how scientists measure the speed of light.
2. On a map of your state, pick a city and find the shortest highway route from your city to the one you selected. Find the distance of that route by holding a string along the road on the map and measuring the string's length. Then determine the displacement involved.
3. Mark off a 100-meter length on a sidewalk. Calculate your running speed by timing how many seconds it takes you to run the 100 meters. Check a book of records to see how your speed compares to the world record.

Chapter Test

A. Vocabulary Write the numbers 1–10 on a piece of paper. Match the definition in Column I with the term it defines in Column II.

Column I

1. the rate of change in velocity
2. the length between two points
3. a change in position
4. affects how an observer views motion
5. distance divided by time
6. unchanging
7. the rate of decrease in velocity
8. a fixed point used to determine movement
9. speed in a certain direction
10. distance and direction from one point straight to another

Column II

a. acceleration
b. constant
c. deceleration
d. displacement
e. distance
f. frame of reference
g. motion
h. reference point
i. speed
j. velocity

B. Multiple Choice Write the numbers 1–10 on your paper. Choose the letter that best completes the statement or answers the question.

1. As you read at your desk at school, you move in relation to   a) the sun.   b) the earth.   c) your house.   d) your school.

2. What is the displacement of a jogger who runs once around an 800-m track?   a) 0 km in any direction   b) 80 km to the north   c) 400 km   d) 200 km

3. A truck that travels 100 km in 2 hrs has an average speed of   a) 200 km/hr.   b) 0.02 km/hr.   c) 50 km/hr.   d) 102 km/hr.

4. In 5 hrs, a car traveling at 80 km/hr can cover   a) 16 km.   b) 400 km.   c) 580 km.   d) 800 km.

5. An object traveling at a constant velocity   a) changes speed.   b) changes direction.   c) covers equal distances in equal times.   d) covers unequal distances in equal times.

6. A graph can be used to describe an object's   a) motion.   b) speed.   c) acceleration.   d) a, b, and c.

7. A ball rolling down a hill has   a) displacement.   b) velocity.   c) acceleration.   d) a, b, and c.

8. To accelerate, an object must change   a) speed.   b) direction.   c) both speed and direction.   d) either speed or direction.

9. The acceleration of a car that increases from 0 km/hr to 100 km/hr in 10 s is   a) 50 km/hr/s.   b) 10 km/hr/s.   c) 20 km/hr/s.   d) 1 km/hr/s.

10. A car traveling in a circle at a constant speed of 100 km/hr   a) is accelerating.   b) is decelerating.   c) has a constant velocity.   d) a and c.

37

Chapter 3
The Laws of Motion

Launching the space shuttle pictured is a complex project. But scientists can send this huge vehicle into orbit partly because they understand the natural laws that describe how objects move. Scientists discovered these laws hundreds of years ago. Yet the laws are still fundamental to every space launch.

This chapter deals with three basic laws of motion that describe how all objects move. It compares movement in a straight line to movement along a curved path. The laws of motion discussed here apply to your movements and those of everything around you.

Chapter Objectives

1. Describe Newton's first law of motion.
2. Describe Newton's second law of motion.
3. Describe Newton's third law of motion.
4. Explain the law of conservation of momentum.
5. Discuss how Newton's laws explain circular motion.

3-1
Newton's First Law of Motion

If you slide a hockey puck on a sidewalk, it does not go as far as it does if you slide it on an ice rink. As you learn why these motions happen, consider these questions:

a. How does friction affect a moving object?
b. What is inertia?

Friction Slows Down Moving Objects

The sliding hockey puck in the picture slows down only when something interferes with its movement. For instance, the puck might hit a player's stick or crash into the side of the rink. Or it might slow down because of friction, which occurs when one object scrapes against another. Less friction occurs on smooth ice than on rough pavement. So the puck glides farther on ice than on a bumpy sidewalk.

In the 1600s, the physicist Isaac Newton observed moving objects and thought about the nature of motion. He noticed that moving objects tend to keep moving unless something stops them. Before Newton realized this fact, people thought that objects tended to slow down and stop by themselves. People had not realized that friction, which results from rubbing, rolling, or sliding along a surface, slows objects down.

Newton Defined Inertia

Newton realized that a moving object keeps moving as long as friction does not slow it enough to stop it. He also observed that a stationary object remains still unless something makes it move. This tendency to resist changes in motion is **inertia** (in er′shə). A hockey puck lying still on a rink stays in place because of inertia. A sliding puck moves at a constant velocity until something slows or stops it because of its inertia.

The path of *Voyager*

Anything that changes the velocity of a moving object is a **force.** The side of a rink exerts a force on a puck that crashes into it. **Friction** is a force between surfaces that opposes other forces on an object.

Newton's first law of motion states that, if no force acts on a moving object, the object will keep going at the same speed and in the same direction. If the object is stopped, it will remain still. This law is sometimes called the law of inertia.

Newton's first law explains much about the motion of spacecraft. The *Voyager* rocket, shown above, was launched from earth. Once *Voyager* escaped the friction of earth's atmosphere, no more power was needed to keep the craft moving. Outer space has little matter and, therefore, offers almost no friction to slow an object. Only gravity or firing the rocket engines would change the spacecraft's motion.

Review It

1. What effect does friction have on a moving object?
2. What is Newton's first law of motion?

Activity

Newton's First Law

Purpose
To illustrate inertia.

Materials
- index card
- cup
- sandpaper
- penny or other 1-cm disk
- 3 cylinders of the same size —one metal, one wood, and one polystyrene
- clay
- 3 rulers or books

Procedure

Part A
1. Rest the index card on top of the cup. Place the penny on the card so the penny is centered over the cup.
2. Slowly slide the card to one side, with the penny sliding along with the card. Friction is holding the penny to the card. Continue sliding the card until the penny falls off the card.
3. Repeat step 1.
4. Flick the card off the top of the cup with a quick motion of your fingers, as in a.
5. Record what happened to the penny in both cases.
6. Repeat steps 3–5 with a piece of sandpaper, rough side up, replacing the card.

a

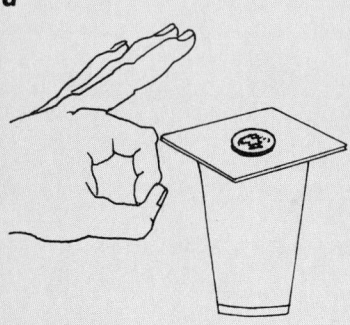

b

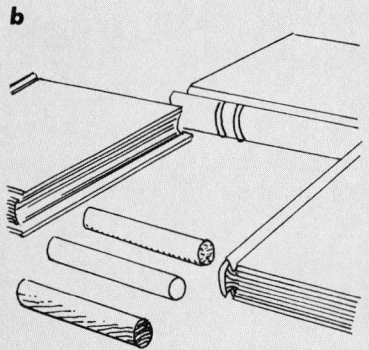

Part B
1. Use the rulers or books to make an alley, as in b.
2. Roll the cylinders at about the same slow rate toward the closed end of the alley.
3. Record whether a big difference occurs in the time each takes to reach the other end.
4. With clay, build a very low ridge 4 cm in from the far end of the alley. This ridge will exert a small force on the cylinders.
5. Repeat steps 2–3.
6. Make a slightly greater ridge. Repeat steps 2–3.
7. Repeat step 6 until two cylinders are stopped, but the third goes over.
8. Record the order in which the cylinders stop passing the ridge.

Analysis
1. What made the penny drop into the glass?
2. What had to be overcome before the card moved?
3. Which cylinder needed the largest force to stop it? Which had the largest inertia? Why?

Breakthrough

Special Relativity

How would you like the secret of long life? Would you like to live for 2,000 years? You do not need any magic potions. Instead, you need *special relativity*.

Albert Einstein became interested in motion in the early 1900s. "What happens to objects that travel very fast?" he asked. By *very fast* he meant traveling at speeds close to the speed of light. Light travels at 300,000 kilometers per second. At this speed, a beam of light can go farther than seven times around the earth in one second.

Scientists cannot make any real objects move close to the speed of light. Even a fast rocket ship travels less than 50,000 kilometers per hour. This speed is only 0.004 percent of the speed of light.

Einstein predicted that time for a fast-moving object slows down. Suppose you make a round trip from earth to a nearby star at 99 percent of the speed of light. The clock inside your space ship would slow down. About 62 years would pass on earth. But only 8.7 years would pass in the space ship. A twin you said good-bye to at the start of your trip would be very old when you returned. But you would have aged very little.

An object traveling near the speed of light should also gain mass. A person with a mass of 70 kilograms on earth would have a mass of about 496 kilograms at 99 percent of the speed of light.

Any measurement taken in the same direction as the ship's velocity should also become shorter. A space ship 50 meters long at rest would appear only 7.1 meters long when it passed you at 99 percent of the speed of light.

These predictions are surprising, but are they true? At first, Einstein's theory was difficult to test. But now it is tested in laboratories on tiny particles that travel faster than 99 percent of the speed of light. And a clock in a moving jet plane really did run slow.

So far, all the tests have shown that Einstein is correct. Someday we might use Einstein's theory to travel to the stars in a human lifetime.

For Discussion
1. What three changes did Einstein predict about objects traveling near the speed of light?
2. How would Einstein's theory affect space travel to the stars?

3-2 Newton's Second Law and Momentum

Because of inertia, a pitched baseball keeps sailing toward the catcher unless the batter hits it. But how does the force of the player shown affect the ball? Think about this question and the following questions as you read:

a. How does Newton's second law link force, acceleration, and mass?
b. What is momentum?

Mass, Force, and Acceleration

In his first law, Newton described inertia. He went on to discover that different objects have different amounts of inertia. Newton realized that the amount of inertia an object has depends on its mass. The more mass the object has, the more force it takes to change the object's motion. Newton also determined that an object's motion changes more—it accelerates more—if you apply a stronger force. The direction of the resulting acceleration is always the same as that of the force applied.

Newton combined these ideas in his **second law of motion,** which states that an object accelerates because a force acts on it. The larger the force, the greater the acceleration. The larger the mass of an object is, the greater the force must be to accelerate it. Newton defined this law by the equation $F = ma$. F means force, m means mass, and a means acceleration.

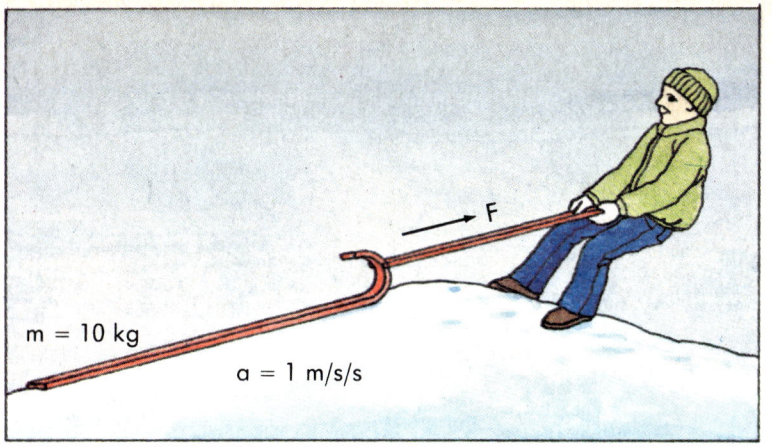

According to Newton's second law, the force of a bat hitting a ball causes the ball to accelerate. Newton's second law also applies to the racer in the picture. The car's engine exerts a strong force during the whole race. Therefore, the racer accelerates all the time.

Consider the example of a boy pulling a 10-kilogram toboggan by its rope. In the drawing, the boy pulls on the rope with a constant force. Newton's second law tells us that the toboggan will accelerate as long as the boy keeps pulling. The force needed to accelerate the toboggan by 1 meter per second each second is

F = 10 kilograms × 1 meter/second/second
= 10 kilogram-meter/second/second.

The unit to measure force is the **newton** (N). One newton is the force required to accelerate 1 kilogram by 1 meter per second every second.

If the boy exerts the same force (10 newtons) on a toboggan that has twice as much mass, it will accelerate at

10 newtons = 20 kilograms × a.

$$a = \frac{10 \text{ newtons}}{20 \text{ kilograms}}$$

= 0.5 meters/second/second.

Doubling the mass when the force is the same cuts the acceleration in half.

If the boy wants the toboggan to accelerate at 2 meters per second each second, he must pull with a force of

F = 10 kilograms × 2 meter/second/second
= 20 N.

Have You Heard?

A space shuttle and its rocket boosters produce 30,800,000 N of thrust (upward force) at launch.

For Practice

Use F = ma to solve these problems.
• How much force does it take to accelerate a 40-kg bundle by 5 m/s/s?
• How many times more force does it take to accelerate the bundle twice as much?

Challenge!

Find out what the Coriolis force is, how it is related to Newton's laws of motion, and how it affects storms on earth.

Moving Objects Have Momentum

The strength of an object's motion is its **momentum** (mō men′təm). The faster you throw a ball, the more momentum it has, so the more damage it can do when it hits something. Or, if the ball moves at half the velocity, it will do just as much damage if its mass is doubled.

A bullet has a small mass, but it has a great momentum because of its high velocity. A slow-moving train has a great momentum because of its large mass. The momentum of an object depends on both its mass and its velocity.

To change an object's momentum, you must apply a force. The amount of force needed depends on how quickly you want to change the momentum. You have probably experienced this if you ride a bicycle. You must brake more to stop quickly than to slow down gradually. You stop in both cases, because a small force acting over a long time can have the same result as a large force acting quickly.

The train in the picture crashed into the station. The walls of the station had to exert a force for a long time to stop the train. A much larger force would have been needed to stop the train more quickly.

Review It

1. Explain Newton's second law of motion.
2. What are two ways to increase momentum?

Activity

Newton's Second Law

Purpose
To show the relationship among force, acceleration, and mass.

Materials
- wood block, 4 cm × 8 cm × 30 cm
- 2 wood blocks about 2 cm × 2 cm × 1 cm
- screw eye
- piece of light rubber band at least 5 cm long
- tape or glue
- thin cardboard, 10 cm × 50 cm
- scissors

Procedure

Part A
1. Screw the screw eye into one of the smallest sides of the large block.
2. Tie one end of the rubber band to the screw eye.
3. Stretch the band just enough to move the block.
4. Practice step 3. Keep the band stretched the same amount to keep the force on the block constant.
5. Pull the block for about 5 seconds. Note what happens to the velocity of the block as you keep pulling with a constant force.

a

b
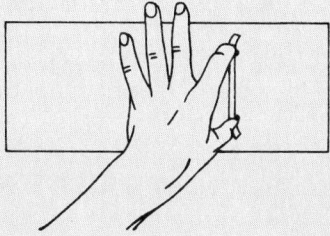

c
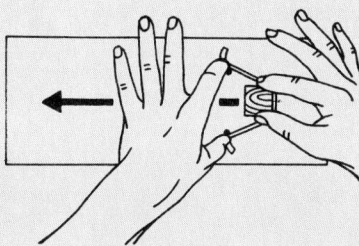

Part B
1. About 10 cm from one short edge of the cardboard, mark two dots about 5 cm apart, as in *a*.
2. Hold the band between the dots, as in *b*.
3. With your other hand, place a small block against the middle of the band, as in *c*. Pull back on the block until the band is stretched almost as far as it will go.
4. Ask your partner to mark with a pencil how far the band stretches.
5. Release the block so the band shoots it forward.
6. Mark on the cardboard how far the block travels. The more it accelerates, the farther it will go.
7. Tape or glue the two small blocks together. This block has twice the mass of either of the small ones.
8. Repeat steps 5–8 with this new block.

Analysis
1. In Part A, what happened to the block's velocity as you exerted a constant force on it?
2. In Part B, how did doubling the mass affect the acceleration of the blocks?

3–3
Newton's Third Law of Motion

The same law of motion that explains how hitting a tennis ball makes it go faster also tells how rockets are launched. As you read about this law, ask yourself these questions:

a. How does Newton's third law of motion relate action and reaction?
b. What does conservation of momentum mean?

Action and Reaction Are Equal

Suppose you are on a raft on a lake, as shown below. As you dive into the water, you exert a force on the raft, and it moves away from you. At the same time, the raft exerts a force against you, which propels you off the raft. The force of your body against the raft is called the **action** force. The force of the raft on your body is the **reaction** force.

Newton's third law of motion states that for every action there is an equal and opposite reaction. Whenever one object applies a force on another object, the second object applies a force of equal strength on the first object. The reaction force acts in the opposite direction to the action force.

Examine the picture at the left. Notice how both the racquet and the ball are affected. The ball's force on the racquet is equal in strength and opposite in direction to the racquet's force on the ball.

Action and reaction

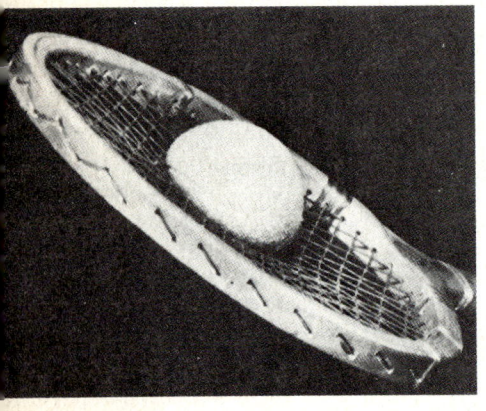

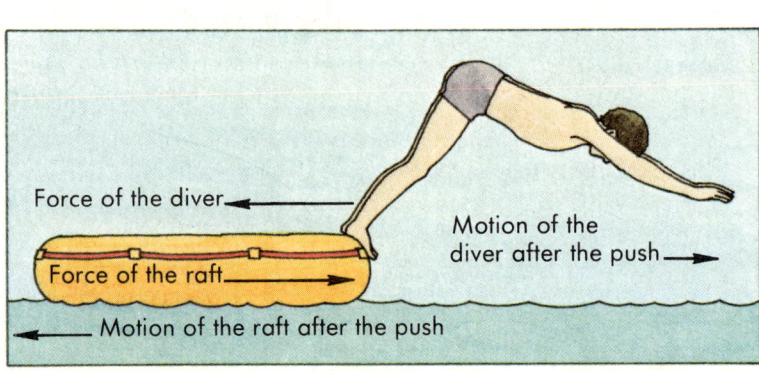

Newton's third law also applies to rockets. A rocket gets its lift from the gases gushing out of its tail. The force of the rocket pushing on these gases is the action. As a reaction, the gases exert an equal and opposite force, which pushes the rocket up in the upper picture. Newton's third law tells us that this reaction force always exists.

The rocket gases do not have to push against anything, such as the ground. The reaction force exists even in outer space, which has no air for the gases to act on. When astronauts need to change a rocket's path slightly, they rely on the action of the gases. A rocket expels gas in one direction, creating a reaction force that pushes the rocket in the opposite direction. The rocket accelerates.

The astronaut in the picture is hovering in the middle of *Skylab*. He can reach the side by throwing anything he is carrying. His throwing action causes a reaction that will push him the other way. A similar thing happens on earth when the friction between your body and the ground is small. Suppose you are ice skating. If you throw a heavy snowball forward very hard, the reaction force will push you backward.

The law of conservation of momentum in action

Momentum Is Conserved

The pictures of a billiard game illustrate another law of motion that is closely related to Newton's third law. In this game, the balls have equal masses. The player hits the yellow ball, which strikes the red ball. The red ball moves forward with some of the momentum the yellow ball had. Then the red ball gives some of its momentum to the blue ball. Before the collisions, the yellow ball had all the momentum. After the collisions, each ball has some of this momentum. The total amount of momentum is the same before and after the collisions. In this example, the action of one ball accelerates another. The reaction of the other slows the first.

The total momentum of the three billiard balls together remains the same—it is "conserved." If outside forces do not act on a group of objects, the total amount of momentum the objects have will not change. This statement is the **law of conservation of momentum.** The momentum of any one object in a group can change. But whatever momentum is lost by one object must be gained by another. The *total* momentum remains the same.

Review It

1. What is Newton's third law of motion?
2. Explain the law of conservation of momentum.

Activity

Newton's Third Law

Purpose
To study action and reaction.

Materials
- 2 long balloons
- 3-m-long string
- 6 strips of paper, 1 cm × 3 cm
- tape
- thin cardboard, 3 cm × 3 cm
- table tennis ball

Procedure

Part A
1. Stretch the string across the room. Tie the ends to the backs of two chairs.
2. Blow up the balloon and hold the end closed.
3. Ask your partner to tape each of two paper strips together and attach them as in *a*.
4. Release the balloon and note what happens.
5. Repeat step 2 and replace the balloon on the string.
6. Repeat step 3 for the table tennis ball so that any air rushing out of the balloon will strike it.
7. Repeat step 4.

Part B
1. Remove the balloon from the string. Roll and tape the cardboard to form a cylinder.
2. Repeat Part A, step 2.
3. Place one end of the cylinder in the neck of the balloon. Repeat Part A, step 3. Keep the balloon inflated.
4. Place about 2 cm of the closed end of the second balloon into the cylinder. To seal the first balloon, repeat Part A, step 2.
5. Repeat Part A, step 3, as shown in *b*.
6. Release the second balloon. Note what happens to the two balloons and when it happens.
7. Repeat Part B, steps 3–7 three times.

Analysis
1. In Part A, compare the directions of the escaping air and the moving balloon.
2. In Part A, did the escaping air have to hit anything to make the balloon move? (Compare your results from steps 4 and 7.)
3. In Part B, what force makes the second balloon move away from the first? Why was there a delay?

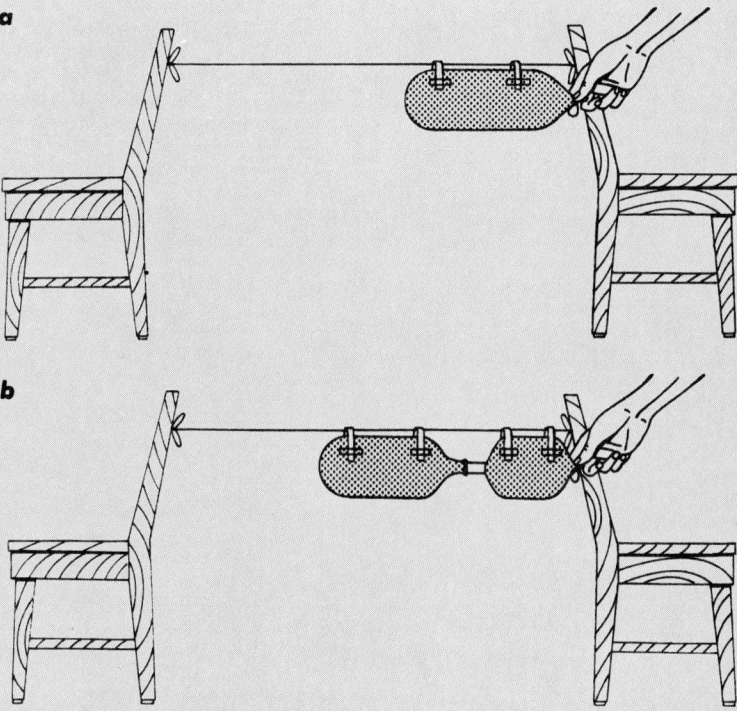

51

3-4 Circular Motion

When you are in a car going around a curve too fast, you feel thrown to one side. A grocery bag may also slide or fall over. Newton's laws explain why these events happen. As you study the explanation, consider these questions:

a. How is circular motion similar to straight motion?
b. What is centripetal force?

How Newton's Laws Explain Circular Motion

If you are in a car moving on a straight road, both you and the car go straight ahead. If the car stops suddenly, you continue to move ahead, as Newton's first law predicts. However, you are soon stopped by the friction between the seat and your clothing, by the seatbelt, by the dashboard, or even by the windshield.

Imagine yourself in the car in the diagram. When the car turns to the left, you again tend to keep going straight ahead. You feel as if you are sliding to the right. Actually the car is turning to the left under you. The faster the car turns, the more you will slide. You move in a straight line, but the car follows a curved path. Eventually, friction or some other force (such as the car door pushing on you) links you to the car's curving motion.

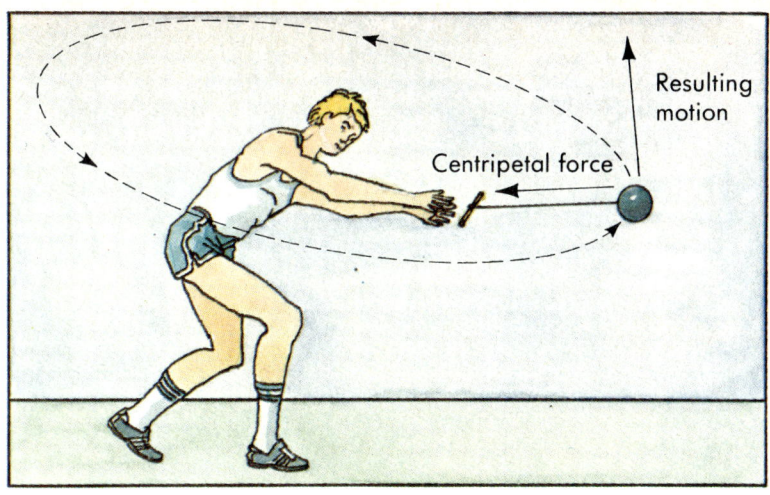

Centripetal Force Causes Circular Motion

The Olympic hammer thrower in the picture is whirling a sphere at the end of a rope. At the instant he releases the sphere, it goes straight ahead, as the arrow shows. But until he releases it, the rope pulls the sphere inward and around in a circle. Newton's first law explains this motion.

The hammer thrower has to exert a force to keep the sphere from going straight ahead. The force he exerts as he pulls inward on the rope is **centripetal (sen tri′pə təl) force.** This force is the inward force needed to keep an object moving in a circle.

The diagram shows a ball moving through a spiral. The walls of the spiral keep the ball circling by exerting a force on it. When the ball comes out of the spiral, the force of the wall stops acting on the ball. Thus the ball goes straight ahead. It follows the path shown by the arrow.

A force is needed to keep objects moving in a curve. Without a force, objects would go straight ahead.

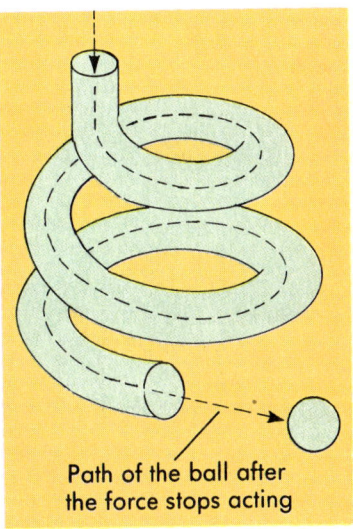

Path of the ball after the force stops acting

Review It

1. Why do you slide to the right inside a car that is turning left?
2. What force keeps an object moving in a circle?

Chapter Summary

- Friction is a force that slows objects down. (3–1)
- Newton's first law of motion states that, as long as no outside force acts on an object, the object will keep going at the same speed and in the same direction or the object will remain still. (3–1)
- Newton's second law of motion states that a force is required to accelerate an object. (3–2)
- The second law also states that the more mass an object has, the more force is needed to change its acceleration. (3–2)
- The momentum of an object depends on its mass and speed. (3–2)
- Newton's third law of motion states that every action is accompanied by an equal and opposite reaction. (3–3)
- According to the law of conservation of momentum, if outside forces do not act on a group of objects, the total momentum the objects have does not change. (3–3)
- Centripetal force is an inward force that keeps an object moving in a circle. (3–4)

Interesting Reading

Ross, Frank. *The Space Shuttle*. Lothrop, 1979. Learn what a space shuttle mission is like and what it might do for future space travel.

Smith, Norman F. *Gliding, Soaring, and Skysailing*. Messner, 1980. Learn how a glider flies and how a pilot uses these forces to stay in the air longer.

Questions/Problems

1. If Newton's first law of motion is correct, why do moving objects on earth eventually stop?
2. How does Newton's first law of motion explain the need to wear seatbelts in an automobile or airplane?
3. How much force is needed to accelerate a 50-kg object by 3 m/s/s?
4. How fast can you accelerate a 60-kg object if you push on it with 120 N of force? (Assume no friction opposes you.)
5. Which has greater momentum: a fullback with a mass of 100 kg or one with a mass of 90 kg, if both are running at the same velocity?
6. Use Newton's third law of motion to explain the direction a crew works the oars to move a boat forward.
7. When a cowhand whirls a lasso overhead, what force is pulling the lasso inward? In what direction would the lasso travel if the cowhand released the rope?

Extra Research

1. Cut out magazine pictures showing examples of friction, Newton's three laws of motion, and centripetal force. Explain how each picture shows forces at work.
2. Try the following with a partner. Both of you should wear roller skates or ice skates and stand on the appropriate surface. Stand facing each other so that your palms are flat against your partner's and about shoulder high. Then push away from each other. Try again, exerting a different amount of force. Explain what you observe in terms of Newton's three laws and the law of conservation of momentum.

Chapter Test

A. Vocabulary Write the numbers 1–10 on a piece of paper. Match the definition in Column I with the term it defines in Column II.

Column I

1. the momentum lost by one object is gained by another
2. acts between surfaces to oppose other forces on an object
3. sometimes called the law of inertia
4. the law that describes action and reaction
5. the law that includes the equation $F = ma$
6. a force equal in strength and opposite in direction to a force you apply
7. what must be exerted to change an object's motion
8. causes an object to move in a circle
9. the strength of an object's motion
10. the tendency of an object to resist changes in motion

Column II

a. centripetal force
b. conservation of momentum
c. force
d. friction
e. inertia
f. momentum
g. Newton's first law
h. Newton's second law
i. Newton's third law
j. reaction

B. Multiple Choice Write the numbers 1–10 on your paper. Choose the letter that best completes the statement or answers the question.

1. Objects traveling along a rough surface slow down because of a) friction. b) reaction. c) action. d) inertia.

2. When a bus stops suddenly, the passengers a) stop immediately. b) keep moving ahead. c) move to the left. d) move to the right.

3. The mass of a body affects a) the force needed to accelerate it. b) its momentum. c) its inertia. d) a, b, and c.

4. You can increase your momentum as you run by a) stopping suddenly. b) slowing down. c) running faster. d) sitting down.

5. How does the floor push on you when you do pushups on it? a) not at all b) downward c) upward d) sideways

6. A rocket in space going to the moon a) gains momentum from earth's gravity. b) fires its engines to change its speed or path. c) must keep its engines firing. d) a, b, and c.

7. When a bat hits a baseball, a) only the bat exerts a force. b) the ball and bat exert an equal force on each other. c) only the ball exerts a force. d) no reaction occurs.

8. If you use similar car engines to exert the same force in two cars with different masses, the more massive car will a) accelerate more. b) accelerate the same amount. c) accelerate less. d) not accelerate at all.

9. If you swing a ball on a string in a circle and the string breaks, the ball will a) keep going in a circle. b) go up. c) go down. d) go straight forward in the direction it was moving when the string broke.

10. An ice skater whirling another skater in a circle is an example of a) centripetal force. b) friction. c) a loss of momentum. d) a lack of inertia.

Chapter 4
Balanced Forces, Friction, and Rotation

The circus performers in the picture are doing a balancing act. The long poles help keep the forces equal on the two sides of the rope. If a performer begins to tip to one side, he instantly moves the pole to regain his balance. The rope provides another kind of balance. It holds the performer up by balancing the downward pull of gravity.

This chapter discusses how two forces often balance each other. It will also help you learn more about friction and its effects. In addition, this chapter explains the forces that cause objects to rotate.

Chapter Objectives
1. Contrast balanced and unbalanced forces.
2. Compare the three kinds of friction.
3. Identify the two quantities that affect the rotation of an object.

4–1 Balanced Forces

Have you ever tried opening a stuck door? You pull and pull, but nothing happens. The door pulls in as hard as you pull out. Suddenly, the door gives way, and you fall backward. For a moment, two forces that balanced each other become unbalanced. As you read more about such forces, think about these questions:

a. When are forces balanced?
b. What do Newton's laws tell us about balanced forces?

Balanced Forces on Stationary Objects

All the stationary objects around you have at least two forces acting on them. These two forces are the downward pull of gravity and an upward supporting force. Hold this book while you read. The force of gravity pulls the book down against your hand. Your hand pushes up on the book with an equal force. Place the book on your desk. Now the desk is providing the supporting force, just as your hand did before. Without this upward force, the book would fall to the floor.

All forces have both strength and direction. In the case of the book in your hand, the forces acting on it were equal in strength but opposite in direction. The arrows in the diagram represent these forces. Forces that cancel each other, as these do, are **balanced.** They keep the book staying still.

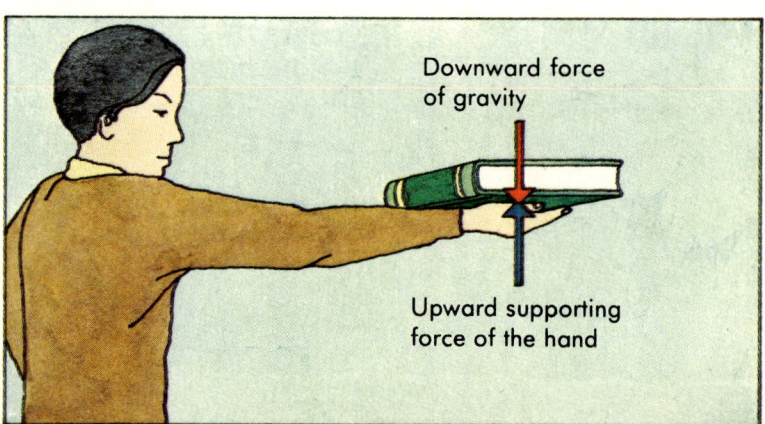

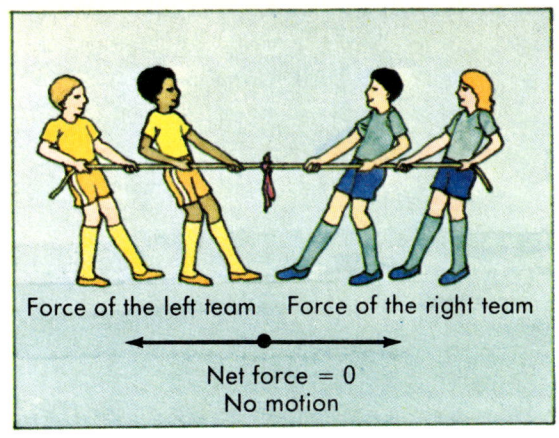

Force of the left team   Force of the right team
Net force = 0
No motion

Force of the left team   Force of the right team
Net force →
Accelerated motion to the right

In addition to vertical forces, horizontal forces often act on objects. The diagrams above show the horizontal forces in a tug-of-war. On the left, the people on both sides are pushing with the same strength but in opposite directions. Because the forces are balanced, the teams are not moving, and the flag in the middle remains still. On the right, the team on the right pushes harder than the team on the left. The forces are now unbalanced. The flag moves to the right, as do the two teams.

The force that results when all the forces acting on an object are combined is called the **net force.** In the tug-of-war, the net force was zero when the two teams pushed equally hard. The extra part of the arrow in the diagram on the right represents the net force when the team on the right pushed harder.

The picture shows a boy and his two dogs in a three-way tug-of-war. Three forces are acting equally in three directions. The net force is zero because the three forces balance one another. If the boy pulled harder than the dogs, the net force would pull the dogs toward him. If the dogs pulled harder than the boy, the net force would pull him toward them. (The tree would soon stop him, though.)

When the opposing forces acting on an object balance each other, the object is in **equilibrium** (ē kwə lib′rē əm). When the opposing forces do not balance each other, the object will accelerate, as Newton's second law states. The object accelerates in the direction of the net force.

59

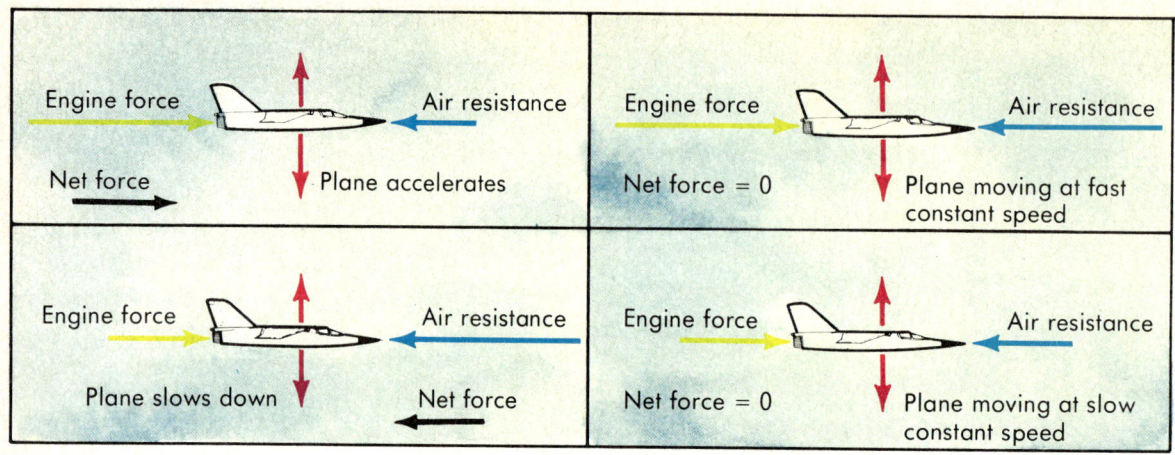

Balanced Forces on Moving Objects

Forces on a moving object can be balanced. In the diagrams of the jet plane, the downward force of gravity is balanced by the upward force of the air pushing against the plane's wings. The horizontal forces are the forward thrust from the engine and the friction of the air pushing back on the plane. This friction is called **air resistance** (ri zis′təns).

At first, the force of the engine is greater than the force of air resistance. As long as the net force is greater than zero, the plane goes faster and faster. But the faster the plane flies, the greater the air resistance becomes. Eventually, the air resistance equals the force of the jet's engine. When the two forces balance each other, the plane moves at a constant speed in a straight line, as Newton's first law predicts.

Now suppose the pilot decreases the force of the jet's engine. Then the force of air resistance becomes greater than the engine's force. The resulting net force causes the plane to slow down. As the plane's speed decreases, the air resistance decreases. Then the two forces again balance each other, and the plane is in equilibrium. The plane's speed again becomes constant.

Review It

1. What is net force?
2. How can the forces on a moving object be balanced?

Did You Know?

Center of Gravity

Football tackle wanted. Must have degree in physics from college or university.

Sports have not quite reached this point. However, science in sports is more important than most people realize. For example, the center of gravity is just one scientific idea used in sports.

Consider the football tackle in the above advertisement. A football tackle does not want to get knocked over. Knowledge about his center of gravity would help make him a better football player. If he gets as close to the ground as possible, his low center of gravity will make him very stable and difficult to tip over.

Race car builders also think about center of gravity. Each year, the speed of race cars increases. Increasing the speed also increases the chance of accidents. That a car might roll over at 250 kilometers per hour is a horrible thought for a driver. One way to keep a car from rolling over is to build it closer to the ground. With a lower center of gravity, the car is less likely to turn over.

Tightrope walkers need to stay up in the air. A low center of gravity helps them too. The poles they hold are heavy and sag on both ends. The center of gravity for a walker-plus-pole is lower than for a walker alone. The center of gravity is directly over the tightrope. In other words, a walker with a pole is "closer" to the tightrope and, therefore, more stable than a walker alone.

Pole vaulters and high jumpers need to think about their centers of gravity too. To clear the bar, the jumper shown wants to keep his center of gravity as low as possible. Then he does not have to lift his weight so far. Notice that the jumper arches his body as he goes over the pole. His center of gravity is low. It is actually *outside* his body and *below* the pole.

For Discussion
1. How is the idea of center of gravity important in building safer cars?
2. Explain how the idea of center of gravity is important for a gymnast to understand and use.

4-2
Friction As a Force

The man in the photograph has placed wheels under the cabinet to lower friction. This heavy object is now easier to move. When he loads the cabinet on a truck, he will remove the wheels to increase friction, which helps hold the cabinet in place. Movers deal with three kinds of friction every day. As you read on to find out about the kinds of friction, answer these questions:

a. What is the strongest type of friction?
b. What are some effects of friction?
c. How does friction affect a moving automobile?

Kinds of Friction

Friction is a force that occurs in three forms. **Static** (stat′ik) **friction** occurs between two surfaces that touch but do not move against each other. Static friction holds a piano in place on a floor. To move an object, you must exert enough force to overcome static friction.

Sliding friction occurs when one object slides over another. If you try to push a piano across a floor, you encounter sliding friction. This type of friction is weaker than static friction and is lowest between two smooth surfaces.

Rolling friction occurs between the surface of a wheel or other round object and another surface. It is usually weaker than sliding friction. When you place a piano on wheels, you can move it more easily because rolling friction is so weak. The diagram shows what causes rolling friction. The surface of both the wheel and the floor are depressed a small amount. The wheel loses more motion from making the depression than it regains when it moves on. The harder the surfaces, the less they are depressed and the lower the friction. Very little friction occurs when a steel wheel rolls on a smooth steel surface, such as the train wheel on the track in the picture. On the other hand, the bike tire in the last picture has a high rolling friction because the tire flattens against the pavement and must regain its shape to move forward.

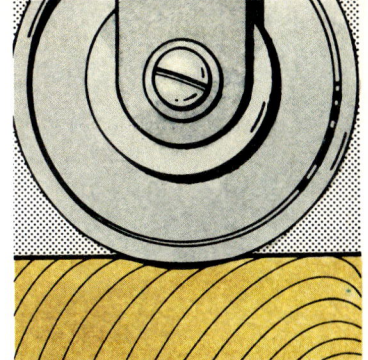

Rolling friction

Some Effects of Friction

Besides affecting motion, friction has some other important effects. For example, sliding friction causes surfaces in contact to heat up. When you rub your hands together, friction creates heat, which warms your skin. Striking a match against sandpaper creates heat, which ignites the match.

Rolling friction also produces heat. When a car travels at high speeds, the rolling friction on the tires can cause overheating, which wears down the tires.

Sliding and rolling friction also cause the surfaces in contact to wear down. Static friction neither creates heat nor produces wear.

In many cases, we want to lessen friction to promote movement and reduce heating and wear. One way to lessen friction between surfaces is to apply a material that allows one surface to slide or roll over another more easily. Such a material is a **lubricant** (lü′brə kənt). Oil and grease are lubricants used in automobiles.

Low rolling friction

High rolling friction

How Friction Acts on a Moving Car

The diagram shows the effect of friction acting on a car moving along a highway at a constant speed. The car's engine produces power to turn the rear wheels. The rear tires push backward against the road. The opposing force is the road pushing forward against the tires. This force propels the car forward. Without friction, movement would be impossible because these forces would not occur. For example, on an icy road, little friction exists between the ice and the car's tires. The tires spin, but the car does not move.

Forces on a moving car

Have You Heard?

As a car goes faster, the force of air resistance increases rapidly. Because the engine must overcome this force pushing backward, going too fast wastes gas.

At first, the force of a road pushing on the rear tires of a car makes the car accelerate. But this force is eventually balanced by both air resistance and the rolling friction between the tires and the road. When the opposing forces are balanced, the car's speed becomes constant.

Review It

1. Give an example of each of the three types of friction.
2. Which kind of friction does not cause heat?
3. Why is friction necessary to make a car move?

Activity

Studying Friction

Purpose
To observe the differences between the kinds of friction.

Materials
- rubber band
- paper clip
- large index card
- scissors
- glue bottle
- tape dispenser that is heavier than the glue bottle
- 2 pencils
- sheet of thick plastic
- grease

Procedure

Part A
1. Fold the card in half and slit the fold about 1 cm, as in *a*. Unfold the paper.
2. Slip the rubber band on the paper clip and place the clip in the slit, as in *b*.
3. Place the glue bottle on the end of the card opposite the band.
4. Gently pull the band to move the card and bottle about 10 cm. Note how much the band stretches just before the card and bottle start moving. You are measuring static friction.

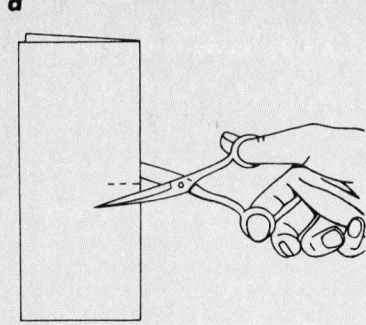

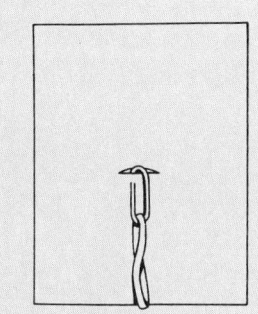

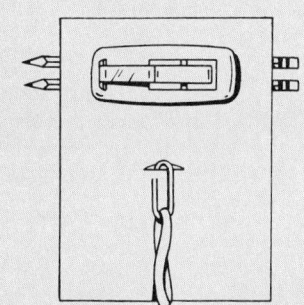

5. Note whether and by how much the stretching changes as the card and bottle move. You are measuring sliding friction.
6. The more the band is stretched, the greater the force. When you stop pulling, mark the point reached by the stretched end of the band on the card. Record your observations.

Part B
1. Replace the glue bottle with the tape dispenser.
2. Try to move the dispenser by pulling the band.
3. Place the pencils so that the dispenser can roll, as shown in *c*.
4. Try to move the dispenser by pulling on the band.
5. Record what happens.

Part C
1. Place the plastic sheet on your desk or table. Coat the sheet lightly with grease.
2. Repeat Part A, steps 3–7.
3. Compare the pencil mark you made in Part A with the mark from Part B.

Analysis
1. Compare the strengths of static, sliding, and rolling friction.
2. What effect does grease have on static and sliding friction?

4–3 Forces on Rotating Objects

You can vary the spin of the bicycle wheel shown by pushing on it at different spots. If you grasp the axle and push down, the wheel does not move. If you grip the middle of a spoke and press down, the wheel spins slowly. If you push down on the rim of the wheel with the same force, the wheel spins faster. Where you apply a force on an object affects its spin. As you read about forces on rotating objects, think about these questions:

1. What determines how fast an object rotates?
2. What is an object's center of gravity?

How Distance from the Axis Affects Rotation

The mechanic in the picture below is tightening a bolt with a special wrench. A pointer on the wrench tells him how tight the bolt is. The mechanic is measuring **torque** (tôrk), which is how strongly an object is turning.

An object turns around an imaginary line called an **axis** (ak′sis). Torque depends partly on the distance between the axis and the point at which you apply the force. It also depends on the amount of force exerted. Torque increases as you move the force farther from the axis or as you apply a greater force. For this reason, a long-handled wrench produces more torque than a short-handled one.

You can produce a torque without turning an object. For example, if you try to remove a rusted bolt, you create a torque even if the bolt does not turn.

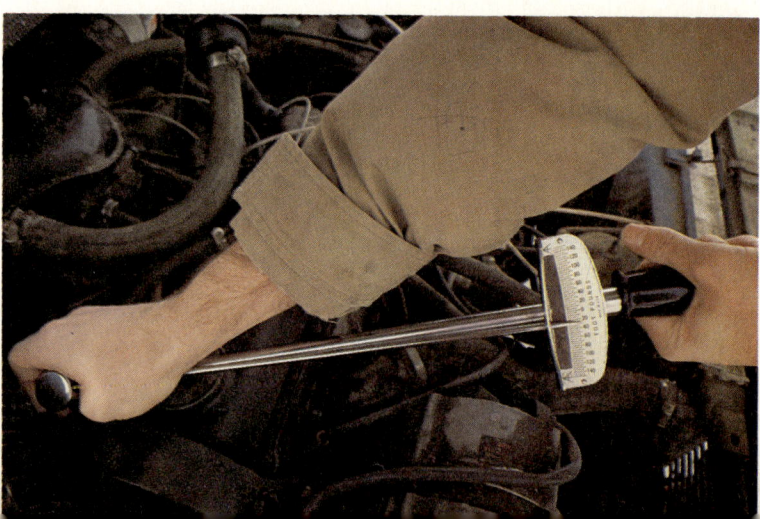

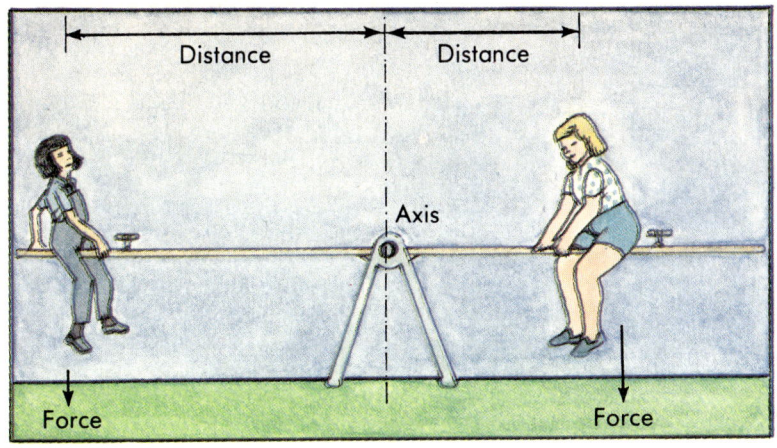
Balancing torques on a seesaw

Think of the seesaw shown above as a wheel spoke that can turn only a certain amount. Here torque depends partly on the pull of gravity. Because gravity pulls more on the girl with more mass, she exerts more force on her side of the seesaw. But torque also depends on each girl's distance from the axis. The girl with less mass exerts a weaker force, but she applies the force farther from the axis. Therefore, the torques on the seesaw are balanced. Either girl could create more torque, and so move the seesaw, by varying her distance from the axis.

Bicycles operate by using torques. The top diagram on the right shows two of the gears on the rear wheel of a 10-speed bicycle. Note how the distance from the axle to each gear varies. When you switch from the small to the large gear, the torque increases. The larger gear makes pedaling easier because you can cause a larger torque with the same force you used on the smaller gear.

The large torque and easy pedaling result in a slower ride, however. The lower diagram on the right shows that the same length of chain turns the large gear one complete turn and the small gear two complete turns. The amount of pedaling required is the same in both cases. But you must pedal harder to turn the small gear. For the same amount of pedaling, the small gear moves the bicycle twice as far. So to attain high speeds, you need small gears that produce less torque.

Torques on bicycle gears

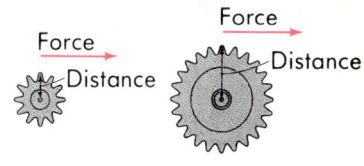

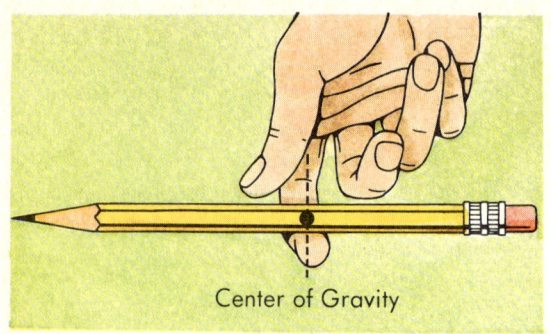
Balancing at the center of gravity

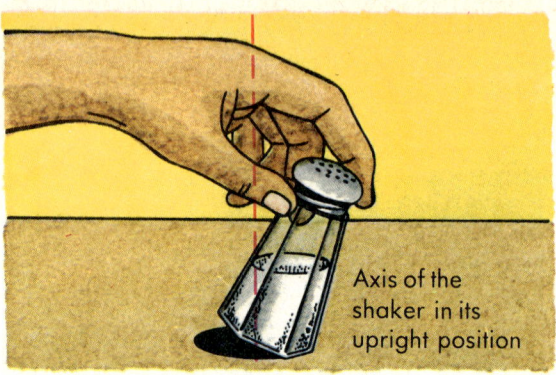
Axis of the shaker in its upright position

How Center of Gravity Affects Rotation

Compare the length of pencil on either side of the finger shown above. The eraser side is shorter than the point side. The eraser makes one side of the pencil heavier. The torque produced is balanced by making the other side longer. Because the torques are balanced, the pencil does not tip. Your finger is under the pencil's **center of gravity,** the point around which the torques on an object are evenly balanced.

Gravity seems to act on an object at its center of gravity. If you gently hold a pencil near one end, gravity will pull the pencil's center of gravity down. The same principle applies to the tilted salt shaker in the picture. The shaker will tip because its center of gravity is to the right of its axis. If the center of gravity were to the left of the axis, the shaker would return to its normal, upright position. If the center of gravity could stay directly above the axis, the shaker would be balanced in that position.

When an object moves freely through the air, its center of gravity always follows a smooth path. In the photograph, the blue tape marks the spoon's center of gravity. The path the tape follows is a straight line.

Review It

1. What two factors determine torque?
2. Where in or near an object does gravity seem to act?

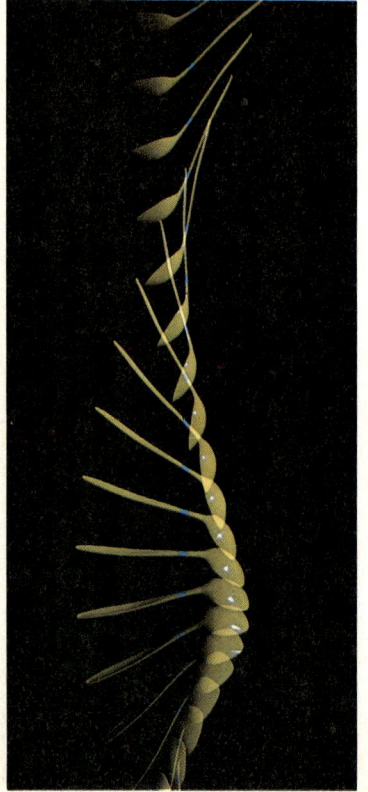
The path of the center of gravity

Activity

Studying Center of Gravity

Purpose
To observe how an object balances at its center of gravity.

Materials
- sheet of heavy cardboard
- scissors
- unsharpened pencil
- masking tape
- 20-cm-long string
- heavy washer
- straightedge

Procedure

Part A
1. Cut a circle, a rectangle, and an irregularly shaped figure from the cardboard. Make each at least 10 cm across.
2. Use the scissors to make two widely spaced holes on the outside edge of each figure.
3. Tape the pencil to the table so that 1 cm extends over the edge.
4. Tie one end of the string to the washer and make a loop at the other end.
5. Hang the circle on the pencil by one hole.
6. Hang the string on the pencil by the loop.
7. Hold the string against the bottom edge of the circle, as in *a*. Mark the point where the string crosses the circle's edge.

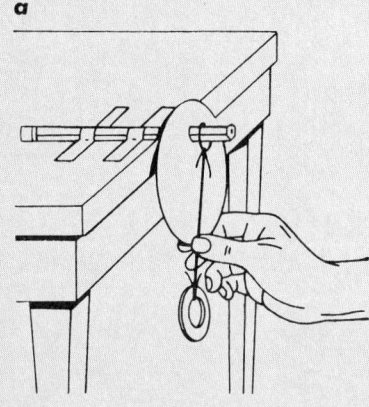
a

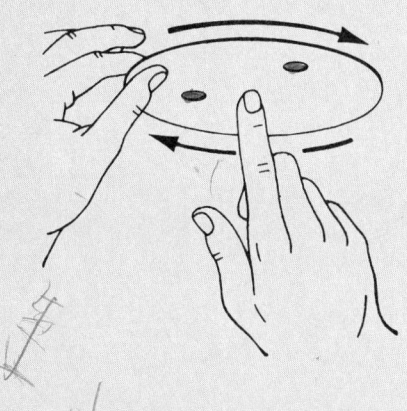
b

8. Remove the string and circle from the pencil. Use the straightedge to draw a line from the mark to the hole you hung the circle from. The circle's center of gravity must have been directly below the pencil and so must be on this line.
9. Repeat steps 5–8, hanging the circle from the other hole. Where your two lines meet, mark "CG" for "center of gravity."
10. Repeat steps 5–10 with the other shapes.
11. Slowly spin each figure on one finger, as in *b*. Note where each balances.

Part B
1. Punch a hole at the center of gravity of the circle and at two other spots nearby.
2. Place the circle on the pencil through a hole.
3. Spin the circle. Note whether gravity pulls any particular part of the circle downward.
4. Repeat steps 1–3, hanging the circle from the two other holes.

Analysis
1. Where does an object's center of gravity always lie?
2. What is the effect of gravity on an object when it is rotating on its center of gravity and when it is rotating on another axis?

Chapter Summary

- The net force is the combination of all forces acting on an object. (4–1)
- When the opposing forces acting on an object balance each other, the object is in equilibrium. (4–1)
- Unbalanced forces acting on an object cause an acceleration. (4–1)
- The three types of friction are static, sliding, and rolling friction. (4–2)
- Static friction is the strongest type of friction. (4–2)
- Rolling friction occurs because the surfaces of both objects in contact are depressed a small amount. (4–2)
- Sliding and rolling friction produce heat and cause the surfaces in contact to wear. (4–2)
- Torque depends on the strength of the force exerted and on the distance between the axis and the point where the force is applied. (4–3)
- Gravity acts on an object as though it were pulling on one point only—the object's center of gravity. (4–3)
- At its center of gravity, the torques on an object are evenly balanced. (4–3)

Interesting Reading

Lewis, Alun. *Super Structures*. Viking, 1980. Descriptions and illustrations of everything that goes into building complicated structures.

Salvadori, Mario. *Building: The Fight Against Gravity*. Atheneum, 1979. Explains the basic principles of architecture and how designs fight the forces of wind, gravity, earthquakes, and temperature change.

Questions/Problems

1. Are the forces acting on a bicycle moving at a constant speed balanced or unbalanced? What are some of the forces acting on the bicycle?
2. Give some examples of situations in which you would try to lessen friction and situations in which you would try to increase friction.
3. Using the terms *axis, force,* and *torque,* explain how a heavier person and a lighter person can maintain balance in a canoe.
4. Using the term *center of gravity,* explain how a tightrope walker keeps from falling.
5. How do you adjust your center of gravity to balance yourself when riding a bicycle?

Extra Research

1. Use the library to find information about three kinds of bridges and how engineers balance the forces on them. Try to build your own bridge with straws or toothpicks and explain how the forces on your bridge are balanced.
2. Slide a book across your desk and then roll it on two pencils. Describe what you observe about static, sliding, and rolling friction.
3. Spin a bicycle wheel from a spot near the axle and from a spot near the rim, using the same force. Observe and explain how distance affects torque. Spin the wheel several times at the same spot, applying different amounts of force. Observe and explain how force affects torque.
4. Cut out five magazine pictures showing motion. Explain the forces acting on the objects in each picture. Indicate which forces are balanced and which are unbalanced.
5. Balance a book, a dustpan, and a roll of aluminum foil on top of a coffee cup to find the center of gravity of each object.

Chapter Test

A. Vocabulary Write the numbers 1–10 on a sheet of paper.
Match the definition in Column I with the term it defines in Column II.

Column I

1. opposing forces that cancel each other
2. the combination of all the forces acting on an object
3. all forces acting on an object are balanced in this condition
4. friction caused by the air pushing against an object
5. the friction that occurs when one object slides over another
6. the point at which gravity seems to act on an object
7. the amount of turn a force exerts on an object
8. usually the weakest kind of friction
9. an imaginary line around which an object turns
10. the strongest kind of friction

Column II

a. air resistance
b. axis
c. balanced forces
d. center of gravity
e. equilibrium
f. net force
g. rolling friction
h. sliding friction
i. static friction
j. torque

B. Multiple Choice Write the numbers 1–10 on your paper.
Choose the letter that best completes the statement or answers the question.

1. When the forces acting on an object are balanced, a) the object accelerates. b) the net force is greater than zero. c) the object is in equilibrium. d) the net force is less than zero.

2. A moving object a) can never be in equilibrium. b) has only unbalanced forces acting on it. c) always accelerates. d) can be in equilibrium.

3. Static friction a) is the strongest kind of friction. b) must be overcome to move an object. c) occurs between two surfaces that touch. d) a, b, and c.

4. Sliding and rolling friction a) produce heat. b) do not cause wear. c) are the same. d) are stronger than static friction.

5. Holding a baseball bat near its middle rather than near its bottom end makes the torque of its top end a) increase. b) decrease. c) stay the same. d) become zero.

6. Without any friction between the tires and the road, a car a) would move much faster. b) would move much more slowly. c) would not move at all. d) would move at a constant speed only.

7. The small gears on a bicycle a) make pedaling easier. b) produce more torque than the large ones. c) produce less torque than the large ones. d) are used when low speeds are desired.

8. An object held at its center of gravity a) tips over. b) remains steady. c) moves to one side. d) is unbalanced.

9. If you switch a short-handled wrench for a long-handled one, the torque you can apply a) increases. b) decreases. c) remains the same. d) becomes zero.

10. Lubricants are used to a) increase friction. b) reduce friction. c) increase the heat produced by friction. d) eliminate air resistance.

Chapter 5
Gravitation

A few seconds after stepping from an airplane, these skydivers were falling nearly 50 meters each second. Now the balancing force of air resistance keeps them from dropping faster and faster. If they pulled in their arms and legs, the earth's gravity would again accelerate the divers to higher speeds. Only the great air resistance of a parachute can slow them to a safe landing.

This chapter describes the force of gravity between objects and the role of gravity in causing weight. The chapter also explains how the sun's gravity controls the planets. In addition, the chapter describes the basic law that tells how gravity works throughout the universe.

Chapter Objectives

1. Define the acceleration due to gravity.
2. Contrast mass and weight.
3. Describe Kepler's three laws of the planets' motion.
4. Explain Newton's law of gravity.

5–1
The Motion of Falling Objects

Many sports are based on the motion of falling objects. When you leap from a diving board, your motion is like that of a falling object. You may shoot a basketball, throw a baseball, or aim an arrow better if you understand the motion of falling objects. As you read about the acceleration due to gravity, ask yourself:

a. How do falling objects accelerate?
b. What are projectiles and how do they move?

Gravity Causes Falling Objects to Accelerate

We often think of gravity as the force that keeps us on the surface of the earth. But gravity also pulls you toward your desk and your classmates and draws your desk and your classmates toward you. Gravity acts between any two objects. It is a force that acts to pull all masses together.

The more mass two objects have, the stronger the gravity between them. The earth has so much mass that its gravitational pull on people is extremely strong. But the force between one person and another or between two objects on earth is so weak that you do not notice it.

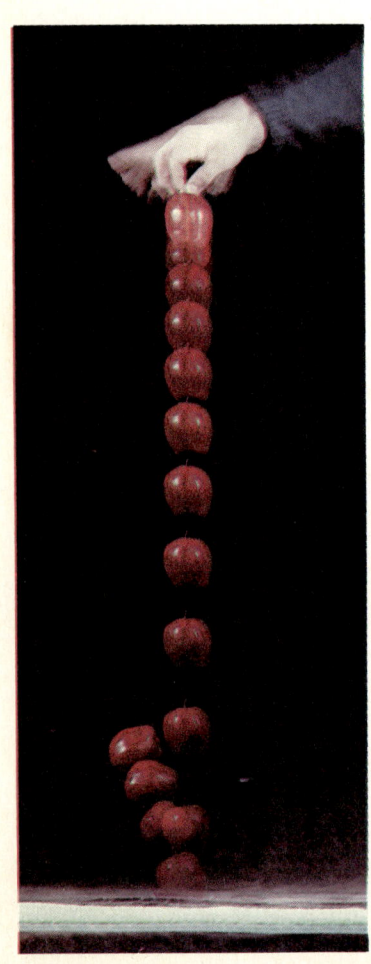

The photograph shows what happens when the earth's gravity attracts an apple. The falling apple was photographed every 0.03 second. Notice how the apple falls farther each time than it did the time before. Thus the apple is accelerating downward. Gravity is the force that makes the apple accelerate.

If nothing slowed falling objects, gravity would make all of them drop with the same acceleration. The **acceleration due to gravity** is the same for both heavy objects and light ones. On or near the surface of the earth, this acceleration is about 10 meters per second each second downward. This means that each second an object is falling freely, its velocity changes in the downward direction by 10 meters per second.

If you drop a crumpled piece of paper and a flat one, the crumpled piece will drop straight down. But the flat piece of paper will probably float around as it drops. Gravity pulls down with equal force on both pieces of paper. But air resistance opposes the acceleration. The flat piece of paper has a greater surface area and so runs into more air than the crumpled piece. Thus the flat paper is slowed more than the crumpled one.

Consider the effects of gravity and air resistance on the group of skydivers shown above. Though gravity pulls them down, air resistance slows their fall. When the divers first jump, they accelerate downward. But for most of their fall, the upward force of air resistance equals the downward force of gravity. Therefore, no net force acts on them, and they do not accelerate. Because they were moving when they stopped accelerating, the sky divers continue to move at a constant velocity. The velocity that any falling object reaches when air resistance prevents it from going faster is the object's **terminal** (tėr′mən nəl) **velocity.**

To prove that the acceleration of gravity is a constant, one of the *Apollo 12* astronauts performed an experiment. On our airless moon, he dropped a hammer and a feather. The hammer and the feather reached the moon's surface at the same time in spite of their different masses and surface areas. This experiment demonstrated that the acceleration of gravity is the same for *any* object.

The Path of Projectiles

The picture above shows the path of a basketball after the student tossed it. The ball rises and then falls, moving forward all the time. The moving ball is a **projectile** (prə jek′təl), which is an object that is thrown, hurled, hit, or shot forward. All projectiles put in motion near the earth are pulled down by gravity.

The photographs compare the motion of a free-falling object—the apple—and a projectile—the ball. The pictures are a series of photographs taken at time intervals of equal length. Notice that the ball falls with the same downward acceleration as the apple. As it falls, the ball moves forward the same distance during each time interval. Therefore, its forward velocity is constant.

Although the ball moves both forward and downward at the same time, scientists can study these motions separately. They can observe how gravity makes the ball accelerate downward. Then they can see that the ball keeps moving forward at a constant velocity because no force acts on it in this direction. Scientists study the motion of such projectiles as baseballs, footballs, and rockets in this way.

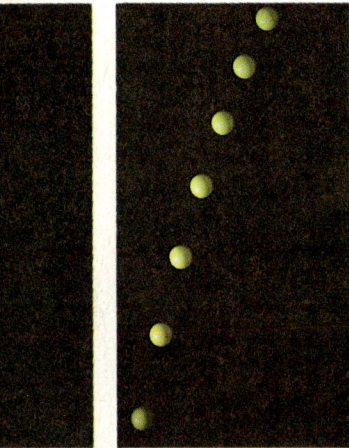

Comparing the motion of two falling objects

Review It

1. What is the acceleration of gravity?
2. Describe the motion of a projectile.

Activity

The Effects of Gravity and Air Resistance

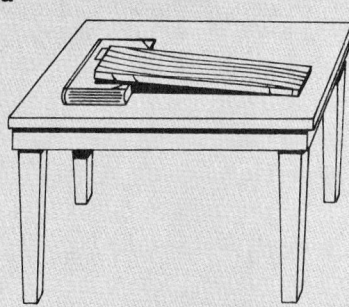

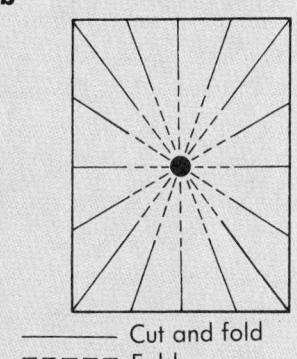

— Cut and fold
- - - - Fold

Purpose
To observe effects of gravity and air resistance.

Materials
- old newspaper
- clay
- golf ball
- thick book
- 1-m-long board
- 2 meter sticks
- 3 pieces of notepaper
- masking tape

Procedure

Part A
1. Spread the newspaper on your desk. Spread the clay on the newspaper so that it covers a 10-cm-square area and is 1 cm thick.
2. Drop the golf ball into the clay from a height of about 30 cm. Sketch the depth of the dent the ball made.
3. Repeat step 2 at heights of 60, 90, and 120 cm.

Part B
1. At one end of a table, set up the plank and book, as in *a*.
2. Make a track on the plank by taping the sticks about 2 cm apart. Release the golf ball from the top of the plank between the sticks. Do not let the ball hit the floor.
3. Your partner should slowly count from 1 to 10 at a constant rate. At each count, mark the ball's position on the plank.
4. Repeat step 3 until you have all ten marks.

Part C
1. Crumple one piece of notepaper and fold the second in fourths.
2. Drop the three pieces of paper in different pairs. For each pair, note which has a larger surface exposed to the air and which hits the floor first.
3. Fold and cut the third piece of paper, as in *b*. Drop it and note how it falls.

Analysis
1. In Part A, the depth of the dent in the clay depended on the ball's velocity as it hit the clay. How did its velocity change in steps 2–3?
2. In Part B, how did the distance traveled and the ball's velocity change each second?
3. Explain how air resistance affects falling objects of the same mass.

5–2 Gravity, Mass, and Weight

If you traveled to Mars, the amount of mass in your body would remain the same. But your weight would change. To make the difference between mass and weight clearer, consider these questions as you read:

a. How is weight defined?
b. What is microgravity?

Contrasting Weight and Mass

Your body has a certain amount of mass. Earth's gravity pulls down on that mass. **Weight** is the force of gravity pulling on a mass.

The picture at the lower left shows a girl weighing herself on a scale on Earth. The girl's mass is 40 kilograms. We can find the amount of force on her by using the equation of newton's second law, $F = ma$. We use the acceleration at which she *would* fall at the earth's surface. The force of gravity on the girl is 40 kg × 10 meters per second per second, which equals 400 newtons. Because weight is a force, we measure it in newtons as we do other forces.

As the girl stands on the scale, a spring in the scale stretches. The amount it stretches depends on how strongly gravity pulls on her.

The picture on the right below shows the same girl and scale on Mars. The force of gravity on Mars is weaker than it is on Earth. Therefore, the girl weighs less, even though her mass is the same.

Microgravity

The boy in the picture above is in an elevator falling freely through space. When he releases the basketball, it seems to hover next to him. The earth's gravity accelerates the boy, the ball, and the elevator by the same amount. However, the boy thinks that gravity has disappeared because he, the ball, and the elevator fall at the same rate.

A spaceship with astronauts aboard acts just like the elevator with the boy in it. Unless they tie themselves down, the astronauts find themselves floating as their ship moves through space. Gravity is still present, but it is accelerating the spaceship and all the objects in it equally. Gravity seems to have disappeared because the spaceship and all the objects in it are falling together. The astronauts could not weigh themselves because any scale they stood on would read zero as it fell with them. This condition is called **microgravity** (mī′krō grav′ə tē). It occurs when objects are falling freely together, as in the photograph.

Challenge!

Find and read magazine articles about how microgravity will be used for experiments on a space shuttle.

Review It

1. What would happen to your mass and weight if you went to the moon?
2. Why does gravity seem to disappear when astronauts are in space?

Activity

Studying Microgravity

Purpose
To observe the effects of gravity on objects falling at the same time.

Materials
- 2 polystyrene cups
- 2 coins
- 2 rubber bands, each slightly shorter than the height of the cups
- tape
- 2 paper clips
- water
- sink or wastebasket

Procedure

Part A
1. Tape each rubber band to a coin, as in *a*.
2. Straighten one end of one clip and use it to make a hole in the bottom of one cup.
3. Fit the rubber bands through the hole and attach them with the unopened paper clip, as in *b*. Hang the coins over the top of the cup, also in *b*. The rubber bands should be stretched slightly.

a
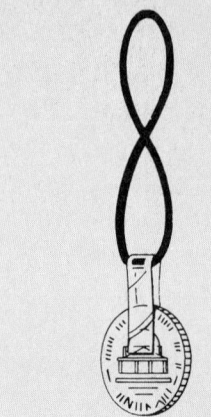

b
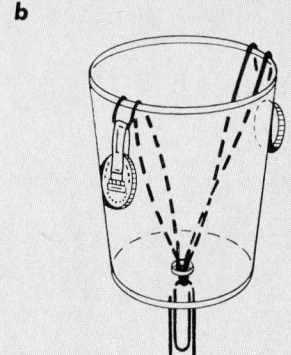

c
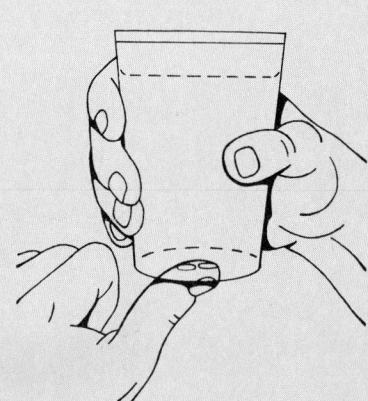

4. Drop the cup from shoulder level to the floor.
5. Note what happens to the coins.

Part B
1. Use the open paper clip to make two small holes near the bottom of the second cup. Or use the first cup with the hole in the bottom.
2. Fill the cup half full with water while holding a finger over the holes, as in *c*.
3. Hold the cup high over a sink or trash can. Uncover the holes and drop the cup.
4. Note what happens to the water as the cup falls.

Analysis
1. Explain why the coins acted as they did in Part A, step 5.
2. Explain why the water acted as it did in Part B, step 4.

Did You Know?

Life Under Different Gravitational Forces

Human beings have lived on the earth a very long time. Our bodies are adapted to the earth's force of gravity. Humans might have problems living where gravitational forces are not the same as on earth.

So far, few humans have had to worry about that problem. Only astronauts have lived and worked in outer space beyond the reach of the earth's gravity. They learned how to get around in a condition of microgravity.

Doctors have observed no serious problems in astronauts returning from space. If you were such an astronaut, you would probably have few health problems resulting from microgravity. Your bones might lose small amounts of calcium and other minerals. You might experience temporary dizziness and difficulty in standing, because blood might collect in the lower parts of your body.

Plans are being made to build large space stations. The space station shown could house 10,000 people. These stations would float in space above the earth or moon. The station's colonists would not feel the earth's or the moon's gravitational pull. A constant condition of microgravity would exist.

How will people adapt to conditions of microgravity? One answer is to spin the space station very slowly. This movement would not create gravity. However, the movement would seem to push people against the floor of the space station. To the colonists, the effect would be the same as gravitation.

Suppose we decide to build colonies on Jupiter, which has a huge mass. The effects of gravity there would be much greater than they are on Earth. A twenty-first century colonist who weighs 705 newtons on Earth would weigh 1,860 newtons on Jupiter. He or she would need help just walking around! Scientists would need to invent ways to fight against gravity. Can you think of any methods to help solve this problem?

For Discussion
1. What effects are caused by forces greater than or less than normal gravity?
2. In what ways might life on a space colony differ from life on earth?

5-3 How Gravity Affects the Planets

Gravity holds us on earth's surface. It also causes the moon to follow a curved path around the earth. The sun's gravity keeps all the planets circling it. As you read about gravity's effects, think about these questions:

1. How does gravity affect the motion of the planets?
2. What is a planet's period of revolution?

Gravity Shapes the Planets' Orbits

The planets, like all moving objects, would move straight ahead if no force acted on them. But the gravity of the sun pulls on each planet. Just as earth's gravity bends the path of a projectile, the sun's gravity bends a planet's path. This force keeps the planet moving around the sun in the same path, called an **orbit** (or′bit).

In the early 1600s, the astronomer Johannes Kepler discovered three laws about the planets' orbits. After years of study, he realized that the planets did not orbit the sun in circles. Instead they orbit in **ellipses** (i lips′ez), which are like the squashed circle below. The sun is at one of the pins. These discoveries are **Kepler's first law.**

Distance from the Sun Affects a Planet's Speed

Because a planet's orbit is an ellipse, the distance between a planet and the sun changes. **Kepler's second law** tells how much faster a planet moves when it is closer to the sun and how much more slowly it moves when it is farther away.

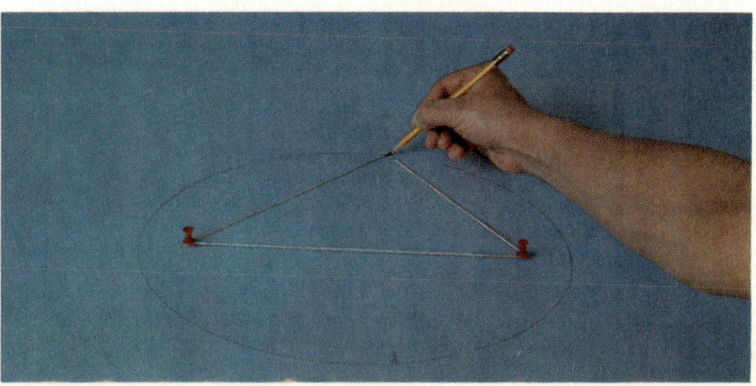

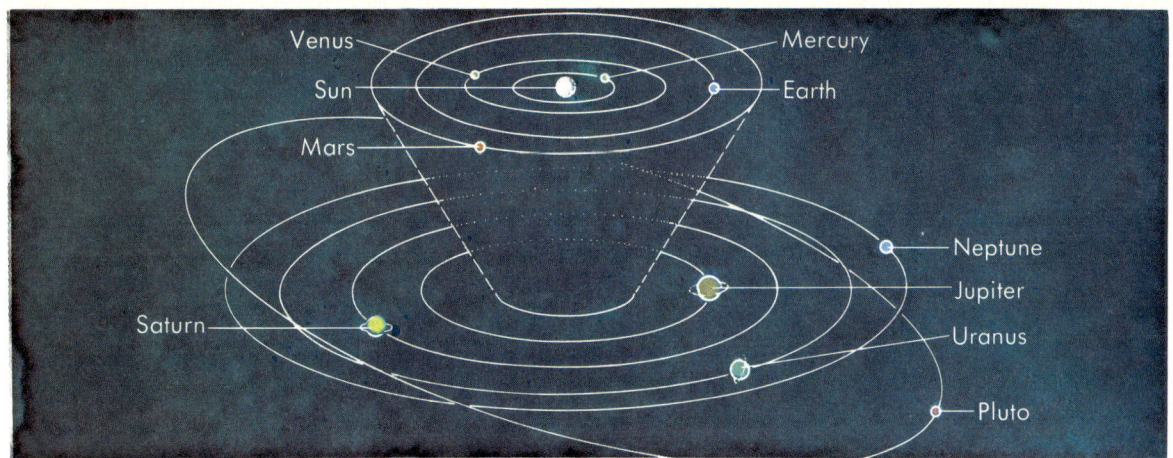

The length of time each planet takes to complete one orbit around the sun is its **period of revolution.** Kepler's third law linked a planet's period of revolution and distance from the sun. The larger its orbit, the longer its period. The more distant planets travel more slowly over longer distances. Compare the sizes of the planets' orbits and their periods in the drawing above and the table below.

Have You Heard?

Scientists take advantage of the planets' gravity when they plan the flight of spacecraft. For example, they planned the *Voyager* flights so that Jupiter's gravity would pull each craft into a new flight path. As a result, the *Voyagers* could fly past Saturn and Uranus.

The Planets' Orbits and Periods of Revolution

Planet	Orbit Compared to Earth's	Period of Revolution
Mercury	0.4	0.2 year (88 days)
Venus	0.7	0.7 year (243 days)
Earth	1.0	1.0 year
Mars	1.5	2.0 years
Jupiter	5.2	12 years
Saturn	9.5	30 years
Uranus	19.2	84 years
Neptune	30.1	165 years
Pluto	39.5	249 years

Review It

1. What holds the planets in their orbits?
2. What happens to a planet's speed as it nears the sun?

5-4
The Law of Gravity

Isaac Newton was one of the greatest scientists of all time. Besides giving us the laws of motion, he also described how gravity worked in the universe. As you read about Newton's ideas, answer these questions:

a. What led to the discovery of the law of gravity?
b. On what does the strength of gravity depend?

Newton Stated the Law of Gravity

In 1543, most people thought that the sun and planets revolved around Earth. Then the astronomer Nicolaus Copernicus proposed a theory that the planets circle the sun. In the early 1600s, the scientist Galileo looked at the night sky with a telescope. He discovered that four moons circled Jupiter—not Earth. His discovery meant that Earth did not have to be the center of the universe.

Kepler made his discoveries about planetary motion at the same time that Galileo was observing the planets with a telescope. Their discoveries supported Copernicus's theory that the sun was the center of the known universe. But no one knew *why* the planets moved around the sun.

In the mid-1600s, Isaac Newton began investigating this problem. He said that the idea of gravity struck him one day when he saw an apple fall. He realized that the force that causes apples to fall also kept the moon and planets in their orbits. The picture points out what an amazing leap in thinking Newton made. He then stated a law that explained Kepler's discoveries.

Gravity Depends on Mass and Distance

Newton's **law of gravity** states that the force of gravity depends on the mass of the objects and on the distance between them. This law enables us to calculate the strength of gravity between any two objects.

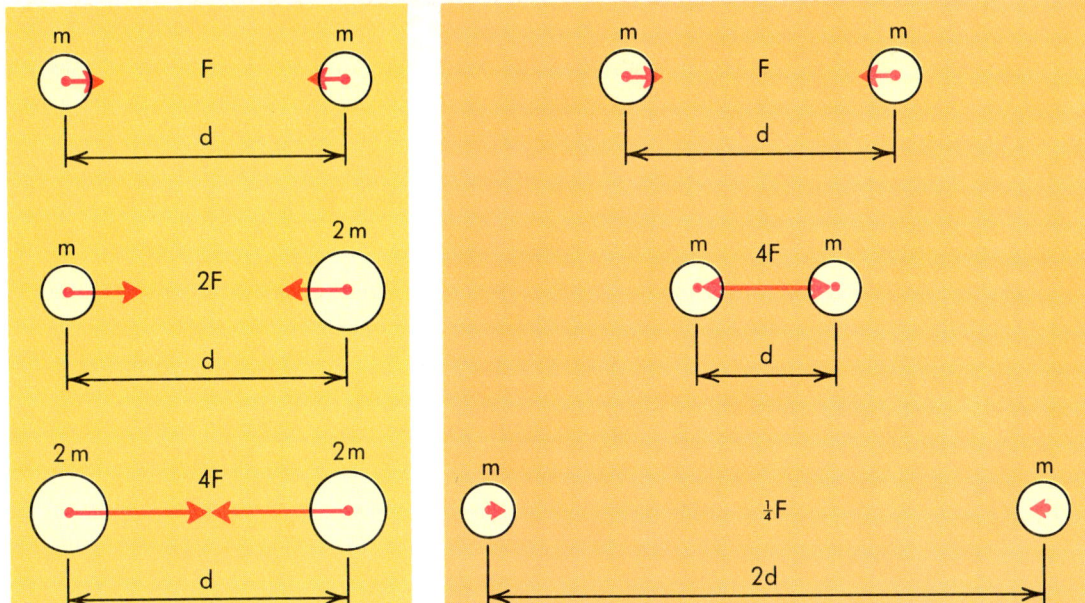

The relationship between mass, distance, and force

The diagrams show the relationships among mass, distance, and the strength of gravity. The larger the mass of either object, the stronger the attraction between them. Doubling the mass of one object doubles the force of gravity. Doubling the mass of both objects multiplies the force of gravity by four. Also, the closer the objects are, the stronger the force of gravity between them. Reducing the distance between two objects by one-half multiplies the force of gravity by four. Doubling the distance reduces the force of gravity to one-fourth.

Newton's law of gravity enables us to determine the strength of gravity, but it does not tell us why gravity exists. In 1916, Albert Einstein explained gravity as the result of a warping of space. A large mass warps space a lot, and approaching objects tend to fall toward the mass, just as a golf ball rolling on a warped green tends to curve.

For Practice

Use Newton's ideas about gravity to answer the following questions.
• What would happen to the force of gravity between you and a box if you added two identical boxes to the first one?
• What would happen to the force of gravity between the earth and the sun if the earth were three times as far away?

Review It

1. How did Newton say he discovered gravity?
2. What does Newton's law of gravity state?

85

Chapter Summary

- Gravity is a force that acts to pull all masses together. (5–1)
- The acceleration due to gravity is the same for both heavy and light objects. (5–1)
- Gravity gives a projectile a downward acceleration. (5–1)
- Weight is the force of gravity pulling on a mass. (5–2)
- Microgravity is a condition in which objects are falling freely together. (5–2)
- The sun's force of gravity bends the paths of the planets. (5–3)
- A planet moves faster closer to the sun and slower farther away from the sun. (5–3)
- The farther a planet is from the sun, the larger its orbit and the longer its period of revolution. (5–3)
- The ideas of Copernicus, Galileo, and Kepler led Newton to realize the law of gravity. (5–4)
- Newton discovered that the force of gravity increases if the mass of either object increases or if the distance between the two objects decreases. (5–4)

Interesting Reading

Cobb, Vicki. *Truth on Trial: The Story of Galileo Galilei.* Coward, 1979. Tells of Galileo's use of the telescope and of his struggle with the Church.

Moore, Patrick. *Astronomy Facts and Feats.* Sterling, 1980. Includes many facts about the solar system and the great astronomers who helped us understand it.

Veglahn, Nancy. *Dance of the Planets: The Universe of Nicolaus Copernicus.* Coward, 1979. A fascinating biography of Copernicus and how he came to believe in the sun-centered solar system.

Questions/Problems

1. Why does a hammer drop more quickly than a feather on the earth but not on the moon?

2. You hurl a basketball down through a hoop and then drop a basketball through the hoop. In each case, you photograph the ball every 1/20 second as it falls from the hoop to the floor. Now suppose you examine the distance between the first and second image in the photograph of the hurled ball. Compare this with the distance between the same two images of the dropped ball. Continue doing this for all the images. Which, if either, ball accelerated faster? Why?

3. Suppose you drop a brick and a small piece of cloth so that you release them at the same time. Explain what you observe as they fall and hit the ground.

4. The acceleration of gravity on the moon's surface is 1/6 of that on the earth's surface. How much does a 30-kg person weigh on the moon? How much mass does the person have?

5. How much stronger is the force of gravity between a 30-kg person and the earth than it is between a 60-kg person and the earth?

Extra Research

1. Push a ball off a table, and at the same time, drop another ball from the same height. Determine which, if either, hits the floor first. Explain what you observe.

2. Observe and consider how birds take off, fly, and land. Use reference books to learn how planes and gliders use these ideas.

3. Observe what happens to your weight as you stand on a bathroom scale in a moving elevator. Explain any differences in weight in terms of microgravity.

Chapter Test

A. Vocabulary Write the numbers 1–10 on a piece of paper. Match the definition in Column I with the term it defines in Column II.

Column I

1. the fastest velocity that a falling object can reach
2. an object that is thrown, hurled, hit, or shot forward
3. the force of gravity pulling on a mass
4. the path of a planet
5. condition in which gravity seems to disappear
6. the shape of the orbit of a planet
7. the length of time a planet takes to complete one orbit
8. how fast a planet moves nearer to or farther from the sun
9. how mass and distance affect the strength of gravity
10. the rate at which the velocity of a falling object changes

Column II

a. acceleration of gravity
b. ellipse
c. Kepler's second law
d. law of gravity
e. microgravity
f. orbit
g. period of revolution
h. projectile
i. terminal velocity
j. weight

B. Multiple Choice Write the numbers 1–10 on your paper. Choose the letter that best completes the statement or answers the question.

1. The acceleration of gravity is   a) a constant.   b) greater for heavy objects.   c) less for heavy objects.   d) impossible to observe.

2. If an acorn and a leaf are dropped at the same time, the acorn hits the ground first because   a) air resistance speeds the acorn's fall.   b) air resistance slows the leaf's fall.   c) gravity pulls harder on the acorn.   d) the acorn weighs more than the leaf.

3. At its terminal velocity, a falling object   a) speeds up.   b) overcomes air resistance.   c) stops.   d) stops accelerating.

4. Projectiles accelerate   a) downward only.   b) forward only.   c) both downward and forward.   d) neither downward nor forward.

5. Astronauts may float in an orbiting spaceship because   a) gravity does not exist in outer space.   b) they fall freely with their ship.   c) the astronauts fall faster than their ship.   d) their masses are less.

6. The weight of an object   a) is the same as its mass.   b) is always the same.   c) depends on the force of gravity on it.   d) is greater on the moon than it is on earth.

7. A planet moves   a) along a circular path.   b) faster when it is closer to the sun.   c) slower when it is closer to the sun.   d) at a constant speed throughout its orbit.

8. The larger a planet's orbit,   a) the shorter its period.   b) the longer its period.   c) the closer it is to the sun.   d) the faster it rotates.

9. The force of gravity affects   a) only large objects.   b) only objects on earth.   c) only nearby objects.   d) all objects.

10. The force of gravity between two objects grows stronger if   a) the objects move closer together.   b) the mass of one object is reduced.   c) the masses of both objects are reduced.   d) the objects move farther apart.

Careers

Surveyor
The next time you cross a city limit, picture the land around you before people built on it. Surveyors were probably there then, doing their job. Surveyors measure and map an area before construction begins.

Surveyors work outdoors with teams of people who help take measurements of the land. From the data they collect, the surveyor draws a map and writes a report about the area. Highway builders, city planners, and boundary makers all use the surveyor's information to better understand the usefulness and safety of building sites.

After high school, surveyors spend two years in technical or junior college. They also get on-the-job training and take a surveyor's license exam.
Career Information:
American Congress on Surveying and Mapping, 210 Little Falls St., Falls Church, VA 22046

Carpenter
Birds build only nests. Beavers build only dams. But carpenters build everything from cabinets to hardwood floors.

All carpenters work with wood. Two kinds of carpenters are "rough" and "finish." Rough carpenters work at construction sites. They put up temporary walls, make forms for cement, and install flooring. Without their accurate measurements, houses would fall down from gravity or other stress forces.

Finish carpenters hang doors, cabinets, and paneling in completed buildings.

Carpenters begin as apprentices. They learn to read blueprints and use tools. As they study, they choose the kind of carpentry they will do.
Career Information:
United Brotherhood of Carpenters and Joiners of America, 101 Constitution Ave., NW, Washington, DC 20005

Pilot
The take-off and landing are the most difficult parts of flying a plane. Slippery runways and bad weather can make the job even more difficult. A pilot must be able to operate a plane under any condition.

Airplane pilots prepare for each trip by making a flight plan. They use weather information to choose the best flight route. A pilot's cockpit controls report on safety conditions, plane position, and control tower data. A plane needs speed to take off against gravity and stay in the air, so pilots watch their velocity closely.

All pilots are licensed. Commercial pilots must have flown for at least 250 hours, pass a written exam, and be in top physical condition to do their job.
Career Information:
Airline Pilots Association, International, 1625 Massachusetts Ave., NW, Washington, DC 20036

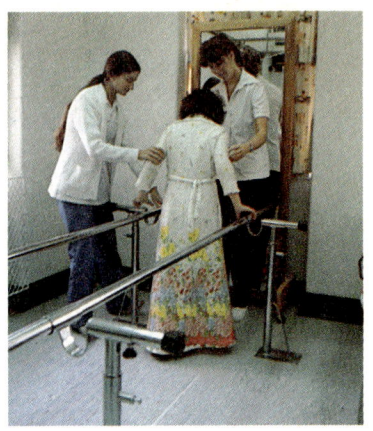  

Physical therapist

What if you could not use your hands or legs? Injured or handicapped people face these problems daily. Physical therapists help them overcome their disabilities.

A physical therapist works to restore movement to people's muscles and joints. The therapist first tests a patient's strength and coordination. Then the therapist designs an exercise and treatment program. This program retrains the body to respond to the forces and stresses of movement. If a disability is permanent, the therapist will help the patient adjust physically to living with the handicap.

After a four-year college degree program, student therapists must pass a licensing exam.

Career Information:
American Physical Therapy Association, 1156 15th St., NW, Washington, DC 20005

Architect

If someone asked you how to build a house, what would you do? An architect would draw the plans for that house.

Architects take people's ideas for buildings and put them on paper. Before your school was built, an architect decided where to put each classroom, locker, and stairway. By understanding the strength of building materials and the stresses and forces acting on them, an architect is able to produce a safe, efficient design. The architect also checks on the school during construction. When architects design projects, they oversee them from beginning to end.

Usually five years of college and two years on the job prepare an architect to become licensed.

Career Information:
The American Institute of Architects, 1735 New York Ave., NW, Washington, DC 20006

Machinist

Henry Ford's Model T was one of the first machines built on an assembly line. Ford made many copies of each part. With them he built cars that were exactly alike. Today, machinists use Ford's idea to do their job.

A machinist works with the equipment that makes metal parts for mass production. First, a machinist reads the blueprint for a needed part. Next, machinery is set up to cut that part. Last, machinists check the accuracy of their work. Since each piece they make must be exactly the same to fit on the assembly line, their measurements must be exactly the same each time.

Most machinists have a high-school diploma. They spend four years as apprentices taking classes and getting on-the-job experience.

Career Information:
National Machine Tool Builders Association, 7901 West Park Dr., McLean, VA 22101

UNIT TWO
ENERGY

What do you think of when you look at this photograph? The object is certainly large. It looks somewhat like the dish of a telescope. But what is the smaller object held above the dish?

This photograph shows an experimental solar oven. The large metal dish reflects sunlight to the small cylinder held above it. Sunlight concentrated in that small area produces quite a high temperature. The experiment may lead to a new method of getting energy from the sun. This unit deals with energy—an important part of physical science.

Chapter 6 Work and Machines
Machines make our lives easier. No matter how simple or complex a machine seems, its purpose is to change the force or motion we put into it.

Chapter 7 Energy and Power
We use and produce energy in many ways. Studying changes in energy is important to understanding many physical processes.

Chapter 8 Heat and Temperature
Using machines to change energy results in the output of heat. Temperature and heat both depend on the energy of objects or processes.

Chapter 6
Work and Machines

Moving sixty metric tons of dirt is not an easy task. But the shovel in the picture can do the job easily. The shovel looks complicated, but it is just a combination of simple tools. Many machines, such as automobiles, typewriters, cash registers, and printing presses, use combinations of simple devices to make our jobs easier.

As you investigate machines, you will learn that your understanding of the word "work" probably differs from the scientific meaning of the word. This chapter will explain the difference. It will also give examples of how simple tools do work and show how simple tools combine to form more complicated machines.

Chapter Objectives
1. Define work and state its unit of measurement.
2. List six types of simple machines and explain how each operates.
3. Explain how a compound machine is made from simple machines.
4. Explain how to measure the efficiency of a machine.

6-1
Doing Work

What do you think of when you hear the word *work*? This word may suggest activities that you must do, but do not necessarily enjoy. But in science, work has a special meaning. As you read, answer these questions about work:

a. What is the scientific definition of work?
b. How is the amount of work calculated?

Work Requires Force and Displacement

In the drawing, the girl is pushing on a large rock. She has been pushing on it for an hour, but the rock has not moved. The girl looks as if she had "worked hard." Scientifically, however, the girl has not done any work!

According to the scientific definition, **work** is done only when the force applied to an object actually moves the object. A force can only cause motion in its own direction. Therefore, only the displacement parallel to the force's direction counts in doing work. So work depends on two things: force and displacement. Force is anything that causes a change in motion. Displacement is the distance and direction through which an object moves. Only the part of the displacement parallel to the force's direction is important.

Calculating an Amount of Work

The equation below shows how to calculate the amount of work done when force and displacement are completely parallel:

$$\text{work} = \text{force} \times \text{distance}.$$

Suppose the girl pushes the rock with 5 newtons of force to the right. She moves the rock a distance of 2 meters to the right. The work done is:

$$\begin{aligned}\text{work} &= 5 \text{ newtons} \times 2 \text{ meters} \\ &= 10 \text{ newton-meters}.\end{aligned}$$

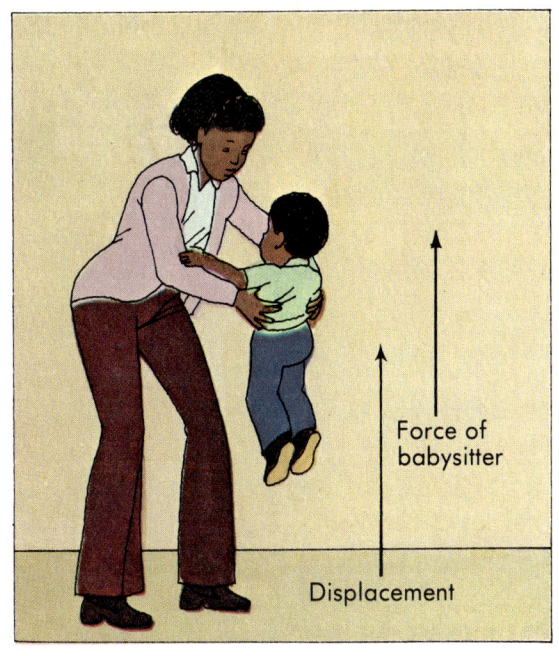 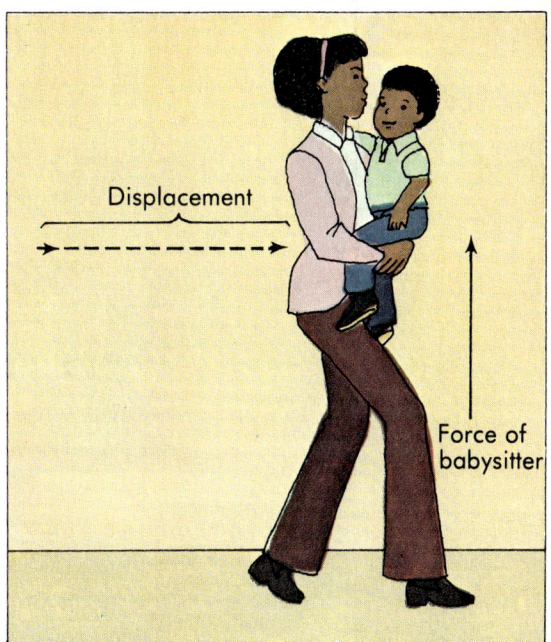

Notice that the **newton-meter** (N•m) is the unit of measurement of work.

In the picture above, the babysitter lifts the child from the floor. Again, the work is the force times the distance—the weight of the baby times the height the child is raised:

$$\text{work} = 80 \text{ newtons} \times 0.5 \text{ meter}$$
$$= 40 \text{ newton-meters.}$$

Then the babysitter carries the child across the room. Their displacement is in the horizontal direction. But the force, gravity, acts in the vertical direction. According to the scientific definition, no work is done in carrying the child because no part of the force is parallel to the displacement.

For Practice

Use the equation for work to answer these questions.
- You push a box with 3 N of force across the floor for 12 m. How much work do you do?
- Kevin pulls a wagon 3 m with a force of 40 N. How much work does he do?

Review It

1. How does the scientific meaning of work differ from the everyday meaning of work?
2. Show how to calculate the amount of work done when an object is moved.

6-2 Some Simple Machines

Machines make our jobs easier. Some machines multiply the force you exert. Other machines change the direction of the movement. As you read, think about:

a. What is a simple machine?
b. How can we calculate mechanical advantage?
c. What is a pulley?
d. How does a wheel and axle work?

Simple Machines Make Tasks Easier

Tools with one or two parts are **simple machines.** One kind of simple machine is just a long, rigid bar with a support. We call this bar a **lever** (lev′ər). Crowbars and seesaws are examples of levers. A lever can help you move objects by multiplying the force you exert.

A lever has a **fulcrum** (ful′krəm), which supports the bar, and two arms. On a seesaw, the fulcrum is under the middle of the bar.

The picture shows how a lever works. The man pushes down on one side of the bar. The opposite side of the bar pushes up on the tree stump and lifts the stump. The distance from the man's force to the fulcrum is the **effort arm.** The distance from the resistance (the tree stump) to the fulcrum is the **resistance arm.**

Calculating the Mechanical Advantage of a Lever

Machines change the force we exert but not the amount of work done. In the picture, examine the work done by the man with the lever. He pushes the lever down a distance of two meters with a force of 10 newtons. The man has done 20 newton-meters of work (10 newtons × 2 meters). The lever can do only that amount of work.

Look at the opposite end, the resistance arm, of the lever. The resistance arm has moved only one meter. The following equation tells us how much force the resistance arm of the lever can exert:

$$\text{work} = \text{force} \times \text{distance}.$$
$$20 \text{ newton-meters} = \text{force} \times 1 \text{ meter}.$$
$$\text{force} = \frac{20 \text{ newton-meters}}{1 \text{ meter}}$$
$$= 20 \text{ newtons}.$$

Since the distance the resistance arm moves is less, the force it exerts is more than the force applied to the effort arm.

The ability of a machine to multiply a force is its **mechanical advantage (M.A.).** A lever's mechanical advantage is the length of the effort arm divided by the length of the resistance arm. In the picture, the effort arm is 120 centimeters long and the resistance arm is 20 centimeters long. The mechanical advantage of the lever is

$$\text{M.A.} = \frac{\text{effort arm}}{\text{resistance arm}}$$
$$= \frac{120 \text{ centimeters}}{20 \text{ centimeters}}$$
$$= 6.$$

A mechanical advantage of six tells us that this lever multiplies a force six times.

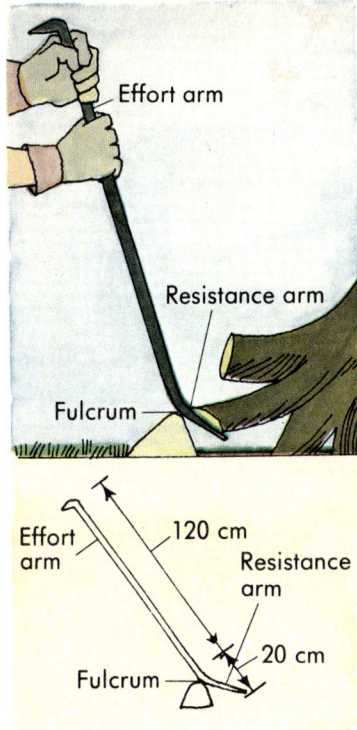

How a lever works

For Practice

Use the equation for a lever's M. A. to solve these problems.
- A crowbar 8 m long is used to pry up some boards. The fulcrum of the crowbar is 1 m from the boards. What is the M. A. of the crowbar?
- A seesaw 10 m long is placed on a fulcrum that is 2 m from the resistance. What is the M. A. of this seesaw?

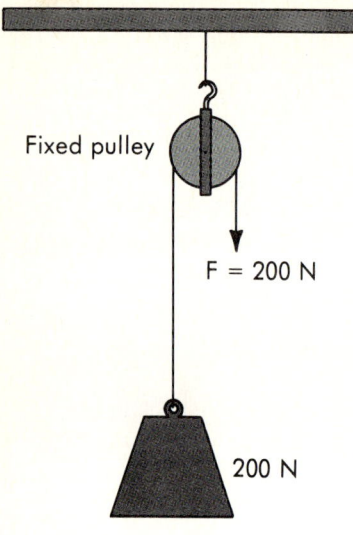
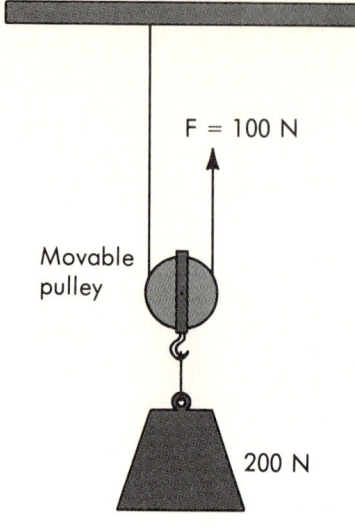
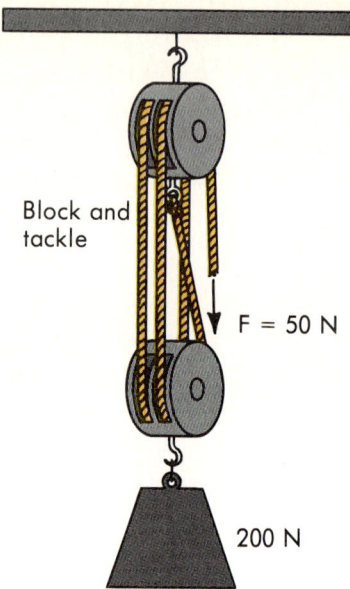

Types of pulleys

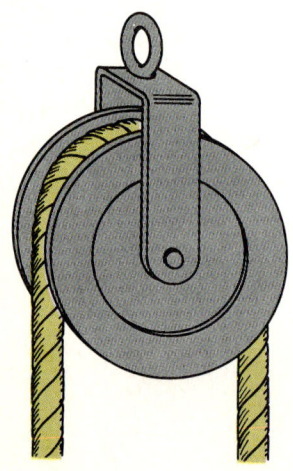

Pulleys Are Simple Machines

A **pulley** is a simple machine that consists of a grooved wheel over which a rope passes. The picture at the left shows a common pulley. Some pulleys change the direction of the force. The pulley on a flagpole raises and lowers the flag. If you pull down on the rope, the flag goes up. The pulley on the flagpole is an example of a fixed pulley. It can turn, but it cannot move up or down. A fixed pulley is always used to change the direction of a force. A fixed pulley helps us to do work even though it has a mechanical advantage of 1.

A movable pulley, such as the one shown above, increases the amount of force. As the pulley moves along a rope, it moves with the object that is the resistance. Using more movable pulleys increases the mechanical advantage.

Sometimes a combination of pulleys is used to do work. A block and tackle, also shown above, is a pulley system that consists of both fixed and movable pulleys. A block and tackle can lift an engine from a car or raise a scaffold along the side of a building. A block and tackle allows a single person to lift very heavy objects.

Wheel and Axle Machines

Doorknobs, eggbeaters, pencil sharpeners, and screw drivers are another kind of simple machine. Because each consists of a wheel attached to an axle, each is called a **wheel and axle.** The diagram shows how the wheel and axle move together. If you push on the wheel, the axle also turns. If you rotate the axle, the wheel rotates too.

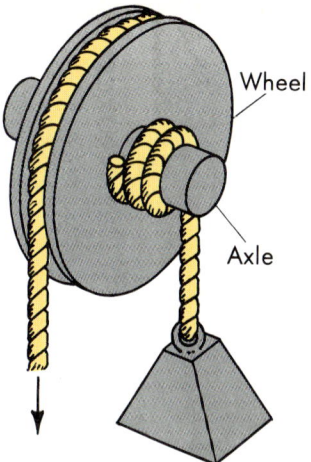

You can think of the wheel and axle as a special kind of lever. The distance around the wheel is equal to the length of the effort arm. The distance around the axle is equal to the length of the resistance arm. The bigger the wheel, as compared to the axle, the larger the mechanical advantage.

The drawing below shows how a screwdriver acts as a wheel and axle. The handle is the wheel, and the blade is the axle. When you turn the handle, the blade turns too. The force you use on the handle is increased. The blade exerts this increased force to tighten the screw.

Review It

1. What are the main parts of a lever?
2. How do you find the mechanical advantage of a lever?
3. How do fixed and movable pulleys change a force?
4. Explain how a wheel and axle increases force.

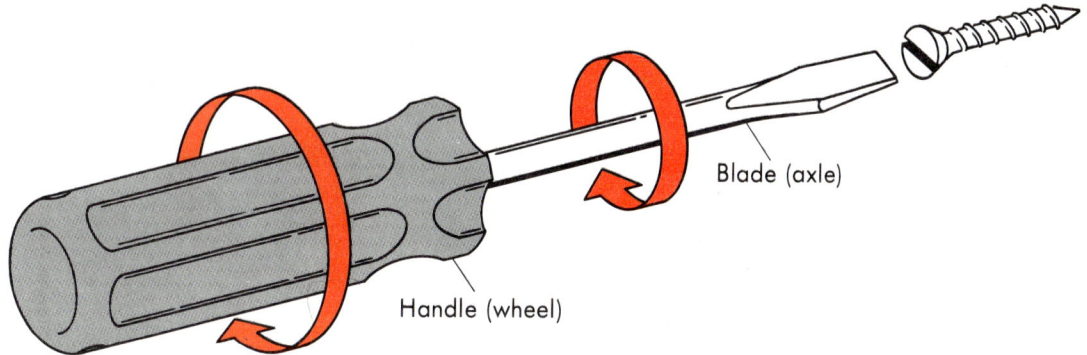

Activity

Using Levers and Pulleys

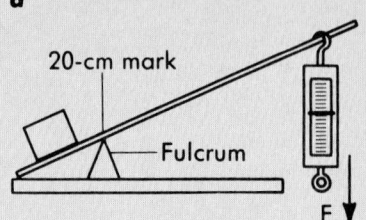

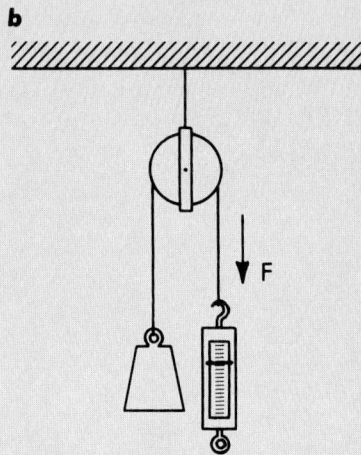

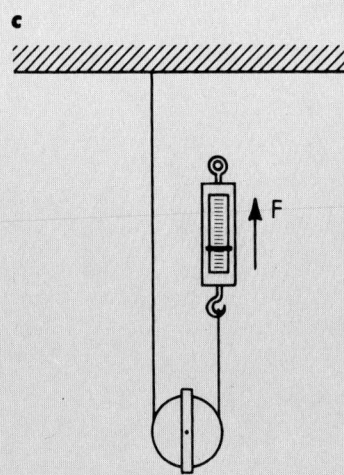

Purpose
To experiment with different kinds of levers and systems of pulleys.

Materials
- meter stick
- wedge-shaped block of wood to use as a fulcrum
- spring scale
- 50-g, 100-g, and 200-g masses
- pulley
- string
- scissors

Procedure

Part A
1. Set up the lever system, as in *a*. Place the fulcrum under the 20-cm mark in the meter stick.
2. Fasten the 50-g mass to the 0-cm mark on the stick. Pull down on the 100-cm mark with the spring scale until the meter stick is level. Record the force needed to lift the 50-g mass.
3. Repeat step 2, using the 100-g mass and then the 200-g mass. Record the spring scale reading.
4. Place the fulcrum under the middle of the meter stick.
5. Place the 200-g mass on one side of the fulcrum and the 50-g mass on the other side.
6. Move the two masses until the system is balanced. Record the distance of each mass from the fulcrum.

Part B
1. Set up a fixed pulley, as in *b*.
2. Hang the 200-g mass from one end of the string.
3. Pull down on the other end with the spring scale. Record the scale reading.
4. Set up a movable pulley, as in *c*.
5. Use the 200-g mass as the resistance.
6. Pull up on the loose end of the string with the spring scale.
7. Record the scale reading.

Analysis
1. Where is it best to place the fulcrum of a lever when the mass is heavy?
2. What kind of pulley changes the direction of the force exerted?

Did You Know?

The Bicycle—A Combination of Simple Machines

You can go for a ride on today's science lesson. We are talking about your bicycle, of course. Many simple machines make up a bicycle.

Look at the two sprockets (gears) on the bicycle shown. They are connected by a chain. One sprocket is in the front, attached to the pedals. The other is in the back, attached to the rear wheel. Both sprockets are wheel and axle machines. Notice the path made by a bicycle pedal as it turns once. This path is the wheel part of the machine. Look at the sprocket itself. This disk is attached to the axle part of the machine.

These front and back wheel and axle machines work together. They change the force of your pedaling into the motion of the back wheel to send you forward.

As you pedal the bicycle, the axle turns. The radius of the pedal and the radius of the sprocket are different lengths. So your foot needs to push with less force while the sprocket produces a greater force. This increased force is transferred to the back sprocket through the chain. The force on the back sprocket is transferred to the wheel through the spokes. Then the back wheel turns.

The back wheel and axle can increase speed. As more force is supplied to the back sprocket, the back wheel turns faster.

Your bicycle contains other simple machines. Screws attach the seat to the bicycle. Look at the hand brakes. You use the brakes as levers when you squeeze them. The fulcrum is the place where the two parts of the brake are joined. Each part is another lever.

For Discussion
1. What is the job of the front sprocket on a bicycle? the back sprocket?
2. What other simple machines can you find in the bicycle shown?

101

6–3 Forms of Inclined Planes

A hill and a set of stairs are one kind of simple machine. Cutting tools, such as knives, scissors, and saws, contain a similar kind of simple machine. As you read about such machines, answer the following questions:

a. How do inclined planes help us?
b. Where are wedges used?
c. What do we mean by the pitch of a screw?

Inclined Planes

Loading the heavy box onto the dock, as in the drawing, could be a difficult job. The dock worker could lift the box onto the dock. But sliding the box up the plank would be an easier job.

An **inclined plane** is a flat surface with one end higher than the other. The plank is an inclined plane.

It is easier to push, pull, roll, walk, or drive up a long, gentle slope than a short, steep one. On both slopes, you do the same total amount of work. But you exert less force at any one time on the long slope than on the short one.

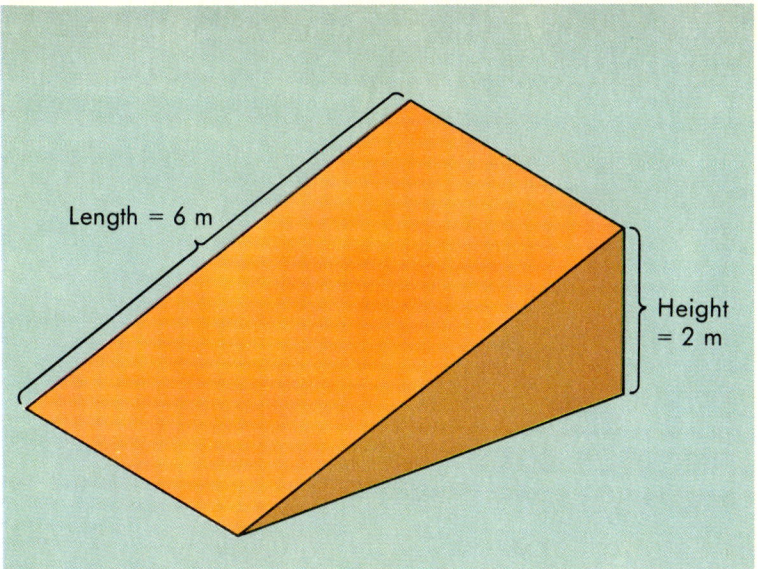

For Practice

Solve these problems by using the equation for an inclined plane's M. A.
• A steep ramp is 20 m long. What is its M. A. if the upper end of the ramp is 5 m higher than its lower end?
• In one 100-m stretch, the upper end of a highway is 2 m higher than the lower end. What is the highway's M. A.?

The longer an inclined plane is in relation to its height, the larger the mechanical advantage is. To find the mechanical advantage of an inclined plane, divide the length of the plane by its height. The mechanical advantage of the inclined plane in the drawing is

$$\text{M.A.} = \frac{\text{length}}{\text{height}}$$
$$= \frac{6 \text{ meters}}{2 \text{ meters}}$$
$$= 3.$$

Wedges Are Inclined Planes

A **wedge** is an inclined plane with either one or two sloping sides. The cutting edge of a wood chisel is a wedge with only one sloping side. A knife blade is a wedge with two sloping sides. Each sloping side of a wedge is an inclined plane.

The ax in the picture is a wedge used to split a log. The person hammers the wedge into the log. The wedge forces the log apart. Saws, scissors, and other cutting tools are machines with wedges.

Tools containing wedges

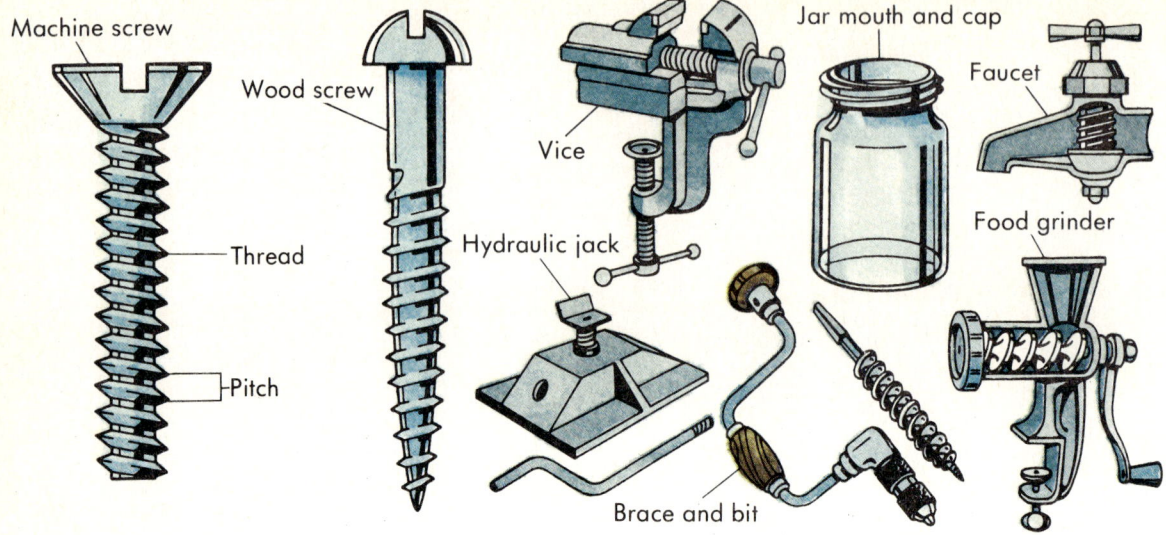

Some uses of screws

Making a screw from an inclined plane

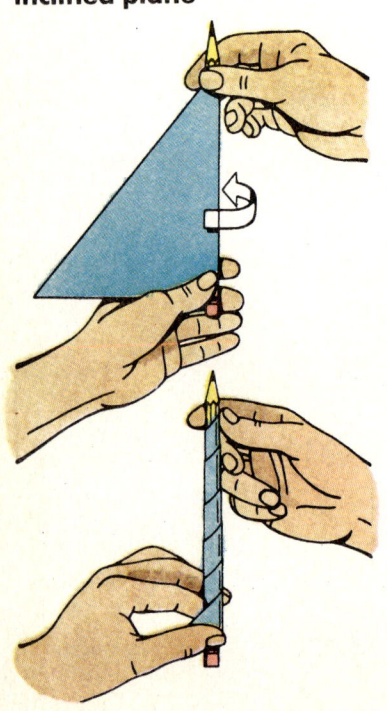

A Screw Is an Inclined Plane

The pictures above show some uses of **screws,** which are spiral inclined planes. You can demonstrate this fact for yourself by wrapping a paper inclined plane around a pencil, as shown at the left. The inclined plane forms a spiral path around the pencil.

The spiral path is the ridge that runs around a screw. The ridge is called the **thread.** The number of threads in a given length is the **pitch** of the screw.

The screw works by increasing and transferring the force that is exerted on the circumference of the screw. When you turn a screw, the direction of the force changes.

The mechanical advantage of a screw depends on its pitch and circumference. If the screw has a large circumference and a large pitch, this simple machine will increase a force many times.

Review It

1. How does an inclined plane help us do work?
2. Give three examples of wedges.
3. How is a screw an inclined plane?

Activity

Experimenting with Inclined Planes

Purpose
To learn how an inclined plane increases force.

Materials
- string
- empty cigar box
- spring scale (1000 g)
- 1-m-long board
- brick or several books
- meter stick
- scissors

Procedure

Part A
1. Support the board so that one end is 10 cm higher than the other end.
2. Carefully put 2 small holes in the smallest side of the cigar box.
3. Thread the string through them, and knot it to make a loop.
4. Place the box as in *a*.
5. Attach a spring scale to the string and slowly pull the box up the ramp. Record the force needed to move the box.
6. Repeat step 3 with the height of the ramp at 20, 30, and 40 cm. Record the force required for each height.

Part B
1. Place the box on the floor. Use the spring scale to lift the box 10 cm off the floor, as in *b*.
2. Record the amount of force needed to lift the box.
3. Repeat steps 1 and 2, raising the box 20, 30, and 40 cm. Record the force needed in each case.

Analysis
1. In Part A, did you exert the most force when the board's slope was more or less steep?
2. How did the force needed to pull the cigar box up the board compare to the force needed to lift it straight up?

a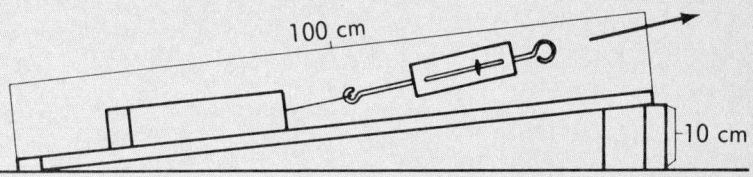

b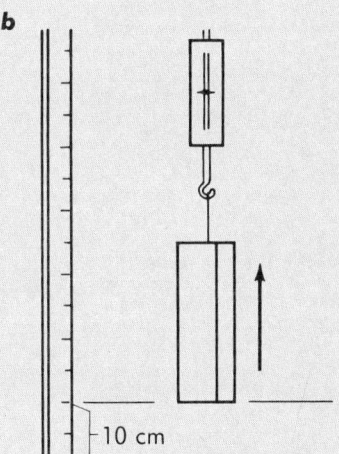

6–4
Compound Machines and Efficiency

After a few turns of a machine, you can have a pencil with a sharp point. A pencil sharpener is a combination of simple machines. Think about the following questions as you read about these tools:

a. What is a compound machine?
b. How do you calculate the efficiency of a machine?

Using Compound Machines

Two or more simple machines are often combined to make one machine. **Compound machines** are combinations of simple machines. The shovel, hoe, ax, scissors, and pencil sharpener shown are examples of compound machines.

In a shovel, hoe, or ax, the handle serves as a lever, and the blade is a wedge. A pair of scissors has two levers with wedge-shaped blades. The pencil sharpener in the picture consists of a wheel and axle and two screws. This type of wheel and axle machine is a crank. When you turn a crank to sharpen a pencil, the wedge-shaped threads of the two screws also turn. The mechanical advantage of a compound machine depends on the mechanical advantages of all its simple machines.

Compound machines that have many parts are complex machines. Automobiles, tractors, and power shovels are just a few of the many complex machines. Each is made of many simple and compound machines.

Have You Heard?

A screw and a wheel and axle make up the compound machine called a jackscrew. The circumference and pitch of the screw are very large. The circumference of the wheel is large compared to that of the axle. So jackscrews are strong enough to lift houses.

Some compound machines

Machines Are Not Completely Efficient

The man in the picture is using a simple machine—an eggbeater. He gets the same amount of work out of the machine as he puts into it. However, not all the work the machine does is *useful* work. Some effort is always wasted by doing useless work, such as overcoming friction.

Complex machines have many parts. All the movable parts produce friction. Because of this friction, less useful work is obtained from large, complex machines.

The **efficiency** of a machine is the amount of useful work obtained compared to the amount of work put into it. No machine is perfectly efficient. To find the efficiency of a machine, divide the useful work by the work put in. To find the percentage of useful work, multiply this number by 100%. For example, if you put 100 newton-meters of work into a machine, and the machine does 15 newton-meters of useful work, its efficiency is

$$\text{Efficiency} = \frac{\text{useful work}}{\text{work put in}} \times 100\%$$
$$= \frac{15 \text{ newton-meters}}{100 \text{ newton meters}} \times 100\%$$
$$= 15\%.$$

An efficiency of 15% means that 85% of the work put into the machine was wasted, probably in overcoming friction.

Many large machines have a very low efficiency. But if it were not for these complex machines, the jobs they perform might be impossible. Automobiles are usually less than 10% efficient, but they help us travel great distances on the ground quickly.

For Practice

Solve these problems using the efficiency equation.
• A complicated new machine produces 6 N•m of useful work. It takes 60 N•m of work to make the machine run. What is its efficiency?
• Riding your new bicycle requires 48 N•m of work. You can get 12 N•m of useful work out of the bicycle. What is its efficiency?

Review It

1. Describe a compound machine.
2. Explain why machines are not 100% efficient.

Chapter Summary

- Work is done when an object moves at least partly in a direction parallel to the direction of the force exerted on it. (6–1)
- Work can be calculated using work = force × distance. (6–1)
- Levers are simple machines with a fulcrum, effort arm, and resistance arm. (6–2)
- The mechanical advantage of a machine tells how much the machine will multiply force. (6–2)
- Pulleys are simple machines that can be grouped as fixed, movable, or block and tackle. (6–2)
- A wheel and axle is a simple machine made of a wheel with a rod through its center. (6–2)
- An inclined plane is a flat, slanting surface. (6–3)
- Wedges and screws are a special kind of inclined plane. (6–3)
- Compound machines are a combination of simple machines. (6–4)
- Machines are not 100% efficient. (6–4)
- Efficiency can be calculated using efficiency = (useful work ÷ work put in) × 100%. (6–4)

Interesting Reading

Adkins, Jan. *Moving Heavy Things*. Houghton Mifflin, 1980. Shows how to lift and move heavy objects, using the ideas behind simple machines.

James, Elizabeth, and Barkin, Carol. *The Simple Facts of Simple Machines*. Lothrop, 1975. Explains the six basic machines and demonstrates their uses.

Leek, Stephen, and Leek, Sybil. *The Bicycle—That Curious Invention*. Nelson, 1973. Traces the history of the bicycle, dealing with its scientific and social aspects.

Questions/Problems

1. Pam and Dan are building a snowman. They lift a block of snow that weighs 200 N 1.8 m off the ground. How much work do they do?
2. Shears for cutting metal have long handles and short blades, while shears for cutting paper have short handles and long blades. Why?
3. Can a machine produce more force than is put into it? Explain. Can it produce more work than is put into it? Explain.
4. A road over a mountain usually winds back and forth instead of going straight up the side. Explain why.
5. A company owns two steel ramps. One is 10 m long, and the other is 15 m long. Which will have the greater mechanical advantage when it is used as an inclined plane? Explain your answer.
6. Would you use a thin wedge or a thick wedge to split a log? Why?
7. What is the efficiency of an engine that produces 12 N•m of useful work from the 120 N•m of work put into it?

Extra Research

1. Collect pictures of these simple machines: a broom, a shovel, a wheelchair ramp, an elevator cable, and a saw. Tell what kind of machine each is and label the important parts.
2. Visit an automobile repair shop or a machine shop to find out how machines are used.
3. Choose a compound or complex machine that you can examine carefully. List all the simple machines you can find in it. Some compound or complex machines you might examine are a typewriter, an alarm clock, a nutcracker, and a corkscrew.

Chapter Test

A. Vocabulary Write the numbers 1–10 on a piece of paper.
Match the definition in Column I with the term it defines in Column II.

Column I

1. simple machine made of a rope over a grooved wheel that changes the amount or direction of a force
2. tools containing only one or two parts
3. can be found using force and displacement
4. compares useful work to the work put into a machine
5. supporting point of a lever
6. number of times a machine increases force
7. unit of measurement for work
8. combination of several simple machines
9. sloping flat surface
10. simple machine consisting of a rigid bar and support

Column II

a. compound machine
b. efficiency
c. fulcrum
d. inclined plane
e. lever
f. mechanical advantage
g. newton-meter
h. pulley
i. simple machine
j. work

B. Multiple Choice Write the numbers 1–10 on your paper.
Choose the letter that best completes the statement or answers the question.

1. A loading ramp is a typical example of a(an)   a) lever.   b) inclined plane.   c) wheel and axle.   d) compound machine.

2. The effort arm of a lever is the distance from the force to the   a) pitch.   b) wheel.   c) resistance.   d) fulcrum.

3. Work is done in all of the following cases *except*   a) pushing on a locked door.   b) raising a window.   c) walking through a revolving door.   d) climbing stairs.

4. A pencil sharpener is an example of a(an)   a) simple machine.   b) inclined plane.   c) compound machine.   d) pulley.

5. The greatest amount of force can be exerted by a wheel and axle if the wheel is 50 cm in circumference and the axle is   a) 5 cm in circumference.   b) 10 cm in circumference.   c) 15 cm in circumference.   d) 100 cm in circumference.

6. A pitcher's arm moves for 1 m before he releases a ball with 5 N of force. How much work is done?   a) 1 N•m   b) 5 N•m   c) 10 N•m   d) 2 N•m

7. A power shovel is an example of a(an)   a) simple machine.   b) efficient machine.   c) complex machine.   d) none of the above.

8. The spiral path of a screw is the   a) space.   b) thread.   c) pitch.   d) circumference.

9. Riding a bicycle requires 44 N•m. You get 11 N•m of useful work out of it. What is its efficiency?   a) 20%   b) 4%   c) 40%   d) 25%

10. A pair of scissors is an example of a(an)   a) wedge.   b) lever.   c) inclined plane.   d) a, b, and c.

Chapter 7
Energy and Power

The picture shows that lights, moving cars and trucks, and having many people around are part of our daily lives. But lights cannot shine, cars and trucks cannot move, and people cannot function without energy. Lights get their energy from electricity. Most cars and trucks get their energy from gasoline or diesel fuel. And people get their energy from food.

This chapter explains the scientific meaning of energy, describes forms energy can take, and explains the difference between energy and power.

Chapter Objectives

1. Describe the relationship between work and energy.
2. Discuss how kinetic energy is related to mass and velocity.
3. Explain potential energy and the meaning of the law of conservation of energy.
4. List six forms of energy.
5. Define power, and distinguish between energy and power.

7–1 Defining Energy

Sometimes you feel as though you do not have enough "energy" to get up in the morning. Just as *work* has a slightly different meaning in science from its meaning in everyday speech, so does the word *energy*. As you read about the scientific meaning of energy, keep in mind:

a. How are energy and work related?
b. What is a joule?

Energy and Work Are Related

The tennis player in the picture hits the ball with her racquet. Because she moves the ball through some distance as she hits it, she does work. To do work, she needs energy. Scientists define **energy** as the ability to do work. The tennis player gets her energy from the food she eats. She uses energy to breathe and move around, and she gives some of her energy to the ball.

Have You Heard?

How much is one joule of energy? When you ride your bicycle one block, you use about 90,000 joules, or 90 kilojoules, of energy.

Energy and work are closely related. We give an object energy by doing work on it. For example, the tennis player did work on the ball. This work gave the ball energy to move. So work is the transfer of energy.

Measuring Energy

The engineers in the picture are designing a new car. They need to know how much energy the engine must supply to move the car at different speeds.

Because work is the transfer of energy, we can measure energy by measuring the work done. For example, if a car engine does 1,000 newton-meters of work, the engine puts out 1,000 newton-meters of energy.

The same units are used for measuring work and energy. One newton-meter is called a **joule** (joul or jül), after the scientist James Joule. The capital letter J is the symbol for the joule.

Review It

1. Define energy in terms of work.
2. What is the unit for measuring energy?

7–2 Kinetic Energy

A speeding car has a lot of energy. If the car hits an object and stops, the car releases that energy and causes a lot of damage. As you read about the energy of moving objects, answer the following questions:

a. What is kinetic energy?
b. How can we calculate kinetic energy?

Moving Objects Have Kinetic Energy

Wind, running water, a moving car, and a swinging hammer all have energy. An object has energy because of its motion. **Kinetic** (ki net′ik) **energy** is the energy of motion. If you try to stop a moving object, you will feel its kinetic energy. All the objects shown have kinetic energy due to their motion.

Examples of objects having kinetic energy

Calculating Kinetic Energy

Kinetic energy depends on both the mass and the speed of an object. Look at the picture of the hammer about to hit the nail. The hammer does work when it drives the nail into the wood. Swinging the hammer faster or using a heavier hammer at the same speed would push the nail farther and do more work. Increasing either speed or mass increases the hammer's kinetic energy.

We can calculate an object's kinetic energy by multiplying the object's mass by its velocity squared:

$$\text{Kinetic energy} = \tfrac{1}{2} \times \text{mass} \times \text{velocity} \times \text{velocity}.$$
$$\text{K.E.} = \tfrac{1}{2} mv^2.$$

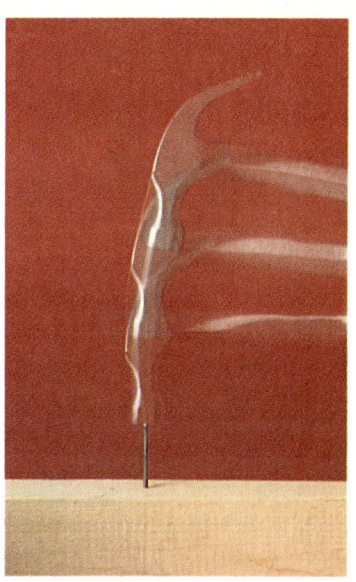

This equation shows that doubling an object's mass doubles its kinetic energy. Doubling the velocity makes the kinetic energy four times as great. So swinging a hammer twice as fast uses more energy and does more work than swinging a hammer with twice as much mass at the same velocity.

Suppose you pedal a 20-kilogram bicycle at 2 meters per second. The bicycle's kinetic energy is

$$\begin{aligned}
\text{K.E.} &= \tfrac{1}{2} mv^2 \\
&= \tfrac{1}{2} \times 20 \text{ kilograms} \times (2 \text{ meters/second})^2 \\
&= 40 \text{ kilogram-meter}^2/\text{second}^2 \\
&= 40 \text{ joules.}
\end{aligned}$$

You can see that pedaling at 1 meter per second (half as fast) would produce a kinetic energy of only 10 joules (one-fourth as much). Notice also that the unit from the kinetic energy equation is the same as the joule:

$$\begin{aligned}
1 \text{ joule} &= 1 \text{ newton-meter} \\
&= (1 \text{ kilogram-meter/second/second}) \times 1 \text{ meter} \\
&= 1 \text{ kilogram-meter}^2/\text{second}^2.
\end{aligned}$$

Review It

1. When does an object have kinetic energy?
2. How much does the kinetic energy of a bicycle increase if you triple its velocity?

For Practice

Use the equation for kinetic energy to solve the following problems.
- A girl whose mass is 40 kg skates around a rink at a speed of 12 m/s. How much kinetic energy does she have?
- A baseball of mass 0.08 kg is thrown at a speed of 20 m/s. How much kinetic energy does the ball have?

7-3 Energy Is Conserved

Energy might change its form, but it does not disappear. We can add energy to an object or take energy away from it, but the total amount of energy does not change. As you read about the total amount of energy, keep the following questions in mind:

a. What is the law of conservation of energy?
b. How does friction affect changes in energy?

The Law of Conservation of Energy

If you consider all the objects in a group of objects that are giving off or taking in energy, you find that the total amount of energy stays the same. This fact has been observed so often that it is called the **law of conservation of energy.**

The skier below hiked up a hill. She gained energy because she did work against the force of gravity. The skier stopped at the top of the hill. Because she was not moving, she had no kinetic energy. But she still had the energy she gained from walking up the hill. This energy, called **potential** (pə ten′shəl) **energy,** is the energy an object has because of its position. When the skier moves down the hill, as shown in the photograph, this potential energy changes into kinetic energy.

The energy the skier gains is the work she does to reach the top of the hill. So the work is equal to her potential energy at the top of the hill. The force she acts against is gravity pulling on her—her weight. The distance she moves is the height of the hill, 10 meters. If her mass is 50 kilograms, you can calculate her potential energy using

$$\text{P.E.} = \text{work done} = F \times d = (m \times g) \times h$$
$$= 50 \text{ kilograms} \times 10 \frac{\text{meters/second}}{\text{second}} \times 10 \text{ meters}$$
$$= 5{,}000 \text{ joules}.$$

Look at the drawings of the skier above. At the top of the hill, she has potential energy. So 5,000 joules is her total energy gain. As she skies down the hill, some of her potential energy is changed to kinetic energy. Since energy is conserved, the skier has 5,000 joules of energy at any place down the hill. Some of the energy is kinetic. The rest is potential. When she reaches the bottom of the hill, she has no potential energy, because h = 0 meters. Because energy is conserved, the energy she gained in climbing the hill must now be kinetic energy. She will travel at whatever speed is needed to give her 5,000 joules of kinetic energy.

For Practice

Use the potential energy equation to answer these questions.
- Ron carries a bowling ball weighing 70 N up a flight of stairs 20 m high. How much potential energy has the bowling ball gained?
- A table tennis ball weighs 0.25 N and rests on the window ledge of a building 600 m above the street. How much more potential energy does the ball have than it would have had if it were on the street?

Friction Wastes Energy

The example of the skier and the hill shows that all the skier's potential energy is changed into kinetic energy. This change would be true only if no friction were present. However, friction wastes some of the potential energy. Friction turns some potential energy into heat before it can become kinetic energy. The law of conservation of energy still holds, but we must remember that the total amount of energy can appear in several ways.

The lower the force of friction, the less energy is wasted and the more potential energy becomes kinetic energy. An object has more kinetic energy the faster it goes. So the less friction that is present, the faster an object goes. The roller skates in the picture, with new plastic wheels, go faster than old roller skates with metal wheels because the new skates lose less energy through friction.

Review It

1. What is true about an object's total amount of energy?
2. Does friction affect the law of conservation of energy?

Activity

Investigating Kinetic and Potential Energy

5. Measure the distance the carton moves along the floor, as in c.
6. Repeat steps 4 and 5 three times, raising the ramp to 20 cm, 30 cm, and 40 cm above the floor.
7. Weigh the ball and the milk carton separately.
8. Calculate the potential energy of the ball when it is 10 cm, 20 cm, 30 cm, and 40 cm above the floor.

Analysis
1. As the ball's potential energy increases, what happened to the amount of work done on the milk carton? How do you know?
2. Explain how the ball's kinetic energy changed as it rolled down the ramp.

Purpose
To show how potential and kinetic energy can do work.

Materials
- tennis ball or baseball
- board to use as a ramp (at least 1 meter long)
- meter stick
- books or pieces of wood to elevate the ramp
- empty liter or half-liter milk carton
- chalk
- balance

Procedure
1. Cut the top and one side off the milk carton, as in a.
2. Support one end of the board so that it is 10 cm above the floor, as in b.
3. Place the milk carton 10 cm from the end of the ramp with the open side down and the open end facing the ramp. Draw a chalk line along the closed end of the carton, also as in b.
4. Hold the ball at the top of the ramp. Release the ball and let it roll down the ramp and into the milk carton.

a

b
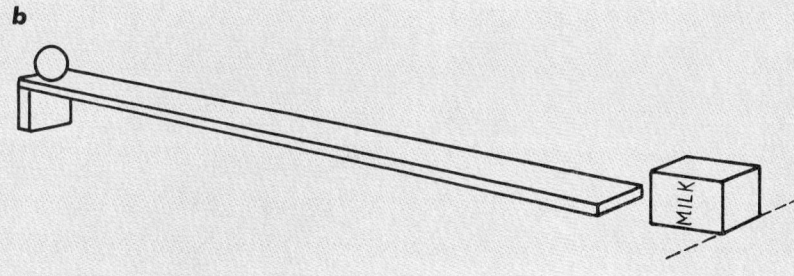

c
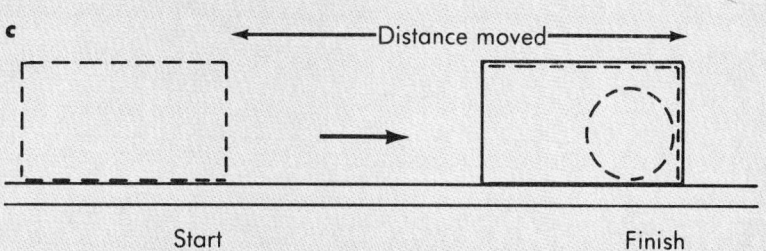

7–4
Forms of Energy

Energy comes into our homes as food, as electricity, and in other forms. Food feeds the people, and electricity runs the lamps and the televisions. As you read about forms of energy, keep the following questions in mind:

a. What are several different forms of energy?
b. How are the forms of energy related?

Energy Comes in Several Forms

A moving person or car has kinetic energy. When you climb or ride up a hill, you increase your potential energy. Both kinetic and potential energy are kinds of **mechanical energy,** which is energy an object has from its motions and the forces acting on it. Mechanical energy is also the energy of machines. The boxer's fist shown has mechanical energy.

If we could see the tiny particles that form objects, we could see that these particles are in motion. So each particle has kinetic energy. Each particle has very little mass, so its kinetic energy is small. But an object has so many particles that the total amount of this energy can be measured. The total amount of energy of all the particles of an object is its **internal energy.**

Electricity is a flow of certain tiny particles of matter. Electricity carries **electrical energy.** This form of energy is very convenient to use.

Energy can also be stored in matter as **chemical energy,** which depends on how the particles of matter are arranged. Therefore, chemical energy depends on the positions of objects and is a form of potential energy.

Energy can also be carried as waves through space in the form of **radiant** (rā′dē ənt) **energy.** Light, X rays, and radio waves are examples of radiant energy.

Nuclear energy is released when very small particles of matter split or combine. After the split or combination, the resulting small particles have a little less mass than the original particles. The rest of the mass is changed into energy. This process of changing mass into energy produces the radiant energy from our sun.

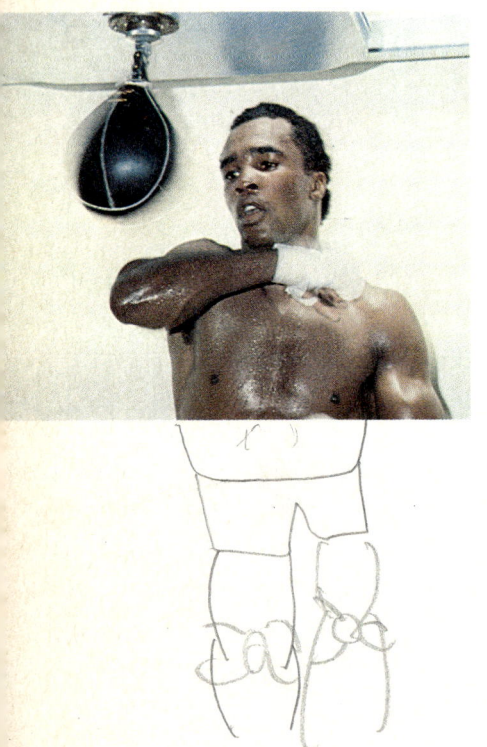

Energy Can Change Form

Nuclear energy is produced when mass changes into energy. Energy can also change into mass and into other forms of energy. For example, the solar oven shown above is cooking food. The sun's energy causes chemical processes to happen in the food. When you eat the food, your body changes the chemical energy stored in the food into the mechanical energy of your motion.

Plants use the sun's radiant energy to grow. Some of the energy is stored in the plants. Plants that died millions of years ago formed the oil, gas, and coal we use every day. The chemical energy we get from these fuels is another form of the sun's energy that was trapped inside plants. Driving a car or heating a house releases this energy.

Burning oil, gas, and coal heats water. Moving steam from the water can produce mechanical energy. Then we can use this energy to produce electricity. In our homes we might change the electricity into light. We change energy from one form to another often during a day.

Review It

1. What are six forms of energy?
2. Explain how energy can change form.

7–5 Measuring Power

Clearing snow from a path can take a long time if you use a shovel. If you use a snow blower, however, moving snow goes more quickly. As you read, keep these questions in mind:

a. What is power?
b. How can we determine energy from power ratings?

Power Involves Work or Energy and Time

In the picture below, Amy and Kate are shown taking care of their lawn. One week, Amy mows the lawn in one hour. The next week, Kate mows the lawn in 45 minutes. Because both girls pushed the lawn mower the same distance, each did the same amount of work. Amy and Kate differed only in the time they spent working.

In science, the term **power** means the rate at which work is done:

$$\text{power} = \frac{\text{work}}{\text{time}}$$

Because work is the transfer of energy, power = energy ÷ time. So power is also the rate at which energy is used.

Challenge!

The unit for power is named after James Watt, a Scottish inventor who lived in the eighteenth century. Find out what he learned about work, energy, and power.

122

If the work done by Amy or Kate is 6,000 joules, we can find the power that each girl used to mow the lawn:

$$\text{For Amy, power} = \frac{6{,}000 \text{ joules}}{3{,}600 \text{ seconds}}$$

$$= 1.7 \text{ joules/second.}$$

$$\text{For Kate, power} = \frac{6{,}000 \text{ joules}}{2{,}700 \text{ seconds}}$$

$$= 2.2 \text{ joules/second.}$$

One joule per second is called a **watt** (W). We measure power in watts.

Calculating Energy from Power

Amy and Kate's parents bought a power lawn mower that supplies an average of 10 watts of power. Now each girl mows the lawn in 10 minutes (600 seconds). We can determine how much work the lawn mower does:

$$\text{work} = \text{power} \times \text{time}$$
$$= 10 \text{ watts} \times 600 \text{ seconds}$$
$$= 6{,}000 \text{ watt-seconds.}$$

We can also say that the lawn mower uses 6,000 watt-seconds of energy.

In this example, the watt-second indicated that power (watt) is used for a certain length of time (second). The watt-second and kilowatt-hour measure energy used. The kilowatt (kW) and the megawatt (MW) are units used to measure large amounts of electrical power. In the picture, the meter shows how many kilowatt-hours of electrical energy a family uses.

Review It

1. How can you calculate power?
2. How is a watt different from a watt-second?

Challenge!

Another unit for power is the horsepower. Although it is not an SI unit, it is often used in measuring the power of engines. Find out where the name *horsepower* comes from. An encyclopedia should have the answer.

For Practice

Use the equation for power to solve these problems.
• A small wind-up truck supplies 5 W of power. If the truck travels for 30 seconds, how much work has it done? How much energy has it used?
• A light bulb uses 60,000 W·s of energy as it burns for 12 hours. How much power does the light bulb use?

Activity

Measuring Human Power

a

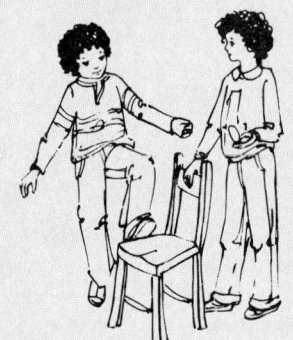

b

Purpose
To measure human power developed during common activities.

Materials
- timing device
- meter stick
- kilogram bathroom scale
- chair
- wastebasket
- spring scale
- masking tape

Procedure

Part A
1. Measure a 2-m length of floor with the meter stick.
2. Mark the length with a piece of tape at each end. Put a few pieces of tape along the length to make a straight-line guide.
3. Put the wastebasket at one end of your taped length and next to the guide line.
4. Pull the wastebasket with the spring scale so that you can read the amount of force you are exerting. You must keep the scale horizontal and move the basket in a straight line, as in a.
5. Practice step 4 until you can keep the scale pointer fairly steady as you pull.
6. Repeat step 4, keeping the pointer steady while your partner times how long you need to pull the basket through the 2-m length.
7. Record the scale reading and the time from step 6.
8. Repeat steps 6–7 twice. Average your force readings and your time readings, and record them.
9. Switch duties with your partner and repeat steps 3–8.

Part B
1. Find your mass by using a bathroom scale.
2. Measure and record the distance from the seat of a chair to the floor.
3. Ask your partner to hold the chair steady so you can step on it.
4. While your partner times you for 10 seconds, count how many times you step on the chair, as in b.
5. Switch duties with your partner, and repeat steps 1–3.

Analysis
1. Use work = force × distance to calculate the work you did in Part A. (Use your average force value.)
2. With the answer to #1 and your average time spent, calculate your power using power = work ÷ time.
3. Calculate the work you did in Part B using work = m × g × h (g = 10 m/s/s).
4. Calculate the power that you exert in Part B while you are stepping up onto the chair. Your time spent is 5 seconds, because you spent the other 5 seconds stepping down from the chair.

Did You Know?

Amusement Park Science

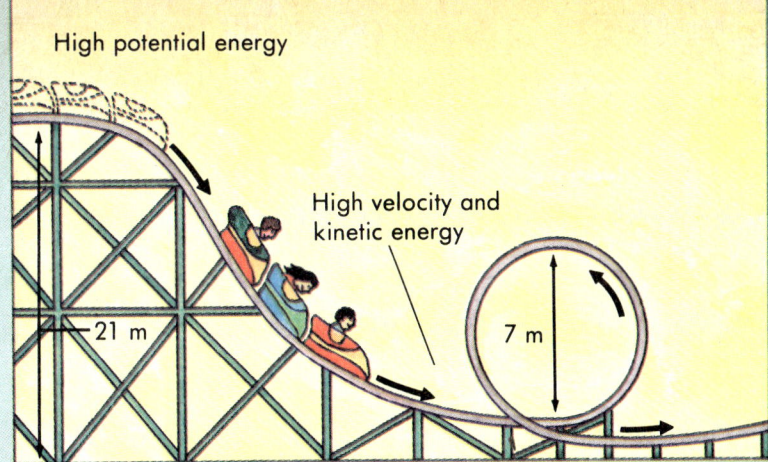

How about a trip to an amusement park for your next science lesson? Roller coasters, Ferris wheels, and merry-go-rounds are fun to ride. But did you know that science is responsible for the fun—and safety—of these rides?

For example, have you ever wondered what keeps a roller coaster going after it reaches the beginning of the first dip? No engine or other device powers it after that point. Yet the roller coaster may reach speeds of 80 kilometers per hour.

The roller coaster gains energy by working against gravity on the way up the hill. This potential energy changes into kinetic energy during the roller coaster's first dip. At the bottom of the dip, a lot of the roller coaster's energy is kinetic. Its energy is great enough to carry it up another hill, or even upside down and around, as the drawing shows.

Friction will waste some of the roller coaster's energy. The roller coaster will not be able to climb as high over later hills and still move fast enough to thrill its passengers. So the first dip must be the deepest.

Sometimes the roller coaster's energy is too great to go around a curve safely. So design engineers build the track with brakes to use up the roller coaster's kinetic energy. The whole system must be safe enough for people to ride.

Bumper cars also waste kinetic energy on purpose. Imagine what would happen to people in the cars if they bounced off each other as billiard balls do. Such collisions result because energy is conserved. So engineers design the cars with big, thick bumpers to use up kinetic energy. You still feel a collision, but not enough to hurt.

Electrical energy operates most amusement park rides. Some of this energy powers the rides' colorful lights. Consider a ride where the floor of a spinning barrel drops out from under the passengers. Such rides use around 20,000 watts to spin the barrel. Another 35,000 watts run the lights!

For Discussion
1. What principles do engineers keep in mind as they design amusement park rides?
2. Some people prefer to ride a roller coaster in which every seat is filled. Others prefer to ride a roller coaster that is almost empty. How do you think the mass of passengers affects the ride?

Chapter Summary

- Energy is the ability to do work. (7–1)
- Work is the transfer of energy. (7–1)
- The unit of measurement for work and energy is the newton-meter or joule. (7–1)
- Kinetic energy is an object's energy due to its motion. (7–2)
- Potential energy is an object's energy due to its position. (7–3)
- The law of conservation of energy states that energy can change form, but the total amount of energy remains the same. (7–3)
- Some forms of energy are electrical, chemical, nuclear, internal, radiant, and mechanical. (7–4)
- Power is the rate at which work is done or energy is used. (7–5)
- Power can be calculated using power = (work or energy) ÷ time. (7–5)
- One watt equals one joule per second. (7–5)

Interesting Reading

Bendick, Jeanne. *Putting the Sun to Work*. Garrard, 1979. Well-illustrated, clear explanations of solar energy principles, including simple experiments.

Kiefer, Irene. *Energy for America*. Atheneum, 1979. A general review of the nation's current energy situation, detailing the use of fossil fuel energy.

Kentzer, Michal. *Collins Young Scientist's Book of Power*. Silver Burdett, 1979. A discussion of electrical power and its effect on our life and environment. Includes simple experiments. (Be sure to ask your parent's permission before you do any experiments.)

Pollard, Michael. *How Things Work*. Larousse, 1978. A detailed description of all kinds of mechanical inventions, with clear pictures.

Questions/Problems

1. What kind of energy exists in each of the following: a) a package on a closet shelf; b) a rock rolling downhill in a rock slide; c) a baseball hit with a bat; and d) a mousetrap with its spring set.

2. An airplane takes off, flies, and lands. Describe any changes in energy during this process. Begin with the use of gasoline in the plane's engine and end with the plane coming to rest after landing.

3. A machine uses 100,000 J of energy.
   a) How much work can the machine do?
   b) If the machine operates with 25% efficiency, how much useful work can it do?
   c) How much energy is useless work?

4. A student throws a ball into the air with a kinetic energy of 100 J. Determine the amount of kinetic energy and potential energy the ball has when it is at 25, 50, and 75 percent of its maximum height.

5. You read under a 100-W bulb for 2 hours. How many kilowatt-hours of electricity do you use?

6. A light bulb uses 1,800,000 J of energy in 10 hours. What is its power?

Extra Research

1. Keep a list of the forms of energy that you use at school and home. After several weeks, determine which form of energy is used most. Which form of energy is most likely to be changed into more useful forms?

2. Record the number of kilowatt-hours registered on the electric meter in your house at the same time each day for the next 7 days. How much energy was used at the end of the week? What is the average amount of energy used in a day?

Chapter Test

A. Vocabulary Write the numbers 1–10 on a piece of paper. Match the definition in Column I with the term it defines in Column II.

Column I

1. energy from machines
2. rate of using energy or doing work
3. ability to do work
4. energy from the sun and stars
5. the total amount of energy remains the same
6. unit of power
7. energy due to an object's motion
8. energy due to an object's position
9. unit of energy and work
10. energy from the splitting or combining of some small particles

Column II

a. conservation of energy
b. energy
c. joule
d. kinetic energy
e. mechanical energy
f. nuclear energy
g. potential energy
h. power
i. radiant energy
j. watt

B. Multiple Choice Write the numbers 1–10 on your paper. Choose the letter that best completes the statement or answers the question.

1. As a rock falls off a cliff, its a) potential energy changes into kinetic energy. b) kinetic energy changes into potential energy. c) total energy decreases. d) total energy increases.

2. A 400-N girl climbs a 2-m high staircase. At the top, the amount of potential energy she has gained is a) 20 J. b) 160 J. c) 120 J. d) 800 J.

3. You can increase the kinetic energy of an object moving at a constant velocity by a) increasing its mass. b) decreasing its mass. c) lowering its height. d) decreasing its velocity.

4. The kinetic energy of a 2-kg object moving at 5 m/s is a) 25 J. b) 50 J. c) 10 J. d) 20 J.

5. Which kind of energy does a light bulb use? a) radiant b) electrical c) nuclear d) chemical

6. One newton-meter is also one a) watt. b) joule-second. c) joule. d) none of the above.

7. Burning coal is an example of a) electrical energy. b) nuclear energy. c) chemical energy. d) mechanical energy.

8. During an energy change, energy is a) increased. b) created. c) conserved. d) destroyed.

9. How much work can an electric pencil sharpener do if it uses 25 W for 5 minutes? a) 750 W b) 7,500 J c) 125 J d) 1250 W

10. What is the power of a record player if it uses 60,000 W·s for 50 minutes? a) 1200 W b) 1200 W·s c) 20 W·s d) 20 W

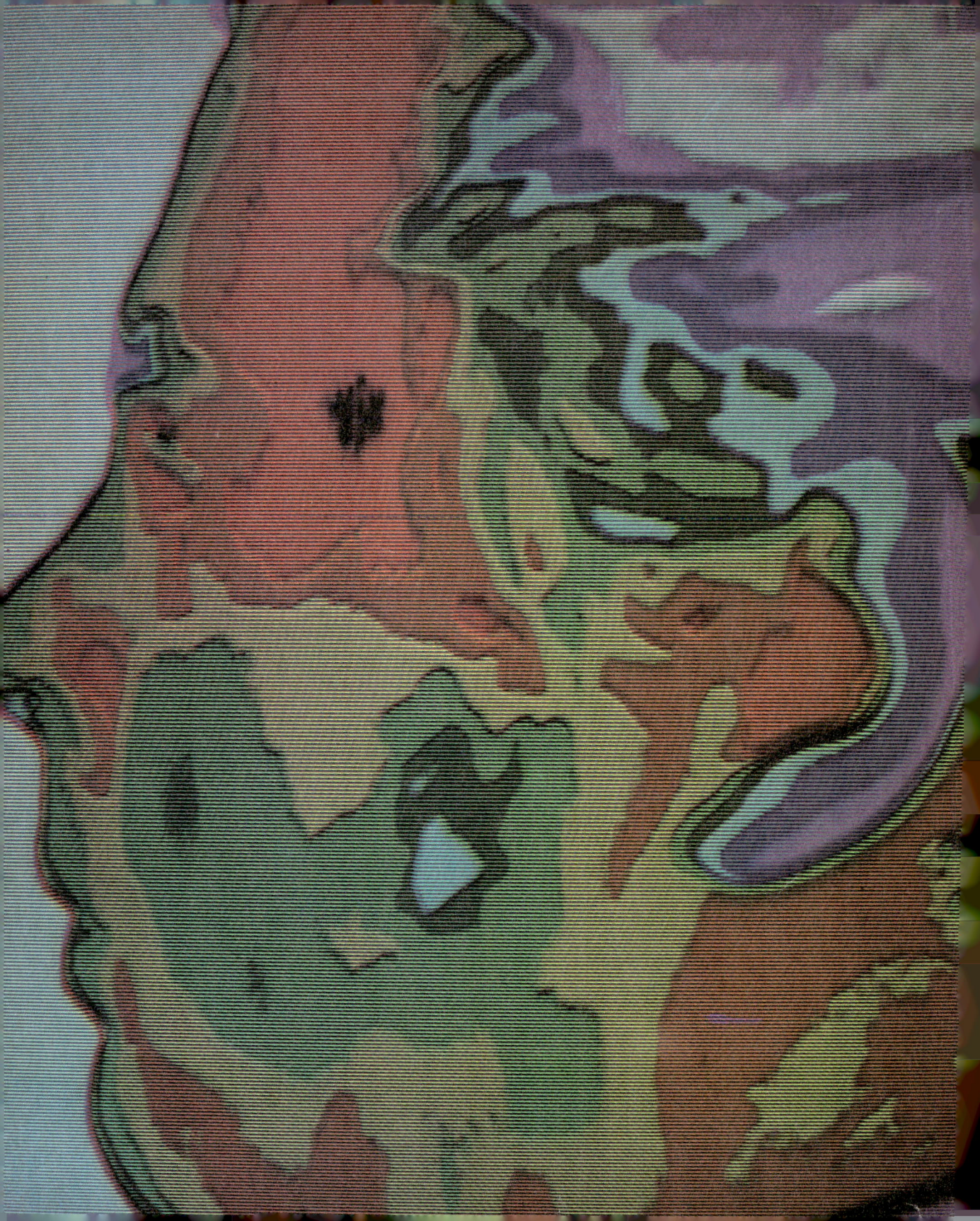

Chapter 8
Heat and Temperature

We usually think of "hot" and "cold" as something that we feel rather than see. But this photograph was taken with a special film that is sensitive to heat. The film records different temperatures and shows them as different colors. Hotter temperatures appear blue or purple. Warm temperatures look green or yellow. Cooler temperatures are red or orange. The picture shows us that different places on the human body give off different amounts of heat.

This chapter describes the nature of heat and temperature. It will describe the sources and explain the movement of heat. Finally, the chapter explains how heat affects different objects.

Chapter Objectives
1. Explain the difference between heat and temperature.
2. Explain how certain instruments detect and measure changes in heat and temperature.
3. Describe some sources of heat.
4. Explain the three methods of energy transfer.
5. Explain how heat affects the size of an object.

8–1
Defining Heat and Temperature

Heat is important in our lives. On cold winter days, the heating system warms your house and school. In the summer, everyone talks about the hot weather and asks for a temperature reading. Heat is also very important for cooking in the restaurant kitchen shown. As you read about heat, keep the following questions in mind:

a. What is the kinetic theory of matter?
b. How can we define heat and temperature?

Moving Particles Make Up Matter

When the temperature rises on a summer day, you feel that it is hot. If you walk into a cool, air-conditioned room, you feel that the temperature is lower. After a brisk winter walk, a warm bathtub of water will heat you better than a spoonful of very hot water. The hot water has a higher temperature, but the large amount of warm water will heat you faster. So heat and temperature are related, but they are not the same. To understand heat and temperature, you need to know more about matter.

All matter is made of tiny particles. These particles are constantly in motion. Both these ideas are part of the **kinetic theory of matter.**

The particles of matter do not all move at the same speed. Water particles move faster than gold particles, and air particles move faster than water particles at the same temperature. Also, the air particles are moving at a wide range of speeds.

Heat Differs from Temperature

Because the particles of matter are in motion, they have kinetic energy. These differences are shown in the drawing at the right. In any object, all the particles do not have exactly the same kinetic energy. We define **temperature** as a measure of the average kinetic energy of all the particles in an object or material. Particles at a high temperature have more kinetic energy. Particles at a low temperature have less kinetic energy. Heating a drop of water makes its particles move faster and increases their kinetic energy and temperature. Putting water in a freezer makes the particles move more slowly and reduces their kinetic energy and temperature.

Suppose you hold an ice cube in your hand. The particles in your warm hand move faster than the particles in the cold ice cube. As you hold the ice cube, energy is transferred. Fast-moving skin particles bump into slowly-moving ice particles and give them energy. The ice particles move faster and become "warmer." Your skin particles move more slowly and become "cooler." **Heat** is the amount of energy transferred from one place or object to another. An object at the same temperature as another object has more energy if it has more particles. So more heat can be transferred from a larger object than from a smaller one at the same temperature.

Temperature is a measure of average K.E.

Review It

1. What does the kinetic theory say about matter?
2. What is the difference between heat and temperature?

8–2
Detecting Heat and Temperature Changes

How can you tell how hot an object is? You could touch it, but touching very hot objects is not safe. You need some kind of measuring instrument. As you read more about measuring heat and temperature, remember the following questions:

a. How do we measure temperature?
b. How do we measure heat?

Measuring Temperature

Thermometers are instruments that measure temperature. All temperature scales have two reference points, one hot and one cold. The temperatures at which water boils and freezes are usually chosen as reference points. A certain number of equal steps, usually called degrees, occurs between the reference points. On the Celsius temperature scale shown, ice melts at 0° and water boils at 100°.

Sometimes the range of temperatures on a common thermometer is too small to measure an object's temperature. For example, a steelworker may need to measure the temperature of steel rods. The photograph shows a worker using an **optical pyrometer** (op′tə kəl pī rom′ə tər) to measure very high temperatures without touching the hot object.

The Celsius temperature scale

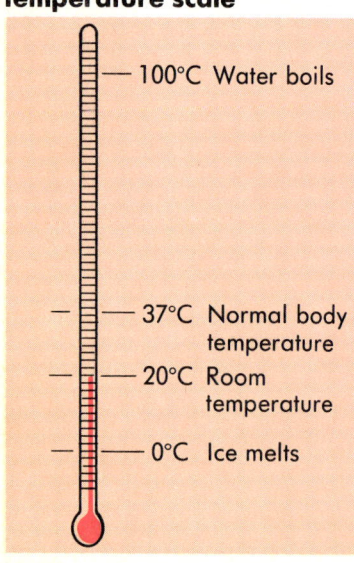

- 100°C Water boils
- 37°C Normal body temperature
- 20°C Room temperature
- 0°C Ice melts

Infrared "photograph" of some stores

An optical pyrometer operates on the idea that the color of light given off by a glowing object depends on its temperature. Warm objects appear reddish, hotter objects are yellow, and even hotter objects give off blue light.

An optical pyrometer measures one kind of radiant energy—light. A kind of radiant energy that is not visible is infrared energy. Hot objects also give off infrared rays. Measuring how much infrared energy an object gives off also measures the object's temperature. Even though we cannot see infrared rays, we can choose a different color to represent each range of temperatures measured. A picture of an object made with such colors is called a **thermogram** (thėr′mə gram). Some industries use thermograms to find weak spots in metal and plastic objects. Thermograms also show where heat leaks out of homes and other buildings. In the thermogram above, hot temperatures appear white. Warm temperatures look orange or red. Cold temperatures look dark blue or black.

Doctors use thermograms of people to detect some kinds of diseases. A higher than normal skin temperature might mean cancer in the tissue below that area of the skin. A lower than normal temperature might indicate poor blood flow in that area.

Have You Heard?

Absolute zero is the lowest temperature we can imagine a substance reaching. At this point, the particles of a substance would not move at all and would have no kinetic energy. We can never cool matter to this point. On the Celsius scale, absolute zero is −273.16°.

Measuring Heat

Thermometers measure the temperature of a material. But they do not measure how much heat a substance gives off or takes in. We need a different unit and instrument for measuring heat.

Heat is related to energy. Just as work is measured in joules, the unit of energy, so also is heat. For example, about 4.2 joules are needed to raise the temperature of 1 gram of water by 1°C.

Scientists use a device for measuring heat called a **calorimeter** (kal′ə rim′ə tər). This device is a closed container. Its outer shell keeps the surroundings from affecting the material or process scientists want to measure.

An important use of the calorimeter is to determine a material's ability to take in and give off heat. This property is called the **specific heat** of the material. Water has a high specific heat. It takes in and gives off heat very slowly. The drawings point out that a lake is much warmer than the air in winter and much cooler than the air in summer. The lake's large mass and water's high specific heat prevent the lake's temperature from changing quickly or by a large number of degrees.

Iron and aluminum have low specific heats. They take in and give off heat quickly. This property of iron and aluminum explains why these metals are used to make cooking utensils.

Have You Heard?

The *calorie* is an older unit for measuring heat. You have probably heard this term in connection with dieting —meaning the number of calories a certain food contains. Actually, the diet calorie, called a Calorie, is a kilocalorie. One calorie is equal to 4.18 joules. A candy bar may supply you with about 250 Calories (250,000 calories). You have to use up that much energy to not gain weight! Think about that the next time you have a sweet tooth!

Review It

1. Describe one instrument for measuring temperature.
2. What is a material's specific heat?

Activity

Investigating Temperature Changes

Purpose
To observe how objects with different temperatures affect each other.

Materials
- balance
- water at room temperature
- hot and cold water
- metal washers
- polystyrene cups
- thermometer
- 100-mL graduated cylinder

Procedure

Part A
1. Copy the chart shown.
2. Place 50 mL of hot water in one cup and 50 mL of cold water in another.
3. Measure and record the temperature of each sample of water.
4. Place the thermometer in the cold water and slowly add the hot water.
5. Wait one minute. Then record the temperature of the water.

Part B
1. Measure 50 mL of cold water into a cup. Record the water's temperature.
2. Measure 100 mL of hot water into another cup. Record the hot water's temperature.
3. Repeat Part A, steps 3–4.
4. Repeat Part B, steps 1–3, with 100 mL of cold water and 50 mL of hot water.

Part C
1. Measure 100 g of water at room temperature into a cup. Record the water's temperature.
2. Repeat step 1 for the hot water.
3. Estimate the temperature you will measure if you mix the water at the 2 different temperatures.
4. Mix the two cups of water. Wait one minute. Then compare the temperature you measured with your estimate in 3.

Part D
1. Measure 100 g of hot water into a cup and record the temperature.
2. Measure 100 g of small metal washers that are at room temperature.
3. Estimate the temperature you will get if the washers are added to the hot water.
4. Mix the washers and hot water. Wait one minute. Then record the temperature of the mixture.

Analysis
1. In Parts A and B, were the temperatures of the mixture closer to the original temperature of the hot water or the cold water? Explain.
2. In Parts C and D, did the liquid or the solid have a greater effect on the final temperature? Explain.

Data table

	Substance	Original Temperature	Final Temperature
Part A	Hot water (50 mL)		
	Cold water (50 mL)		
Part B	Hot water (100 mL)		
	Cold water (50 mL)		
	Hot water (50 mL)		
	Cold water (100 mL)		
Part C	Water (room T)		
	Hot water		
Part D	Hot Water Washers		

8-3 Producing Heat

Rubbing your hands together produces heat and warms your fingers. You are using mechanical energy to produce heat. As you study about ways to produce heat, remember the following questions:

a. What is a heat source?
b. How do the forms of energy produce heat?

Heat Sources

A **heat source** is anything that gives off heat. Most forms of energy produce heat. Directly or indirectly, the sun is the source of our energy. Therefore, the sun is also an important heat source. Any object or place of high temperature is a heat source. Heat moves from regions of higher temperature toward regions of lower temperature.

The Forms of Energy Produce Heat

Electrical energy is an important source of heat. Toasters, electric irons, hot plates, hot water heaters, and electric blankets use electrical energy to produce heat. These appliances contain a device that heats up when electricity passes through it, as the picture shows.

Chemical energy can produce heat. Chemical processes provide much of our energy. During chemical processes, substances release heat as they react with each other. Combustion is the chemical process of burning. The combustion of fuels such as coal, wood, fuel oil, gasoline, and natural gas produces heat and often light.

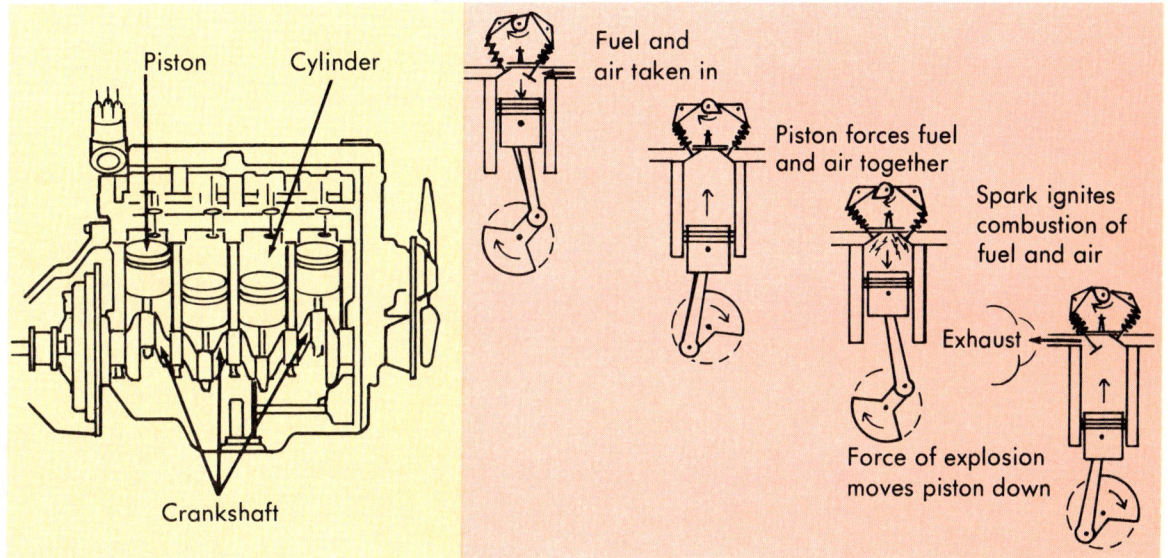

An internal combustion engine   **Piston cycle**

Combustion in a car's engine moves the pistons, as shown. This motion eventually moves the car. But combustion produces too much heat. So a car's cooling system removes this unwanted heat from places in the engine.

Mechanical energy also provides heat, usually from friction. Often friction causes harmful heat effects. For example, oil-well drills produce a lot of heat as they drill through solid rock. The drills must be flushed with water to remove this heat. Cooling the drill with water keeps the drill from melting.

Challenge!
Spontaneous combustion causes millions of dollars in fire damage every year. Use reference books to find out how spontaneous combustion happens.

Review It

1. How does a heat source's temperature compare with the temperature of its surroundings?
2. What are three ways of providing heat?

8-4
Energy Transfer

Energy from a stove reaches a frying pan and cooks the food in the pan. Energy from a furnace moves through all the rooms of a house. Energy from the sun travels through space and warms the earth. How does energy get from one place to another? As you learn about how energy moves, answer the following questions:

a. How does energy travel by conduction?
b. What is convection?
c. How is energy transferred by radiation?

Energy Travels by Conduction

If you put your feet on something warm, such as a heating pad, your feet will become warm. Heat is reaching your body directly. **Conduction** is the movement of energy from a source to an object by direct contact between them.

The kinetic theory of matter explains conduction. The particles in any material move and bump into each other. In the drawing, particles in the hot burner bump into particles in the bottom of the pan. The pan particles begin to move faster. The pan particles bump into and heat (transfer energy to) the particles above them. The top pan particles bump into the food particles and the food becomes hot. In this way, energy travels through the frying pan and cooks the food inside it.

Any material through which heat passes easily is a **conductor.** Most metals, such as aluminum and copper, are good conductors of heat. Other materials, such as cloth and wood, are not good conductors of heat. A poor conductor is an **insulator** (in′sə lā′tər). It prevents the movement of heat. Because wood and plastic are good insulators, we cover the handles of cooking utensils with them. People use insulators in their homes to keep heat inside during the winter and outside during the summer. Storm windows keep heat in the house because they trap air, which is a good insulator.

Cooking a steak with conduction

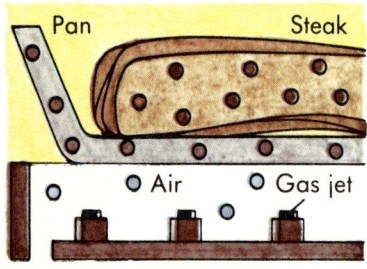
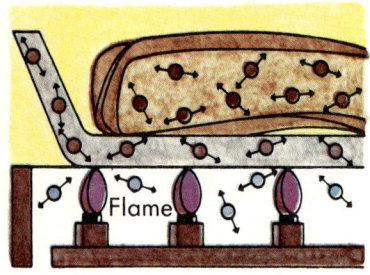

Energy Transfer by Convection

Placing your hands above a stove warms them too. Air moving up from the stove heats your hands. **Convection** is the transfer of energy when a large mass of liquid or gas moves from one place to another. The moving material flows in a **convection current.** Most air and water currents are convection currents. A temperature difference between two places causes convection currents.

The picture shows home convection currents in one kind of heating system. In the furnace, heated air particles begin to move faster, bump into each other, and move farther apart. So the hot air is less dense and rises above the cooler air. The hot air is forced through air ducts into a room. The hot air continues to rise above the cooler air already in the room, as the red arrows show. As the air mixes, the hot air becomes cooler and denser and sinks. The blue area shows this air. The cooler air returns through other air ducts to the furnace. The process of convection then begins again.

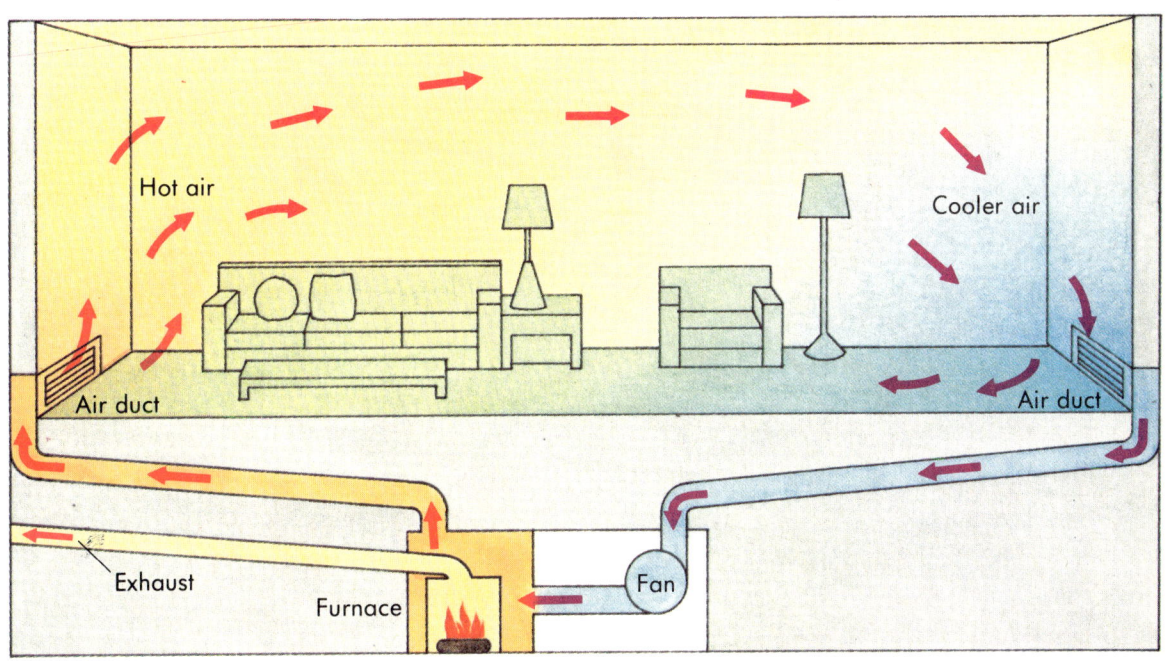

Energy Transfer by Radiation

If you sit a short distance from a fireplace, you take in heat and feel warm. But you did not touch the fireplace, and convection currents did not pass you. Energy reached you in the form of radiant energy, as in the drawing. The heat you feel results from another way to transfer energy. Radiant energy from the fire moves through space in waves. Like sunlight, the waves of radiant energy stream out from a source in all directions—they radiate. All objects radiate some energy. A hot object radiates more energy than a cool one. **Radiation** (rā′dē ā′shən) is the transfer of energy in this special kind of wave.

More energy reaches you through radiation than through conduction or convection. When radiant energy strikes a material, it increases the particles' energy. The particles move faster, and the substance heats up. All objects take in some radiant energy.

Solar panels on a house

Materials with dark, dull surfaces take in a lot of radiation. So the dark-colored solar collectors shown are good for taking in the sun's radiation and changing it into heat. Some materials reflect or allow radiation to pass through them easily. Air, glass, and other clear materials allow radiant energy to pass through them. These materials reflect some radiation and take in very little. Bright, shiny surface materials also reflect radiation.

The sun is a major source of radiation. On a cold winter day, the sun shines through a window and warms you. The outside air and window panes remain cold, because radiant energy from the sun passes through the air and glass without warming them very much. But when radiant energy strikes your clothes and body, it is taken in. The particles in your body move faster and faster, and you begin to feel warm.

Review It

1. Explain how conduction transfers energy.
2. How does convection differ from conduction?
3. Explain how radiation transfers energy.

8–5
Expansion and Contraction

If you hold a tight jar lid under a stream of hot water, the heat from the water causes the lid to increase in size. Soon the lid is loose enough to unscrew. As you read about how heat affects objects, think about the following questions:

a. What makes materials expand?
b. How can we allow for expansion and contraction?

Materials Expand When Heated

Almost all materials take up more space when heated and less space when cooled. **Expansion** is an increase in the size of a material. **Contraction** is a decrease in size. Sometimes expansion and contraction are quite noticeable. For example, the Golden Gate Bridge is about 1.5 meters longer during the summer than it is during the winter.

The kinetic theory of matter explains how expansion and contraction happen. When a steel rod is heated, its particles move around and bump into each other more. Because the particles knock each other farther apart, the space between the particles increases. So the rod expands, as shown. When the rod cools, the particles move closer together and the rod contracts.

Expansion in a metal rod

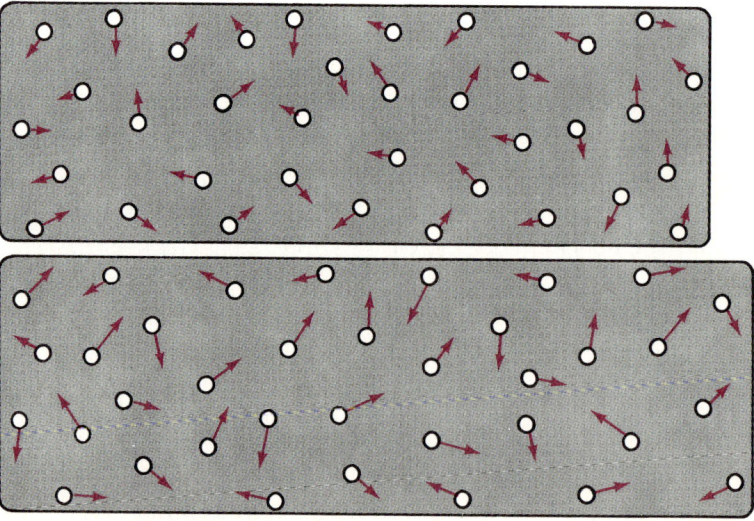

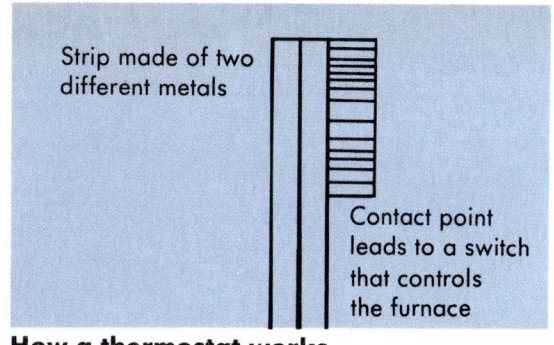
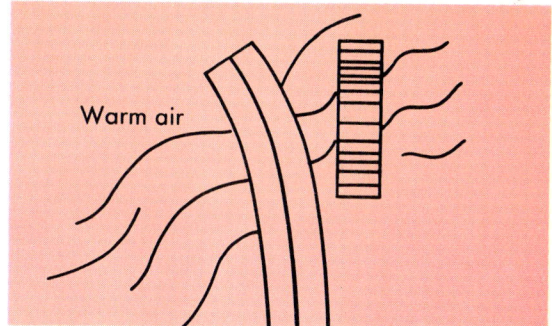

How a thermostat works

Rubber and water do not behave as the rod does. If you heat a stretched rubber band gently, the rubber contracts. As the rubber band cools, it expands. Cooled water contracts until it reaches 4°C. Below this temperature, water expands until it freezes at 0°C. The expansion makes ice less dense than water. So ice floats in water.

Living with Expansion and Contraction

Sometimes expansion and contraction cause problems in structures such as buildings and bridges. Engineers design bridges with toothlike seams, such as the one shown. In summer, the teeth clamp together as the bridge expands. In winter, the teeth separate as the bridge contracts.

All materials do not expand or contract at the same rate. A **thermostat** is a device that contains two different metal strips welded together, as shown. Each metal expands and contracts by a different amount. So the curve of the strips increases. This bending pulls the strips away from a switch that controls an appliance such as a furnace. The furnace shuts off. When the air in a room cools, each metal contracts by a different amount. The strips bend in the other direction and switch the furnace on. This process occurs over and over to control the room's temperature.

Expansion joint in a steel bridge

Review It

1. Explain how expansion and contraction happen.
2. How does a thermostat work?

Activity

A Material's Effect on Radiation

Purpose
To investigate how an object's surface affects the amount of radiation the object takes in.

Materials
- 3 aluminum cans, one white, one black, and one shiny silver
- 3 thermometers
- hot water
- 100-watt light source
- white construction paper
- black construction paper
- ruler

Procedure

Part A
1. Fold each sheet of construction paper as in *a*. Fold the bottom edge of the paper to within 4 cm of the top.
2. Staple the sides of the paper to form a pouch, as in *b*.
3. Place a thermometer into the pouch of the white paper, as in *c*. Record the thermometer reading.
4. Place the pouch flat on a table. Place the light source 10 cm above the pouch, also as in *c*.

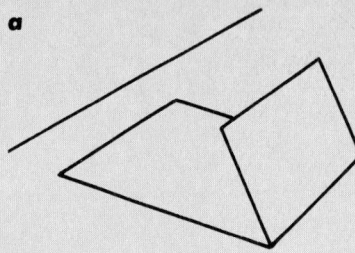

a

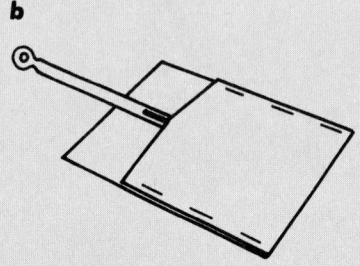

b

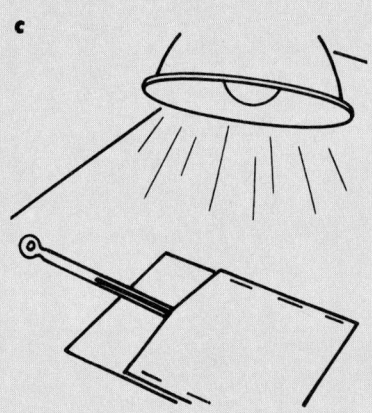

c

5. Turn the light on and record the temperature of the pouch each minute for 6 minutes.
6. Repeat steps 3–5, using the black paper.

Part B
1. Fill each aluminum can three-fourths full with equal amounts of hot water.
2. Place a thermometer in each can and record the temperatures.
3. Predict which can will cool the fastest.
4. Record the water temperature in each can each minute for 10 minutes.
5. Plot your data on a grid and connect the points. Use a different color or kind of line for each can.

Analysis
1. How do you explain the temperature differences you observed in Part A?
2. On what information did you base your prediction in Part B?
3. Explain the differences in the graphs you made in Part B.

Breakthrough

Cryogenics

Some scientists are getting the "cold shoulder" these days. Scientists are learning that materials behave in some strange ways at very low temperatures (between $-100°C$ and $-273°C$).

As materials are cooled, the particles in them begin to slow down. At temperatures near $-273°C$, the particles have little kinetic energy left. Near absolute zero, no motion would occur.

The study of the behavior of materials cooled to near absolute zero is *cryogenics*. Research in cryogenics has turned up some unexpected but useful discoveries. At temperatures near absolute zero, all gases except two are solids. Only hydrogen and helium are liquids.

Normal solids become brittle. Steel, rubber, and plastic shatter like glass. Oxygen turns to a pale blue solid that is magnetic. Liquid helium becomes a superfluid that will not stay inside its container. It flows up the inside of the container and down the outside! The picture shows this unusual behavior.

Cryogenics is now used in dozens of ways. For example, rocket engines use liquid nitrogen as fuel and liquid oxygen to burn the fuel. Some foods are preserved by quick-freezing them with liquid nitrogen.

Another exciting use of cryogenics is in the field of medicine. Red blood cells can be frozen and preserved in liquid nitrogen. They can be stored for more than 15 years without damage. Scientists believe that tissues and organs can be preserved by the same method.

Many forms of surgery now use very cold temperatures too. Severe pain can be treated by injections of liquid nitrogen. Nerves that signal pain are frozen, and pain stops. Similar methods can be used to remove tumors and cataracts. The cold liquid acts as an anesthetic and controls bleeding as it freezes tissue. Unlike normal surgery, no scar is left after using this method.

These medical uses are exciting beginnings. But they are only the "tip of the iceberg." We will find many ways to use cryogenics in our lives.

For Discussion
1. How does cryogenics affect solids and gases?
2. What advantages does cryogenic surgery have over conventional surgery?

Chapter Summary

- The kinetic theory of matter states that tiny particles make up matter, and they are constantly in motion. (8–1)
- Temperature is a measure of the average kinetic energy of the particles of matter. (8–1)
- Heat is the amount of energy transferred from one material or object to another. (8–1)
- On the Celsius temperature scale, water boils at 100° and ice melts at 0°. (8–2)
- An optical pyrometer or a thermometer can be used to measure temperature. (8–2)
- Heat is measured in joules. (8–2)
- Heat capacity is a material's ability to take in and give off heat. (8–2)
- Chemical energy, electrical energy, and mechanical energy are important sources of heat. (8–3)
- Conduction is heating from direct contact between a source and an object. (8–4)
- Convection is heating from movement of large masses of liquid or gas. (8–4)
- Radiation is heating from energy traveling through space in a special kind of wave. (8–4)
- Changes in speed of a material's particles cause expansion and contraction. (8–5)

Interesting Reading

Cobb, Vicki. *Heat*. Watts, 1973. A discussion of the nature of heat and temperature, their scientific explanation, and the role of heat in everyday life.

Kentzer, Michal. *Collins Young Scientist's Book of Cold*. Silver Burdett, 1976. A description of the effects of cold, including simple experiments. (Be sure to ask one of your parent's permission before doing experiments at home.)

Questions/Problems

1. To ventilate a room properly, you should open its windows at both the top and bottom. Explain why.
2. Use the kinetic theory to explain how the end of a spoon you are holding becomes hot when the other end is in a pot of boiling water.
3. In grocery stores, frozen foods are often kept in freezers with open tops. Why are the frozen foods not thawed quickly by warm air in the store?
4. Is an electric light bulb a source of heat, light, or both? Explain your answer.
5. Why would automobile tires appear slightly deflated the morning after a cool night even though the tires were fully inflated the previous warm afternoon?
6. Use the kinetic theory to explain why a thin, helium-filled, rubber balloon might explode if left in the hot sun.

Extra Research

1. Conduct an experiment on the effects of insulation. Fill a large metal can with hot water and find the best way to keep the water hot without adding more heat.
2. Experiment to discover whether equal amounts of hot and cold water have the same weight.
3. Do some research to find out how an insulated vacuum bottle keeps liquids hot or cold for hours.

Chapter Test

A. Vocabulary Write the numbers 1–10 on a piece of paper. Match the definition in Column I with the term it defines in Column II.

Column I

1. chemical change that gives off heat and light
2. measure of the average kinetic energy of particles of matter
3. amount of energy transferred from one object or place to another
4. transfer of energy as waves
5. a material's ability to take in or give off heat
6. an increase in the size of an object
7. energy transfer by contact between particles
8. energy transfer by the movement of a large mass of liquid or gas
9. object or material that prevents conduction
10. the particles of matter are in constant motion

Column II

a. combustion
b. conduction
c. convection
d. expansion
e. heat
f. specific heat
g. insulator
h. kinetic theory
i. radiation
j. temperature

B. Multiple Choice Write the numbers 1–10 on your paper. Choose the letter that best completes the statement or answers the question.

1. Directly or indirectly, most of our heat comes from   a) electricity.   b) chemical energy.   c) the sun.   d) mechanical energy.

2. A thermostat uses the property of
a) radiation.   b) contraction.   c) color.
d) potential energy.

3. An optical pyrometer is most likely to be used with   a) very hot objects.   b) very cold objects.   c) objects at room temperature.
d) all of the above.

4. As an object's temperature decreases, its particles   a) stop moving.   b) gain kinetic energy.   c) slow down.   d) hit other particles harder.

5. Which color indicates the coolest temperature?   a) blue   b) yellow   c) orange   d) red

6. Energy from the sun reaches us through
a) radiation.   b) convection.   c) conduction.
d) circulation.

7. A steel building is taller in summer than in winter because   a) the building has more particles in the summer.   b) the spaces between its particles have become smaller.   c) the spaces between its particles have become greater.   d) its particles move more slowly in the summer.

8. Heat always moves from places of high temperature to places   a) of higher temperature.
b) of lower temperature.   c) with the same temperature.   d) none of the above.

9. The unit to measure heat is the   a) newton.
b) watt.   c) meter.   d) joule.

10. On the Celsius temperature scale, the freezing point of water is   a) 32°.   b) 0°.
c) 100°.   d) −273.15°.

Careers

Mechanical engineer
Mechanical engineers make a lot of power. They are the people who design machines that generate power.

Engines that run trains, planes, cars, and even rockets are developed by mechanical engineers. They concentrate on making efficient machines. The best engines use the least fuel to put out the most power. Mechanical engineers may redesign machines that consume power to use less energy.

In order to do their job properly, mechanical engineers must understand blueprints, machine-building tools, and energy efficiency. They study engineering for at least four years in college.
Career Information:
The American Society of Mechanical Engineers, 345 E. 47th St., New York, NY 10017

Aerospace engineer
The Wright brothers never guessed that their flight would start a new industry. But after Kitty Hawk, the aerospace business began. Aerospace engineers bring new ideas into this business.

Aerospace engineers invent new types of aircraft and equipment for earth and outer space. They design wings, bodies, and engines for flying machines. Their models are tested in wind tunnels, temperature cubicles, and flight simulation chambers. Engineers' models that withstand these experiments may someday get off the ground.

An aerospace engineer studies flight mechanics, design, math, and science during at least four years of college or university.
Career Information:
NASA, Education Services Branch, LCG-9, 400 Maryland Ave., SW, Washington, DC 20546

Automobile mechanic
When you look under the hood of a car, how many parts can you identify? An automobile mechanic could tell you the name and function of almost every part.

Auto mechanics work in gas stations or garages. They repair and tune hundreds of different vehicles. A mechanic replaces tubes, adjusts wires, and rebuilds entire engines. All mechanics must have a good understanding of all the machines and devices in motorized vehicles.

Most auto mechanics learn their trade while working as gas station attendants or mechanics' assistants. They can also take technical classes to learn special skills.
Career Information:
Automotive Information Council, 28333 Telegraph Rd., Southfield, MI 48034

Electronics assembler
Since you woke up this morning, you have probably used at least one of the products built by the largest group of workers in the electronics industry. These assemblers put together all kinds of equipment, including televisions, radios, and calculators.

Thousands of separate parts go into a machine. Each assembler works on just one part. An assembler might follow diagrams, listen to recordings, or watch slides for directions on where to put the part. Assemblers must do their jobs perfectly, or an entire machine might not work properly.

Assemblers learn on the job. After their training they may stay in one position or move from job to job within the same factory.
Career Information:
Electronics Industry Association, 2001 Eye St., NW, Washington, DC 20006

Millwright
Textiles are fabrics, and textile mills are fabric factories. Long ago, a millwright's only job was to set up machines in these mills.

Today's millwrights construct machinery in almost every kind of factory. Millwrights set up very efficient equipment because they pay attention to detail. A good millwright builds machines that run with little friction, which increases production.

Millwrights learn some skills on the job. They may also take special classes on equipment safety, electrical wiring, or the reading of machine blueprints. A millwright's training usually takes four to six years.
Career Information:
United Brotherhood of Carpenters and Joiners of America, 101 Constitution Ave., NW, Washington, DC 20001

Heating and air conditioning system installer
Who makes it cold in summer and hot in winter? A heating and air conditioning system installer lets you change the temperature inside, no matter what the weather is outside.

These installers equip new buildings with heating and cooling systems. The installer learns the path of air flow in a building, then designs a heating or cooling system to match. Installers use their knowledge of convection currents and ventilation requirements in doing their job.

Installers learn their trade in technical school, by apprenticing, or through experience at work.
Career Information:
Air Conditioning and Refrigeration Institute, 1815 N. Fort Meyer Dr., Arlington, VA 22209

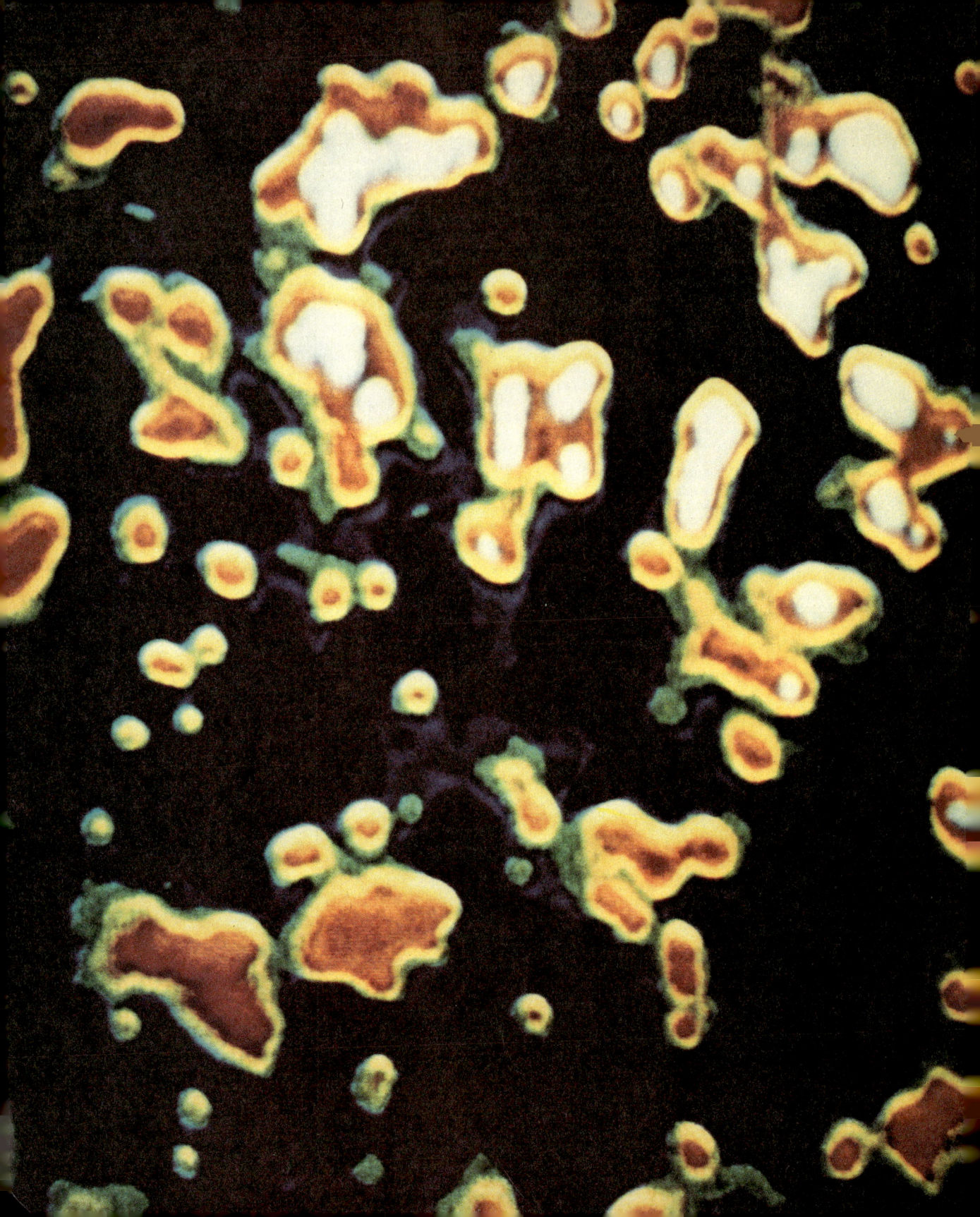

UNIT THREE
THE STRUCTURE OF MATTER

Perhaps you think the photograph shows drops of a liquid. Or are the blobs human cells? But what do the colors show?

The picture is an image of uranium atoms enlarged 10,000,000 times. Atoms are the particles that make up matter. Because we cannot see atoms, scientists shot even smaller particles at them to produce this image. Then a computer added the colors. The red-and-yellow dots are atoms. Larger areas are clusters of atoms. White areas are layers of clusters. This unit describes matter, from large objects down to the basic pieces of matter.

Chapter 9 Matter and Its States
Large amounts of matter exist in one of three conditions. Changes in the surroundings can cause matter to change condition.

Chapter 10 Properties of Matter
We can classify matter according to its observed properties and behavior.

Chapter 11 The Atom
Our understanding of the structure of matter results from years of thought, observation, and experimentation.

Chapter 12 The Atomic Nucleus
Studying the structure of matter led scientists to search for new, more basic particles and new, more basic laws.

Chapter 9
Matter and Its States

Everything you see in this picture of Niagara Falls is made of matter. Air, water, trees, and even people consist of matter in different forms. All of the incredible variety of living and nonliving things on earth are forms of matter.

This chapter describes matter in its different forms. The chapter also explains how matter changes into other forms or into energy.

Chapter Objectives
1. Define the states of matter.
2. Describe the characteristics of gases.
3. Describe the characteristics of solids and liquids.
4. Explain how matter changes from one state to another.

9-1 The States of Matter

What do you have in common with a table, orange juice, and the air? Like these things, you are made of matter. Your body is matter in three forms. As you read about the forms of matter, think about these questions:

a. What is matter?
b. What are the states of matter?
c. What determines the state of a bit of matter?

Defining Matter

In the early 1900s, Albert Einstein discovered that matter can be changed into energy and energy can be changed into matter. In other words, energy and matter are different forms of each other. Inside the sun, shown in the picture, tiny amounts of matter are changed into huge amounts of energy.

In everyday life, however, we think of matter differently. We describe matter as anything that has mass and takes up space. This description applies best to the large amounts of matter that we can see and work with. We can describe a large amount of matter by its properties.

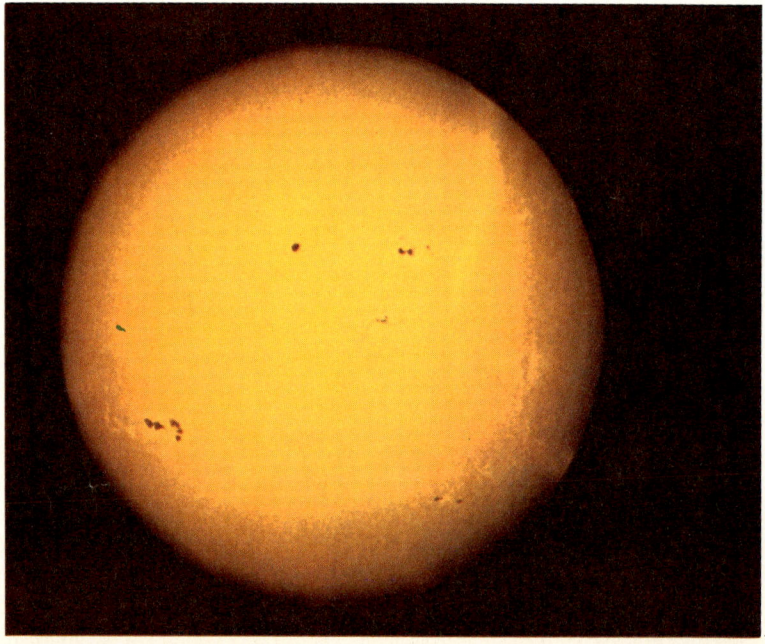

The Three States of Matter

Except for the smallest bits, matter exists as a solid, liquid, or gas. These three different conditions of matter are **physical states**. A **solid** has a definite shape and volume. A book, for example, has a regular shape and a certain size that can be measured.

A **liquid** has a definite volume but no definite shape. A liquid takes on the shape of the container that holds it. Water, for example, flows to fit the shape of the glass into which you pour it. But the volume of water remains the same if you empty it into a larger glass.

A **gas** has neither a definite shape nor a definite volume. It takes on the shape and the volume of its container. For example, a small amount of gas can expand to fill a room, or it can be contained in a balloon. The picture includes examples of the three states of matter. The mountain is a solid, the water is a liquid, and the air is a gas.

In all its states, matter behaves as if it consists of tiny particles too small to see. Scientists call these particles atoms (at′əmz). Forces can hold groups of atoms together in molecules (mol′ə kyülz).

Have You Heard?

About 400 B.C., the philosopher Democritus argued that matter could be divided into smaller and smaller pieces. Eventually, however, a tiny particle would remain that could not be broken apart without changing it. These particles made up all matter. Our word *atom* comes from his name for these particles —*atomos*. In Greek, this word means "something that cannot be cut."

States of Matter Depend on How Particles Are Linked

The state of a substance depends on how strongly its particles are held together. The pictures use tennis balls to show how particles of a solid, a liquid, and a gas are joined.

In a solid, the particles are held together closely and strongly. As a result, a solid keeps a definite shape and volume.

In a liquid, the particles are not held together as tightly as they are in a solid. The particles are held loosely and slide past one another. As a result, a liquid can flow. Thus, even though a liquid has a definite volume, it takes on the shape of its container.

In a gas, the particles can move quickly in all directions. As a result, gases do not have a definite shape or volume. Instead, their container determines both their volume and shape.

Review It

1. What did Einstein discover about matter and energy?
2. Describe the three states of matter.
3. How are particles joined in a solid, a liquid, and a gas?

The arrangement of particles in each physical state

Did You Know?

Plasma

QUESTION: Which of the following terms does not belong with the others?

solid   liquid   gas   plasma

ANSWER: None of them! All four items are examples of states of matter.

You are familiar with solids, liquids, and gases. You see examples of them every day of your life. But what are "plasmas"?

The structure of a gas changes when it is heated to very high temperatures. Heated gas atoms are torn apart into negatively and positively charged particles. This "gas" of very hot charged particles is a plasma. Unlike ordinary gases, plasmas can conduct an electric current.

Plasmas exist in only a few kinds of places on earth. One place is inside a neon or fluorescent light bulb. The gas in these bulbs is heated by electricity. Inside, the atoms break apart into charged particles. The result is a small amount of plasma.

Plasmas are very common in the universe. Over 99 percent of all matter in the universe occurs as plasma. The matter in stars is always very, very hot. The atoms are torn apart into charged particles. So stars contain plasma.

The charged particles in a plasma have so much energy that they often combine when they crash into each other. These collisions release energy. Often the new particles split apart when other particles crash into them. The combining and splitting occur over and over again.

Scientists are now trying to use plasmas to produce energy on earth. To do so, however, they need a temperature of 20,000,000°C. If they can contain such a plasma for long enough, energy will be released from our own little "star."

This method could solve many of our present energy problems. However, the problems of working with hot plasmas are enormous. More time is needed before all the problems will be solved and plasmas can become an energy source on earth.

For Discussion
1. What is plasma?
2. What kinds of problems must be solved to use plasma as an energy source?

9-2 Characteristics of Gases

Why are tires filled with gas? How can the air in tires hold up a vehicle as heavy as a loaded truck? Think about these and the following questions as you read:

a. How do particles of a gas produce pressure?
b. How do the volume and temperature of a gas affect its pressure?

How Gases Produce Pressure

The picture shows the particles of gas in a bicycle tire. The gas particles bump into and bounce off one another as they fly about. When the particles hit the tire wall, they press against it. **Pressure** (presh′ər) is a force acting over a certain area. The pushes of the gas particles against the container make up the pressure of the gas.

The earth's atmosphere consists of the air that surrounds our planet. The atmosphere creates pressure even though it is not in a container. Gravity acts like a container, keeping the gases of the air from escaping outward. The gas particles bump into and bounce off one another as they fly about. The particles produce pressure on everything they hit, including your body.

Gas particles causing pressure in a bicycle tire

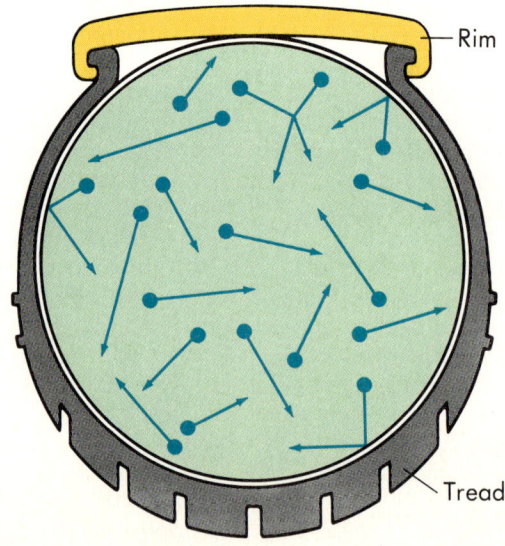

Cross-section of the tire

158

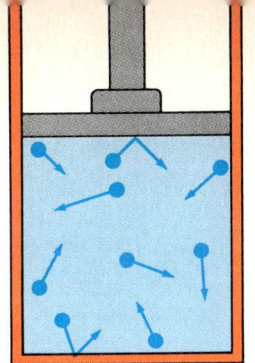 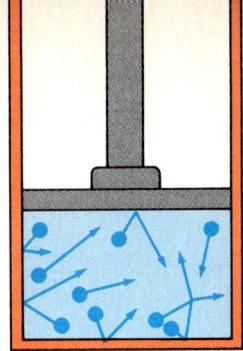

How volume affects gas pressure

Changes in the Volume and Temperature of Gases

The pictures above show how the pressure of a gas changes as its volume changes. When you reduce the volume of a container by half, the same amount of gas takes up only half as much space as before. The change in volume forces the gas particles closer together. The pressure of the gas then doubles because more particles bounce against each square centimeter of the container's walls each second.

If you double the volume of a container, fewer particles bounce against each area of the container's walls each second. As a result, the pressure is cut in half.

We fill tires with a gas at high pressure to hold up heavy objects and to make the ride smoother. Air is the cheapest gas to use. When a car passes over a bump, the volume of gas decreases for a short time, and the pressure rises.

The pictures on the right show how the pressure of a gas changes as its temperature changes. A rise in temperature increases the energy of gas particles. The particles gain energy and move more quickly. They bounce against the container's walls more often and with greater force. Therefore, as you heat a gas, its pressure increases. On the other hand, if you lower the temperature of a gas, the particles move more slowly. The pressure of the gas then drops.

How temperature affects gas pressure

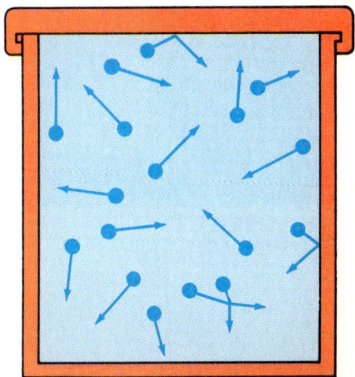
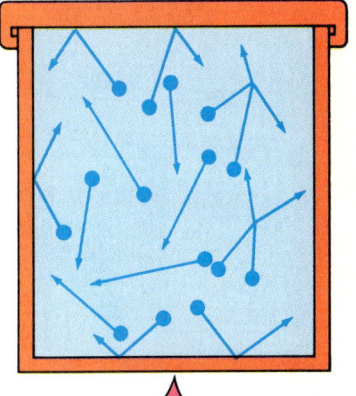

Review It

1. Explain how a gas creates pressure.
2. How do changes in volume and temperature affect the pressure of a gas?

9–3 Characteristics of Solids and Liquids

How are the structures of ice and sugar alike? Why does motor oil flow more slowly than water? Consider these and the following questions as you study solids and liquids:

a. How are particles arranged in a solid?
b. What makes some liquids flow more easily than others?

How Particles Are Arranged in Solids

The particles of a solid vibrate in one place. The particles in most solids form structural units called **crystals** (kris′tlz). In a crystal, the particles are arranged in a definite, repeating pattern. The lower left picture shows the pattern in a sugar crystal.

Some solids, such as diamonds, consist of a large, single crystal. Other solids, such as ice, are made of many smaller crystals.

A crystal has smooth, flat surfaces. The other picture shows the smooth surfaces of a quartz crystal. In most cases, crystals break evenly along their surfaces.

Sugar crystals magnified five times

The shape of a quartz crystal

Crystals form when a melted substance cools. The size of the crystal depends on how quickly the substance cools. If the substance cools slowly, a large crystal may develop. If it cools quickly, many smaller crystals form.

The rock candy in the picture results when hot sugar syrup cools slowly and forms crystals. Notice that the candy breaks evenly along the crystal's smooth surface.

Sometimes a liquid cools so rapidly that crystals do not have time to form. The lollipop in the picture is made by cooling the syrup so quickly that you cannot see any crystals. Notice how unevenly the lollipop breaks.

Compare the pictures below. On the left, liquid sulfur is poured into water. The sulfur cools so quickly that no crystals form. Then look at the orderly arrangement of ice crystals in the snowflake pictured on the right. The sulfur does not have the orderly structure of a crystal. But the ice has a characteristic six-sided structure.

A definite crystal structure

No crystal structure

Comparing the structures of two solids

The Viscosity of Liquids

The particles of a liquid are not bound as tightly as those of a solid. Yet the particles of a liquid do not move about as freely as those of a gas.

In some liquids, the particles are attracted to one another more than in others. The greater the attraction, the less easily the liquid flows. A liquid's resistance to flowing is its **viscosity** (vi skos′ə tē). The more strongly the particles attract each other, the more viscous (vis′kəs) the liquid is. Viscous liquids are thick and sticky and do not pour easily.

Motor oil is a viscous liquid. It is used as a lubricant in automobiles because it forms a slick film between the moving parts of the motor. Some liquids, such as  tar, are so highly viscous and flow so slowly that they appear to be solids.

Heating liquids makes them less viscous. Heat causes the particles in the liquid to move more quickly and slide more easily past one another. The molten rock in the picture is so hot it flows faster than water.

A decrease in temperature makes liquids more viscous. The particles in the warm candle wax shown move more slowly and interact more as the temperature drops.

Review It

1. What is a crystal?
2. How does heating a liquid make it less viscous?

Activity

Liquids and Solids

Purpose
To observe the effect of temperature on the formation of crystals.

Materials
- bowl or tray
- water
- plastic spoon
- sugar
- Bunsen burner
- matches
- potholder
- copper sulfate
- ring stand and ring
- wire gauze
- 100-mL beaker
- small object to use as a weight
- string
- rubber band
- paper towel
- safety goggles

Procedure

Part A
1. Set up your ring stand, ring, wire gauze, and burner as shown.
2. Put about 50 mL of water in the beaker.
3. *Put on your safety goggles.* Heat the water to make it hot. It does not have to boil.
4. Turn off the burner. Stir about 2 teaspoons of copper sulfate into the hot water.
5. Tie the weight to some string. Put the weight in the hot water so the loose end of string hangs over the edge of the beaker.
6. When the beaker is cool enough, cover it with a paper towel held in place by the rubber band.
7. Let the beaker stand undisturbed for 1–2 days.

Part B
1. Put some cold water in the bowl.
2. Light the burner and slowly heat a spoonful of sugar over it. *CAUTION: Hold the spoon with the potholder and keep hair, clothes, and papers away from the flame.*
3. As soon as the sugar melts, remove the spoon from the burner.
4. Put the spoon and the liquid sugar into the cold water so the sugar will cool quickly.
5. Observe and record in what form the sugar becomes solid again.

Analysis
1. Compare the structure of a solid that formed quickly with one that formed slowly.
2. Examine the solid copper sulfate attached to the string. Determine whether or not the copper sulfate formed crystals.

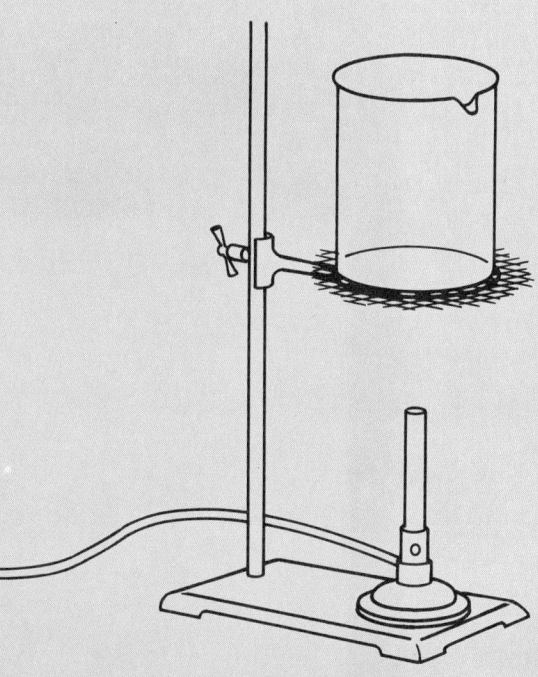

163

9–4
Changes in the State of Matter

How does a rain puddle on a sidewalk disappear? Why does a lake freeze in winter and melt back into water in spring? Remember these and the following questions as you read about the states of matter:

a. How does matter change from one state to another?
b. What is latent heat?

How Matter Changes from One State to Another

The picture shows water changing its state. Like water, most kinds of matter can be a solid, a liquid, or a gas. Matter changes from one state to another when its temperature changes enough. Temperature affects the way particles in a substance are joined. At low temperatures, particles move slowly and become closely joined. At high temperatures, the particles gain energy and move more quickly and more freely.

Ice, for example, melts when its particles gain energy. The temperature at which a solid melts is its **melting point.** Each substance has a different melting point. For example, the melting point of water is 0°C at normal atmospheric pressure. That of iron is 1535°C.

Water freezes into ice. The temperature at which a liquid changes into a solid is its **freezing point.** For each substance, its freezing point is the same as its melting point.

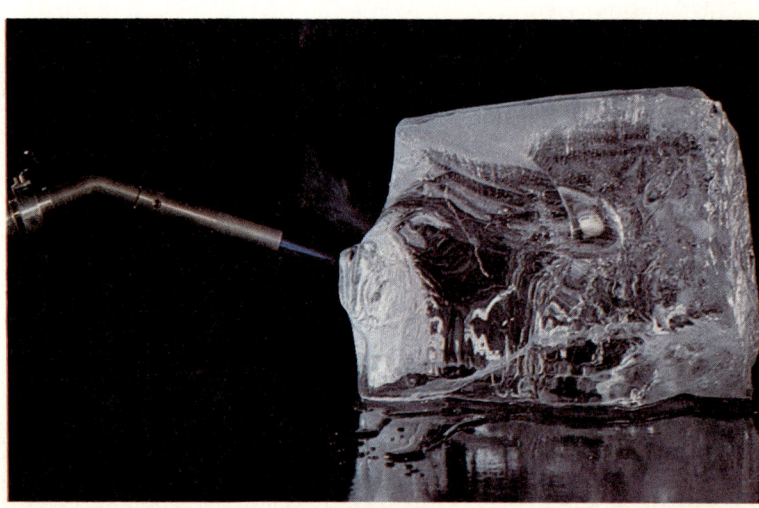

Water boils into steam. The temperature at which a liquid boils is its **boiling point.** At its boiling point, a liquid changes into a gas in a process called **vaporization** (vā′pər ə zā′shən). Water in the gas state is called water vapor.

The air pressure on a substance affects its boiling point. Under the air pressure at sea level, water reaches its boiling point at 100°C. The air pressure at high altitudes is lower than it is at sea level. For this reason, water boils at a lower temperature at high altitudes. As a result, you need to boil food longer at high altitudes than you do at low altitudes to get the same result.

Water can evaporate—change from a liquid to a gas—even when its temperature is below the boiling point. This process is **evaporation** (i vap′ə rā′shən). At any temperature, some particles in a liquid move faster than others. The particles moving fastest can fly out of the liquid. In this way, water evaporates little by little. A puddle on a sidewalk slowly disappears by evaporation.

Water vapor can change into a liquid—or condense—in a process called **condensation** (kon′den sā′shən). If its temperature is lowered enough, a gas can become a liquid. Natural gas for heating is often condensed to a liquid in plants such as the one shown. The liquid form takes up less space, so it is easier to store and transport.

Challenge!

Find out how air pressure affects how we cook. Use reference books to learn how a pressure cooker works, and look at cake or bread mix packages for the high altitude directions.

165

Temperature During Changes of State

The pictures show ice melting in a beaker. A thermometer measures the temperature of the melting water. The temperature does not change as long as the melting continues. It remains at the melting point of water—0°C, under normal atmospheric pressure. While the ice is melting, the heat does not raise the temperature of the water. Instead, the heat makes the ice change its state. It is called **latent** (lāt′nt) **heat,** meaning *hidden heat*. The temperature of the water in the beaker will not start to rise until all the ice has melted.

In melting ice, latent heat provides the energy that breaks the orderly arrangement of particles in the ice crystals. In boiling water, latent heat provides the energy needed to separate one water particle from another.

Review It

1. Explain the processes of evaporation and condensation.
2. Explain the meaning of latent heat.

Have You Heard?

When boiling water condenses, the energy from the latent heat is released. If steam at 100°C touches your skin, the steam gives off a large amount of latent heat. For this reason, steam burns you more seriously than water at the same temperature.

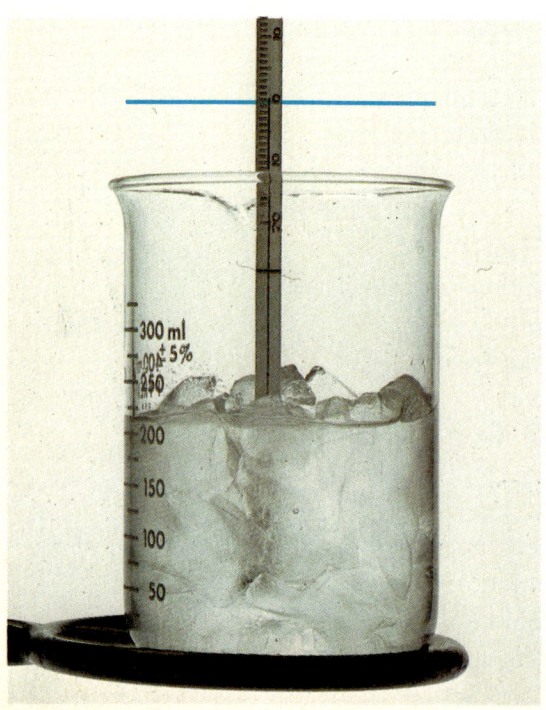

Activity

Changing Physical State

Purpose
To observe the effects of latent heat.

Materials
- 2 beakers
- 2 hot plates, or 2 Bunsen burners, rings and ringstands, wire gauze, and matches
- 2 thermometers calibrated to include 0°C and 100°C
- water
- 2 watches to measure seconds
- safety goggles
- stirring rod

Part A
1. Fill the beaker with ice.
2. Set up your equipment as in *a*. *CAUTION: Mercury is poisonous. Handle the thermometer carefully so you do not break it.*
3. Add enough water to the beaker to cover the bulb of the thermometer.
4. Stir very gently with the stirring rod and check the water temperature every 30 seconds. When the water has remained at 0°C for 2 minutes, light the burner.
5. Record the temperature every 30 seconds and sketch how much ice remains each time.
6. After the ice has completely melted, continue recording the temperature for 5 minutes.
7. Copy the grid in *b* and plot your temperatures. Extend the grid as much as needed. Mark the temperature at the time the ice completed melting.

Part B
1. Fill a beaker half way with water. *Put on your safety goggles.*

a

2. Repeat Part A, step 2. Heat the water and record the water's temperature every 30 seconds.
3. Continue recording the temperature as the water starts to boil. *CAUTION: Steam and hot water can cause painful burns.* Then record the temperature every 30 seconds for 3 more minutes.
4. Repeat Part A, step 7. Mark the temperature at the time the water started to boil.

Analysis
1. What happened to the temperature of the water while the ice was melting? Explain in terms of latent heat.
2. What happened to the temperature of the water while it was boiling? Explain in terms of latent heat.

b. Temperature during changes in state

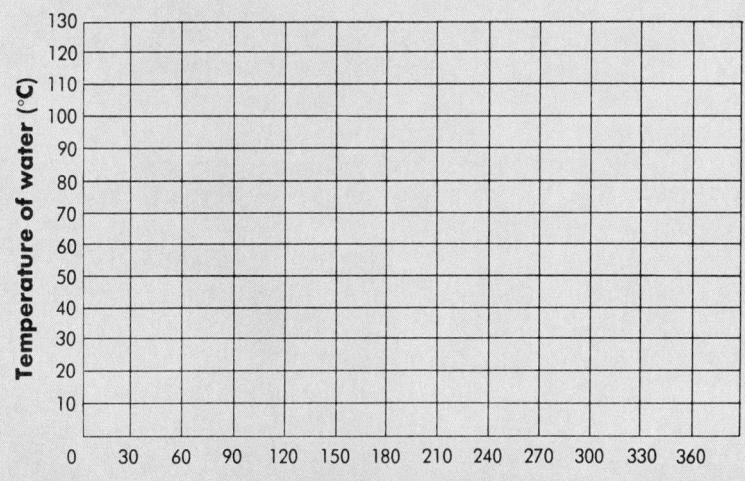

Chapter Summary

- All things are made of matter. (9–1)
- Matter and energy are different forms of each other. (9–1)
- Large amounts of matter exist in one of three physical states. (9–1)
- A solid has a definite shape and volume. (9–1)
- A liquid has a definite volume but takes on the shape of its container. (9–1)
- A gas takes on the shape and volume of its container. (9–1)
- Particles of gas bumping against the walls of a container create pressure. (9–2)
- The pressure of a gas changes as its temperature or volume changes. (9–2)
- The particles of most solids are arranged in crystals. (9–3)
- Viscosity depends on the attractive forces between a liquid's particles. (9–3)
- Matter can change state by melting, freezing, boiling, evaporating, or condensing. (9–4)
- The heat required to produce a change in state is called latent heat. (9–4)

Interesting Reading

Arnov, Boris. *Water: Experiments to Understand It.* Lothrop, Lee, and Shepard, 1980. Demonstrations that point out the properties of water.

Cobb, Vicki. *Supersuits.* Lippincott, 1975. Descriptions of how people have learned to survive under conditions of high pressure (underwater) and low pressure (little or no air).

Zubrowski, Bernie. *Messing Around with Water Pumps and Siphons.* Little, Brown, 1981. Experiments and demonstrations using the principles of suction and compression. (Be sure to ask your parent's permission before doing any experiments at home.)

Questions/Problems

1. If gas expands to fill any size container, why do bicycle tires show the effects of air loss and require periodic filling?
2. Explain what happens inside a closed container of a gas as the temperature is raised.
3. Explain what happens inside a closed container of a gas when the container's volume is tripled.
4. You find a large crystal of quartz in some rocks. Describe the conditions under which that crystal formed.
5. Contrast evaporation and vaporization.
6. Explain what happens to the molecules of a liquid and any changes in temperature as the liquid freezes.
7. In summer, would you use a more viscous motor oil in a car or a less viscous one than in the winter?
8. Why do roofers heat tar before pouring it?

Extra Research

1. Press the sharp edge of a table knife on an ice cube. How does pressure affect the melting point of ice? Find out how this process explains how ice skates and sleds work.
2. Blow on a mirror several times. What process do you observe? Explain this process.
3. Ancient people considered glass a precious substance. Use reference books to learn more about how people have made and used glass. If you can, visit a glass works to see glass being made. Then report what you learned to your class.
4. Use an encyclopedia or earth science text to find out what the dew point is, if it is related to the freezing point of water, and what affects the formation of dew.

Chapter Test

A. Vocabulary Write the numbers 1–10 on a piece of paper.
Match the definition in Column I with the term it defines in Column II.

Column I

1. heat required to produce a change in state
2. state of matter that has a definite shape and volume
3. has a definite volume but no definite shape
4. takes on the shape and volume of its container
5. resistance of a liquid's particles to flowing
6. structure of most solids
7. temperature at which matter changes from a solid to a liquid
8. temperature at which matter changes from a liquid to a gas
9. process by which matter changes from a gas to a liquid
10. process by which matter changes from a liquid to a gas

Column II

a. boiling point
b. condensation
c. crystal
d. evaporation
e. gas
f. latent heat
g. liquid
h. melting point
i. solid
j. viscosity

B. Multiple Choice Write the numbers 1–10 on your paper.
Choose the letter that best completes the statement or answers the question.

1. Matter   a) can exist as a solid, liquid, or gas.   b) is another form of energy.   c) in large amounts has mass and occupies space.   d) a, b, and c.

2. All true solids   a) are viscous.   b) are made of crystals.   c) have particles that move about freely.   d) have particles that stay close to one location.

3. Liquids   a) can be made less viscous by cooling.   b) have particles that are held in place.   c) change into gases by vaporizing.   d) a and c.

4. The particles of gases   a) move slowly at high temperatures.   b) are held closely together.   c) exert more pressure when heated.   d) exert more pressure when cooled.

5. If you double a container's size, the pressure of the gas in it   a) decreases.   b) doubles.   c) remains the same.   d) triples.

6. Which of the following liquids is the least viscous?   a) honey   b) tar   c) maple syrup   d) water

7. The boiling point of water is   a) the same everywhere.   b) lower at high altitudes.   c) lower at low altitudes.   d) unaffected by air pressure.

8. Water droplets appear on a cold glass because water vapor in the air   a) evaporated.   b) condensed.   c) melted.   d) vaporized.

9. As ice is heated and melts, the heat   a) separates the particles in the ice crystals.   b) immediately lowers the temperature of the water.   c) immediately raises the temperature of the ice.   d) binds the particles in the ice crystals.

10. Highly viscous liquids   a) flow easily.   b) flow slowly and are thick.   c) have strong interactions between molecules.   d) b and c.

Chapter 10
Properties of Matter

Once the Statue of Liberty was as shiny as a new penny. Over many years, the air has affected the metal in the statue. Now, the statue is dull and green.

This chapter introduces the basic substances that make up everything on earth and throughout the universe. Then the chapter describes the different properties of these substances and explains how the substances change.

Chapter Objectives

1. Compare and contrast metallic elements with nonmetallic elements.
2. Distinguish between physical and chemical properties of matter.
3. Distinguish between physical and chemical changes.

10–1
The Elements

The ancient Greeks believed that all matter formed from four substances: fire, air, earth, and water. As you read about how we describe substances, you will discover answers to these questions:

a. What is an element?
b. How do we describe the elements?

Elements Are Basic Substances

An **element** is a substance that cannot be broken down into other substances by heat, light, or electricity. Elements are the basic substances that form matter. Gold, silver, mercury, and oxygen are elements. Carbon, shown below, is an element that occurs naturally in more than one form. Graphite is shown on the left, and diamond on the right. All living things contain carbon.

An element contains only one kind of atom. Copper is made of only copper atoms, and iron is made of only iron atoms. We define an **atom** as the smallest particle of an element with that element's chemical characteristics.

Everything in the world is made of a combination of atoms of the elements. For example, atoms of the elements hydrogen and oxygen make water. A combination of hydrogen, oxygen, and carbon atoms makes sugar.

Your body contains many of the elements, mainly oxygen, hydrogen, carbon, and nitrogen. Calcium is important in forming your bones and teeth. Zinc keeps your taste buds working properly. Copper helps your nerves function. Iron helps carry oxygen through your blood.

Gold

Silver

Copper

Iron

Tin

Lead

Mercury

Carbon

Sulfur

We know of 106 elements. Eighty-nine of them are found in nature. Scientists make the other elements in the laboratory by using high-energy machines.

Discovering all the elements took thousands of years. Ancient peoples worked with several of the elements, even though they did not know they were elements. The pictures above show these elements. Gold and silver were used in jewelry. Mercury was used as a medicine. Copper and tin were used for cooking utensils and were combined to make bronze for weapons. As time went by, people learned to use iron for making weapons.

For hundreds of years, only a few elements were known. During the time of the American Revolution, about twenty elements were known. But by the end of the Civil War, more than sixty elements had been discovered. In the century after the Civil War, the remaining naturally occurring elements were discovered. Today most scientists agree that all naturally occurring elements have been found.

Challenge!

Use an encyclopedia or another reference book to find out how and where the element helium was discovered.

Describing Elements

Scientists group elements according to their properties. The elements tin, lead, iron, and aluminum are examples of one group—the **metals.** Most elements are metals. Sulfur, carbon, and helium are examples of another group—the **nonmetals.** The pictures below illustrate the properties of metals.

Most metals are solids at room temperature under ordinary pressure. Mercury is an exception. It is a liquid at room temperature.

Metals have a **luster** (lus′tər), which means how they shine. Manufacturers trim cars with the metal chromium because of its high luster. The ancient Romans polished silver to a high luster and used it as a mirror.

Metals conduct heat well. Some cooking utensils are made of copper, iron, or aluminum. A pot made of one of these elements heats quickly and distributes heat evenly.

Metals also conduct electricity. Even though silver and gold are the best conductors, electricians use copper in electrical wiring because of another property. Copper, platinum, and some other metals are **ductile** (duk′təl), which means that they can be drawn into a wire.

Some metals can also be hammered, rolled, or shaped without being broken. These metals are **malleable** (mal′ē ə bil). Copper, gold, and silver can be pounded into jewelry, and aluminum foil folds easily around your sandwich, because these elements are malleable.

Fewer than two dozen elements are nonmetals. Some nonmetals are shown below. Chlorine, neon, hydrogen, and nitrogen are nonmetals. The properties of metals and nonmetals are opposites. The luster of nonmetals is low. They conduct heat and electricity poorly, and they are not ductile or malleable. At room temperature under ordinary pressure, most nonmetals are solids or gases. Sulfur, iodine, and phosphorus are solid nonmetals. Sulfur is a yellow crystal. Iodine is a dark purple crystal. Phosphorus, usually a soft, white solid, also occurs in a red form. The gaseous nonmetals include nitrogen, oxygen, and fluorine. Bromine is the only liquid nonmetal.

Some elements, such as arsenic and silicon, have properties of both metals and nonmetals. They are called **metalloids** (met′l oidz).

Review It

1. Define the word element.
2. How are metallic and nonmetallic elements different?

Sulfur

Iodine

Red phosphorus

10–2
Physical Properties and Changes

All matter can be described by its properties. Chewing gum, for example, could be described as thick, sticky, and stretchable. Answer these questions to begin learning how scientists describe matter:

a. What are the physical properties of matter?
b. What is a physical change?

Describing Physical Properties

A **physical property** is a property that can be observed without changing the identity of the substance. For example, as it melts, cheese changes its state, but it is still cheese. It still tastes and smells like cheese. If you cut a sheet of paper, you change the shape and appearance of the paper, but it is still paper. The pictures below show several physical properties. The properties of metals described earlier are physical properties.

Distinctive tastes or odors are physical properties of some substances. When blindfolded, you may not be able to taste the difference between two brands of bread, but you can taste the difference between bread and a lemon. Many substances have distinctive odors too. Rotten eggs, for example, get their smell from a substance in them that contains hydrogen and sulfur.

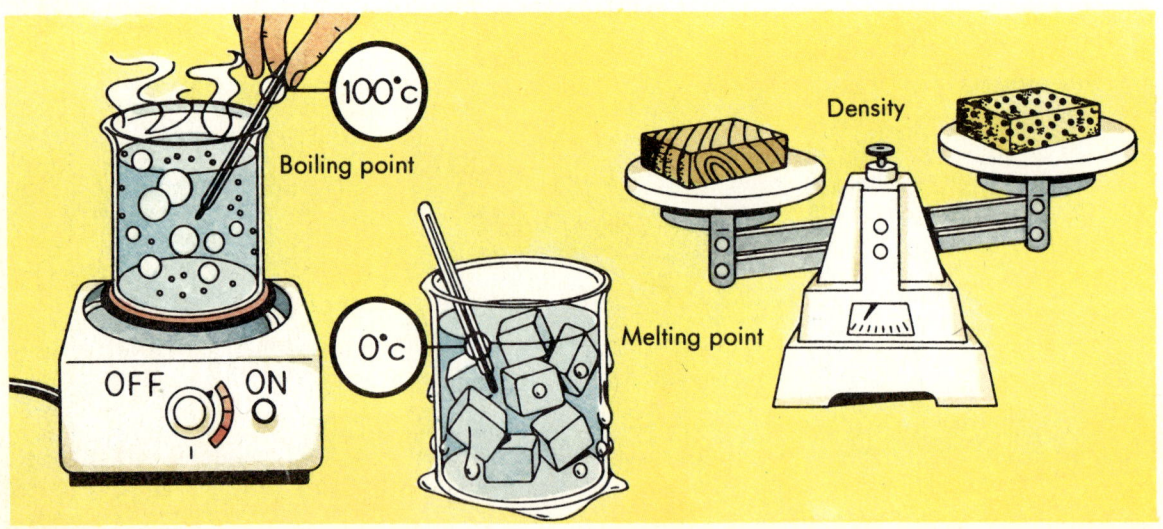

Hardness is another physical property. The harder the substance, the more difficult it is to scratch the surface of the substance. Geologists use this property to identify rocks and minerals. The diamond is the hardest naturally occurring substance. A diamond will scratch any other substance. Gold, on the other hand, is soft and easily scratched. Therefore, gold jewelry is made of a combination of gold and other metals that make the jewelry harder.

If you melt a silver spoon and shape the silver into a bracelet, the silver remains silver. The boiling point and melting point of a substance are physical properties. Each element has its own unique boiling point and melting point. Chlorine, which is a yellowish-green gas at room temperature, boils at −34.6°C and melts at −101.6°C. Iron melts at 1535°C and boils at 3000°C. The metal gallium melts in your hand because its melting point is 29.8°C.

Color, crystal shape, and density are other physical properties that sometimes help identify a substance. For example, sulfur crystals are typically bright yellow. Osmium is the densest element.

Magnetic properties are also physical properties. Iron is the most magnetic element. Nickel and cobalt are somewhat magnetic.

Producing Physical Changes

Iron can be hammered into sheets, molded into a radiator, or bent into a nail. During each change, the iron remains iron. A **physical change** is any change that does not alter the identity of a substance.

Sawing wood produces a physical change. The sawdust in the picture is still wood. Shredding paper and crushing a sugar cube are also physical changes. In each case, the identity of the substance—the wood, the paper, and the sugar—remains unchanged. In a physical change, the size, shape, or state of the substance changes. Boiling and freezing are, therefore, physical changes.

Dissolving sugar in tea is a physical change. The sugar no longer exists in the cube form, but the sugar is still sugar. You know the tea contains sugar because the tea tastes sweet. If you boil the tea, the water in it will boil away, but the sugar will remain.

Evaporation is a physical change from the liquid to the gaseous state. When water evaporates, the liquid changes to a gas. Condensation reverses the process.

At room temperature and under standard pressure, phosphorus is a white solid. If you heat phosphorus to 250°C in the absence of air, it turns red. The element phosphorus is still present, but its color changes. The color change occurs because the atoms move into a different crystal shape. But the new crystals are still phosphorus.

Under most conditions, carbon dioxide is a gas. When it is cooled, however, carbon dioxide becomes the white solid known as dry ice. Although the white solid looks different from the colorless gas, both substances are carbon dioxide.

Review It

1. List three physical properties of an element.
2. Describe three physical changes of elements.

Activity

Physical Properties

Purpose
To observe physical properties of substances.

Materials
- aluminum foil
- copper wire
- sulfur crystals
- zinc strips
- small hammer
- nail
- balance
- 50-mL graduated cylinder
- water
- safety goggles

Procedure
1. Copy the table shown. Put on your safety goggles.
2. Examine each substance. Describe its shape, color, state, and luster.
3. Record your observations in the table.
4. Scratch the surface of each substance with a nail. Record "yes" if the nail scratches it or "no" if the nail does not scratch it.
5. Try to reshape each substance by bending, hammering, or folding it. Record your results.
6. Use the balance to find the mass of each element. Record the masses.
7. Half-fill the graduated cylinder and record the height of the water.
8. Drop the zinc into the water. Record the new height of the water.
9. Subtract the water levels you measured in steps 7 and 8. The difference is the volume of the zinc.
10. Calculate and record zinc's density using: density = mass ÷ volume.
11. Repeat steps 7–10 for the other substances.
12. Classify the substances you tested according to their properties as metals or nonmetals.

Analysis
1. Explain how you can shape some substances and not others.
2. How are properties useful in classifying substances?
3. How have people used knowledge of the properties of elements to design the nail, wire, and foil?

Data chart

Element	Shape	Color	State	Luster	Scratch	Malleability	Density
Aluminum							
Copper							
Sulfur							
Zinc							

10-3
Chemical Properties and Changes

In the old days, prospectors mistook iron pyrite for the element gold. This mistake occurred so often that iron pyrite was called fool's gold. To learn which other properties prospectors consider in identifying gold, keep these questions in mind:

a. What is a chemical property?
b. What is a chemical change?

Identifying Chemical Properties

Two substances can look the same even if they are really quite different. To avoid mistaking them, we identify substances by more than one kind of property. A **chemical property** is a property that describes how a substance reacts with other substances.

You can observe a chemical property by trying to change a substance into a new material. If you add gold to nitric acid, gold remains unchanged. Gold does not react with the acid, so it does not become a new substance. If you add fool's gold to nitric acid, a new substance forms. Iron pyrite has the chemical property of reacting with acid.

If you ignite a piece of paper, the paper burns. It is flammable—at the proper temperature, it burns in the presence of oxygen. If we try to ignite asbestos, however, it does not burn. It is not flammable, as shown below.

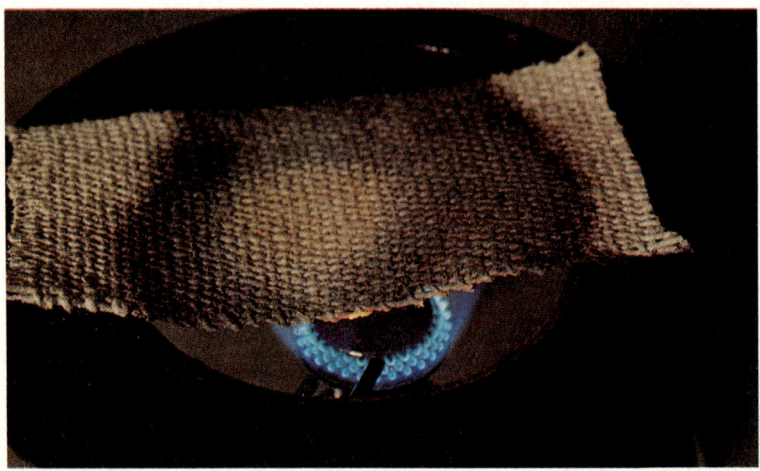

Reacting with oxygen

Reacting with acid

Reacting with oxygen

Reacting with water

Producing Chemical Changes

A nail rusts because the iron in it reacts with oxygen in water. Silver tarnishes because it reacts with sulfur in the air. Reacting with air is a chemical property.

Rust and tarnish are new materials with new properties. Both result from **chemical changes,** which are changes that form new substances with new properties. The pictures show several chemical changes.

Many substances react with water or air. The bubbling or fizzing that sometimes results indicates a chemical change. Lithium, sodium, and potassium react so violently with water that an explosion results.

After a chemical change, the properties of the original substance are no longer present. Burning wood is a chemical change. The powdery, gray ash left behind by a burning log is not at all like wood.

Digesting food causes chemical changes. The substances in the carrot you eat and digest do not look like your skin. Yet your body changes the substances in carrots into the substances of skin—another chemical change.

Have You Heard?

Mixing wood ashes and water produces lye. Pioneers made lye in this way to make soap.

Review It

1. Name two chemical properties.
2. How can you tell if a chemical change has occurred?

Activity

A Chemical Change

Purpose
To observe what happens during a chemical change.

Materials
- several wooden splints
- ring stand with test tube clamp
- test tube
- 1-hole stopper, with glass tubing inserted, for test tube
- Bunsen burner
- matches
- safety goggles

Procedure
1. Break the wooden splints into small pieces.
2. Fill the test tube about one-fourth full with the wood.
3. Stopper the test tube firmly.
4. Clamp the test tube to the ring stand. The test tube should be slanted, as in *a*.
5. CAUTION: *Do not point the test tube toward yourself, another person, or the aisle. Wear your safety goggles.* Heat the lower one-fourth of the test tube slowly for about 10 minutes, as in *b*.
6. Observe and record any changes in the wood.
7. Note and record whether any material escapes from the test tube or appears on the sides of the test tube.
8. Turn off the burner and let everything cool before you put your equipment away.

Analysis
1. What evidence for a chemical change did you see?
2. What can you say about the substances before and after the chemical change took place?

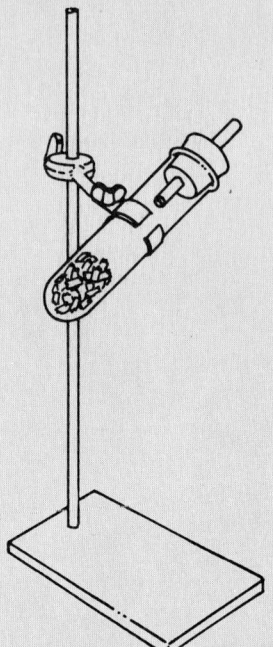
a

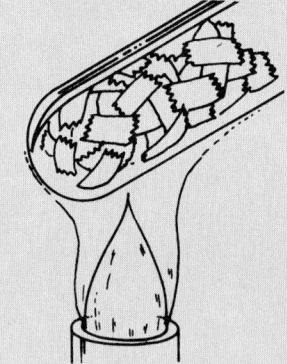
b

Did You Know?

Naming the Elements

What is in a name? If you are talking about the elements, the answer is "many interesting stories!" For example, did you know that one element is named for a city in the United States and another is named for a state? They are element 97, berkelium, and element 98, californium. The two elements were given these names because they were first produced at the University of California at Berkeley.

Another popular way to name elements is after famous people. Samarium, for example, was named after the Russian mining engineer Samarski. Can you guess for whom the elements curium, fermium, einsteinium, and nobelium were named?

Sometimes scientists disagree about the naming of an element. Recently, elements 104, 105, and 106 were discovered by two different groups. The American discoverers wanted to call one element rutherfordium to honor Ernest Rutherford. The Russian discoverers wanted to name the same element kurchatovium after a Russian scientist. However, the International Union of Pure and Applied Chemistry has decided against both of these names. Instead, they suggest that all new elements be named after their atomic numbers. Element 104 would be called unnilquadium (un = 1, nil = 0, quad = 4); element 105, unnilpentium (pent = 5); and element 106, unnilhexium (hex = 6).

Elements that have been known for a long time are named from Latin or Greek words. Mercury's symbol, Hg, comes from *hydrargyrum,* which means "liquid silver." The symbol Au comes from the Latin name for gold, *aurum.* One ancient name may sound familiar to you. Lead, Pb from the Latin *plumbum,* was often used to make water pipes and drains. The people who repaired them were named after the element they used.

Many elements are named after one of their properties. For example, bromine comes from the Greek word *bromos,* which means "stink." How would you describe the nature of argon, knowing that *argos* means "idle"?

For Discussion
1. List some ways the elements were named.
2. What is the system now recommended for naming new elements?

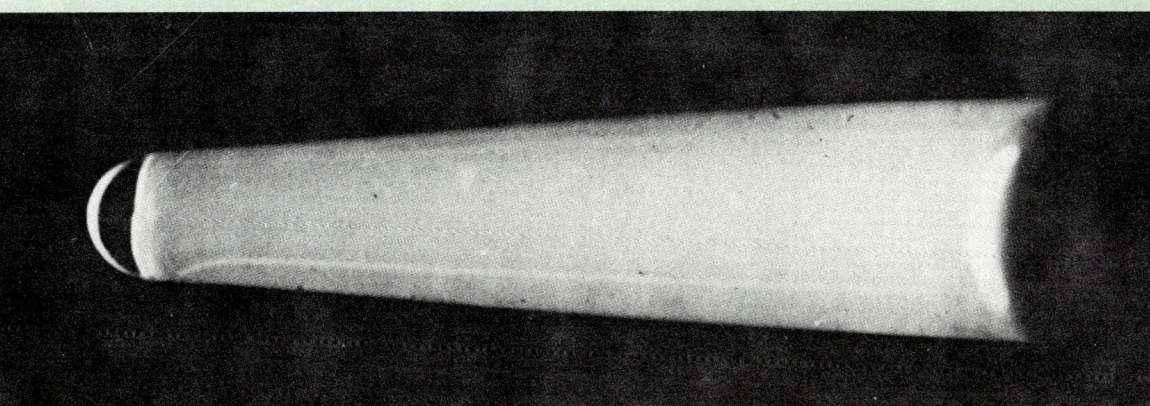

A test tube containing a glowing curium compound

Chapter Summary

- An element is a pure substance that cannot be broken down into another substance by heat, light, or electricity. (10-1)

- An atom is the smallest particle of an element with the chemical characteristics of that element. (10-1)

- Metals have a shiny luster and are good conductors of heat and electricity. Some metals are malleable and ductile. (10-1)

- The properties of metals and nonmetals are generally opposites. (10-1)

- Metalloids are elements with properties of both metals and nonmetals. (10-1)

- A physical property distinguishes a substance without changing its nature. (10-2)

- Taste, odor, hardness, boiling and melting points, physical state, color, crystal shape, density, magnetic properties, luster, and being malleable, ductile, and a good conductor are physical properties. (10-2)

- A physical change is one that does not change the identity of a substance. (10-2)

- A chemical property explains how a substance reacts with other substances. (10-3)

- Reacting with air and acid are chemical properties. (10-3)

- A chemical change is a one that produces new substances with new properties. (10-3)

Interesting Reading

Coombs, Charles. *Gold and Other Precious Metals*. Describes the properties, uses, and refining of gold, silver, and platinum.

Wohlrabe, Raymond A. *Metals*. Lippincott, 1964. A classic history of our uses of metals, with home experiments that could be done with adult supervision. (Be sure to ask your parent's permission before performing any experiments at home.)

Questions/Problems

1. You perform certain tests on an unknown substance. How would you decide if it is an element or some other kind of substance?

2. Which properties of copper allow it to be used in an electrical wire, a cooking utensil, or a bracelet?

3. Which of the following statements about the element sodium refer to physical properties? Which refer to chemical properties? a) reacts violently with water; b) is a shiny metal; c) is so soft that a knife can cut it; d) will tarnish quickly if exposed to air.

4. You have a box of sugar cubes, a blender, some water, a pot, and a hot plate. What physical change or changes can you make in the sugar? What chemical change or changes can you make?

5. What could you do to keep an iron nail from rusting?

Extra Research

1. Gold is a very soft metal. Talk with a local jeweler to find out what is added to gold to make it stronger and less easy to bend.

2. Industry uses a lot of iron. But iron has some undesirable properties. Use an encyclopedia to find out how various industries use and improve iron.

3. Visit a museum to see samples of the elements. Or collect samples of as many elements as you can. Use chemistry or mineralogy books to help you identify each element.

Chapter Test

A. Vocabulary Write the numbers 1–10 on a piece of paper.
Match the definition in Column I with the term it defines in Column II.

Column I

1. a pure substance that cannot be broken down into another substance by heat, light, or electricity
2. can be hammered, rolled, or pounded into different shapes without being broken
3. can be drawn into a wire
4. how a substance shines
5. smallest particle of an element that acts like that element
6. the elements copper, aluminum, and gold
7. a change that maintains the substance's identity
8. a change that produces a new substance
9. the characteristics of color, odor, taste, and hardness
10. describes how a substance reacts with other substances

Column II

a. atom
b. chemical change
c. chemical property
d. ductile
e. element
f. luster
g. malleable
h. metals
i. physical change
j. physical property

B. Multiple Choice Write the numbers 1–10 on your paper.
Choose the letter that best completes the statement or answers the question.

1. An element forms from   a) different kinds of atoms.   b) metals.   c) one kind of atom.   d) nonmetals.

2. Which of the following is not an element?   a) silver   b) rust   c) chlorine   d) calcium

3. Tarnish on silver indicates   a) a physical change.   b) density.   c) a chemical change.   d) hardness.

4. Copper is   a) malleable.   b) a metal.   c) ductile.   d) a, b, and c.

5. The physical properties of gold include the fact that it   a) is shiny.   b) reacts with oxygen.   c) is hard.   d) reacts with acid.

6. Lead is a   a) metal.   b) nonmetal.   c) metalloid.   d) a and c.

7. Choose the correct statement.   a) All elements are found in nature.   b) Eighty-nine elements are found in nature.   c) All elements are made in the laboratory.   d) Eighty-nine elements are made in the laboratory.

8. A dull, yellow solid that conducts electricity is probably a   a) metal.   b) nonmetal.   c) metalloid.   d) none of the above.

9. In a physical change, the change is in   a) size.   b) state.   c) shape.   d) a, b, and c.

10. Iron and oxygen join to produce rust in a(n)   a) chemical change.   b) physical change.   c) element.   d) physical property.

Chapter 11
The Atom

The ancient Greeks knew about and used several substances that we now call elements. They also wondered about the structure of matter that formed these substances. For more than 2,000 years, people continued to wonder about matter. Now, with the help of mathematics, we have an idea of how atoms look. At the left is a model of the element uranium. The model is based on the mathematical equations that describe the uranium atom.

This chapter explains how scientists developed theories about the atoms that make up the elements. It discusses the present theory about the structure of these bits of matter. The chapter also explains how scientists used the atomic theory to classify the known elements.

Chapter Objectives
1. Trace the development of the theory of the atom.
2. Discuss the current theory that matter is made of atoms.
3. Explain how atoms differ from one another.
4. Explain the uses of the periodic table of elements.

11-1
Theories About the Atom

People have thought about the nature of matter for the last few thousand years. However, most of the experiments that tested their ideas were done in the last one hundred years. The results gave us our present understanding of matter and the atom. Consider these questions as you read about the atomic theory:

a. How has the atom been described?
b. How does an atom gain or lose energy?
c. What is the present atomic theory?

Models of the Atom

In 1808, the school teacher John Dalton offered the first scientific theory about the atom. He described atoms as solid particles that could not be divided. He thought each element had its own kind of atom, as the drawings show.

By the early 1900s, the experiments and ideas of several scientists reshaped Dalton's theory. J. J. Thomson discovered a tiny particle with a negative electric charge. He called it an **electron** (i lek′tron).

Thomson knew that matter has no charge—it is neutral. He reasoned that positive charges must exist to balance the negative electrons. But he was unable to detect positive charges. He thought that atoms were electrons scattered through a thin, positive material.

Later, a particle with a positive charge of the same strength as the electron's negative charge was discovered. This particle, called a **proton** (prō′ton), has about 1,800 times as much mass as an electron.

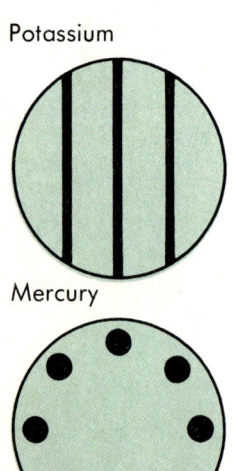

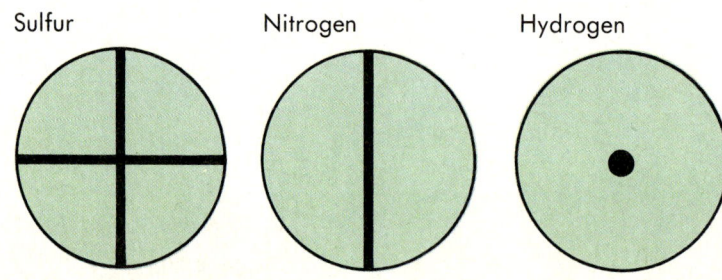

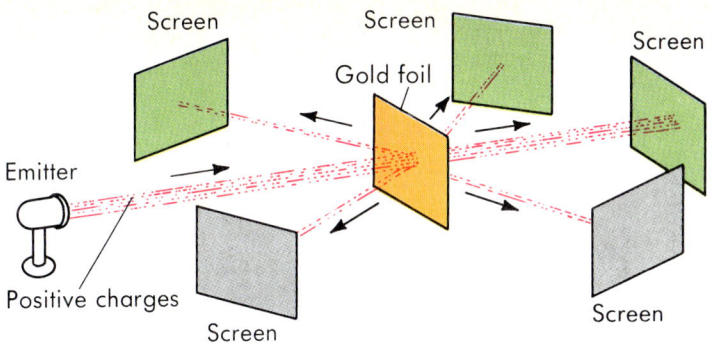

Other scientists also experimented with charged particles. One experiment involved shooting positively charged particles at gold foil. If Thomson's model were correct, the charges would sail through the thin material of an atom. Just in case the model was wrong, Ernest Rutherford suggested that the experimenters place detecting screens as in the drawing above.

The results of this experiment amazed everyone. Even though most of the positive "bullets" passed through the foil, a few did not. Instead, they bounced back. Rutherford used these results to form a new atomic model. Since most charges passed through the foil, the atom must be mainly empty space. Since like charges repel (push away from) each other, whatever caused the positive "bullets" to bounce back must have a positive charge. This positive object is the **nucleus** (nü′klē əs; plural: nü′klē ī). It contains the protons. Compared to the electrons, the nucleus has a large mass.

Since unlike charges attract, the positive nucleus attracts the negative electrons. They whirl around the nucleus. Rutherford compared his model of the atom to the solar system: electrons circle the nucleus as planets orbit the sun.

Rutherford's model raised the question of why the negative electrons in an atom are not drawn directly into the positive nucleus. In 1913, Niels Bohr suggested that perhaps electrons do not fall into the nucleus because they can only move around it in certain fixed paths. Each path is a certain distance from the nucleus, and electrons in those paths have only certain amounts of energy. Each amount of energy is an **energy level.** At the right is Bohr's model of a lithium atom.

Have You Heard?

How many atoms make up a United States penny? A 1973 penny, for instance, is 95% copper and 5% zinc. In its 3.11 g are 2.8×10^{22} copper atoms and 1.4×10^{21} zinc atoms.

The Bohr model of lithium

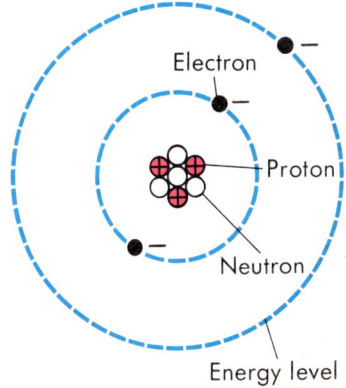

Challenge!

Find out how Aristotle influenced (incorrectly) ideas about the atom for 1,500 years. Use an encyclopedia or a book about the development of the atomic theory.

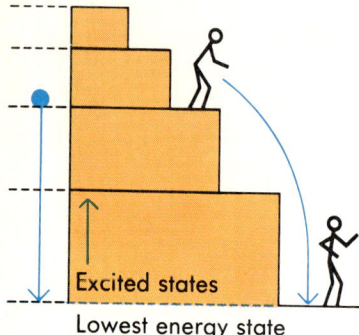

Electrons Have Certain Energies

The drawing compares the energy levels of an atom to a staircase with uneven risers. You can stop on one step or the next, but not in between steps. An atom's energy levels are like the steps. Each electron occupies a certain level, with a certain energy. Other levels and energies in between levels are not allowed. The risers in the model are uneven because the distance and change in energy between levels are also uneven.

Heating an atom or passing electricity through it increases its energy. If an atom gains enough energy, an electron can jump to a higher energy level. The atom is now in an **excited state.** To return the atom to its usual state of lower energy, the electron must release exactly enough energy to move to a lower energy level. Often, we see this released energy as light.

Since each kind of atom has a particular set of energy levels, we can recognize an element by the energy its excited atoms release. This pattern, called a **spectrum** (spek′trəm; plural: spek′trə), is the "fingerprint" of an element. The spectra for calcium, strontium, barium, and zinc, in that order, are shown below.

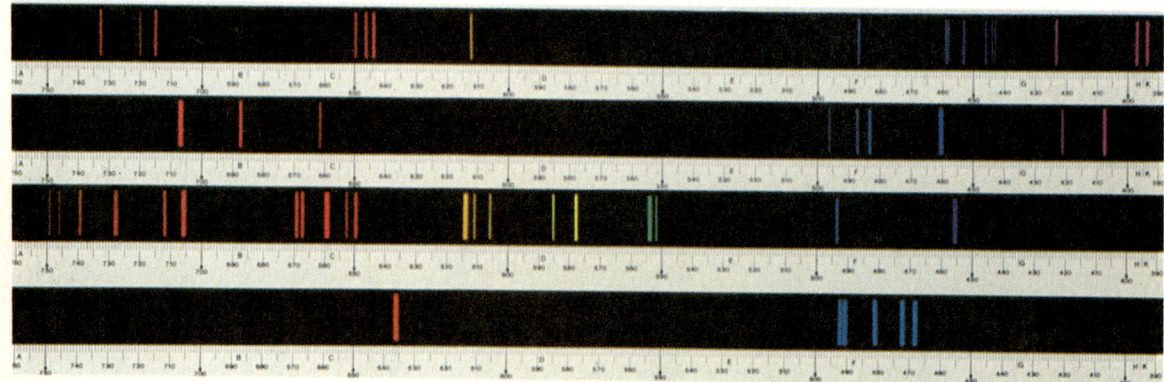

The Atom Today

In 1932, James Chadwick discovered another particle in the nucleus. He called this particle a **neutron** (nü′tron) because it has no charge. It has slightly more mass than a proton.

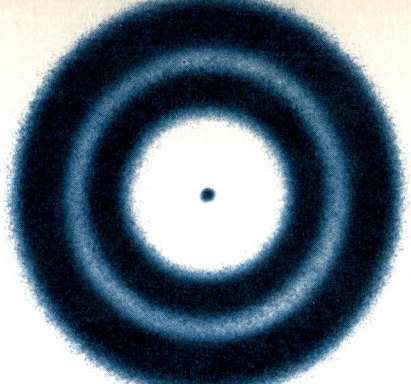

The electron cloud model

By 1926, Erwin Schroedinger, Werner Heisenberg, and others had looked at the evidence about atoms in a new way. They used mathematics to describe the atom. They learned that at any one instant, we cannot know exactly where an electron is. We can only know how likely it is for an electron to be at any place.

The diagram of our present atomic model results from all these ideas. In the electron cloud model, the electrons move rapidly throughout the atom. We describe their position as a cloud. The electrons are somewhere in that cloud. The darker areas show the distances from the nucleus where it is most likely for the electrons to be.

Atoms with electrons close to the nucleus have less energy than atoms with electrons farther away. The atom still has energy levels. Each energy level can hold up to a certain number of electrons, much as a step can hold only a certain number of people at one time. The first energy level holds 2 electrons, the second holds 8, and the third holds 18. The four remaining energy levels each hold 32 electrons.

Even though Bohr's model is not quite correct, scientists often use it to describe the behavior of electrons. For many purposes, scientists find it useful to assume that electrons move in orbits, as the photograph shows.

Have You Heard?

Even though the nucleus accounts for 99.9% of an atom's mass, the diameter of the atom is about 100,000 times larger than the nucleus. If each atom could be pressed into a sphere as small as its nucleus, the Washington Monument could be crammed into a space the size of a pencil eraser.

Review It

1. Explain one early theory about the atom.
2. How does an atom produce a spectrum?
3. Describe the present model of the atom.

11-2 How Atoms Differ

You and your classmates may have a lot in common, but important differences also exist among you. Each of you is a member of this class. Yet your birthdays, physical characteristics, and personalities are different. Similarly, even though the atoms of an element act alike, they come in a few different kinds. To find out about these different kinds, remember these questions as you read:

a. How are the atoms of one element different from those of another?
b. What does an element's mass number tell us?

Atomic Number

During the early 1900s, scientists determined the number of electric charges in the nuclei of many kinds of atoms. In this way, they found that all the atoms of one element have the same number of protons. In fact, the number of protons determines what the element is. For example, every atom of iron has 26 protons. If an atom has 27 protons, it is cobalt. The number of protons in an atom of an element is the **atomic number** of the element. The atomic number of iron is 26. Cobalt is number 27.

All the atoms of one element have the same number of protons and the same atomic number. For example, hydrogen, shown below, has one proton. Its atomic number is 1. Oxygen has eight protons, so its atomic number is 8.

A hydrogen atom

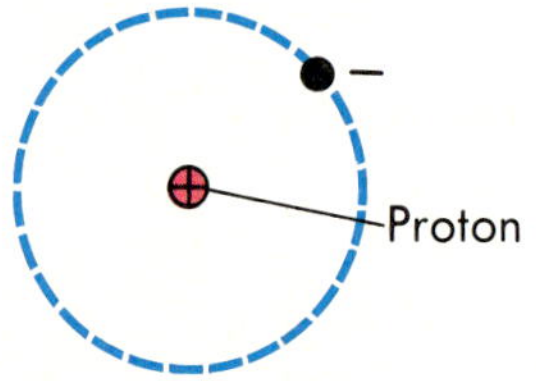

Number of protons = atomic number 1

An oxygen atom

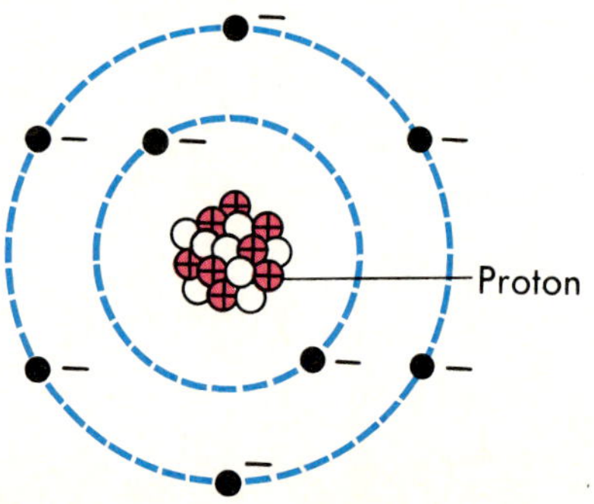

Number of protons = atomic number 8

Even though all atoms of an element have the same number of protons, their masses are not always the same. Some atoms of an element have slightly different masses because they have a different number of neutrons. Atoms of one element that have different numbers of neutrons are **isotopes** (ī′sə tōps) of the element. The mass of each isotope of an element is slightly different.

For example, each of the three isotopes of hydrogen has one proton. The most common form of hydrogen has no neutrons. A form of hydrogen with one neutron is the isotope deuterium (dü tir′ē əm). Another isotope of hydrogen is tritium (trit′ē əm), which has two neutrons. Tritium does not occur naturally on earth, but it can be made artificially.

Though their nuclei and masses differ, all isotopes of an element have similar chemical properties. For example, all three isotopes of hydrogen burn easily.

The number of electrons in an atom is the same as the number of protons. For example, in the drawing, sulfur has 16 protons. It also has 16 electrons. The negative charges of the electrons balance the positive charges of the protons, which makes the atom neutral.

A sulfur atom

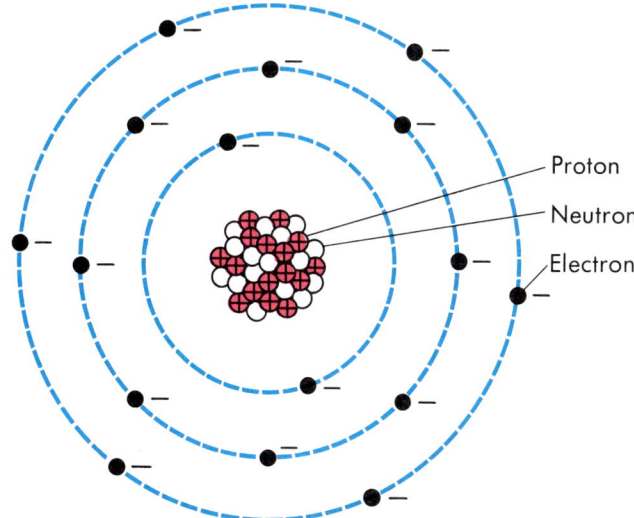

16 protons balance 16 electrons

193

Mass Number

The total number of protons and neutrons in an atom is the **mass number** of the atom. Each isotope of an element has a different number of neutrons. Therefore, each isotope has a different mass number.

The diagrams show the isotopes of hydrogen. The atomic number of all three isotopes is 1. But the mass numbers of the isotopes differ. The mass number of ordinary hydrogen is 1, because it has one proton and no neutrons. Deuterium has one proton and one neutron, so its mass number is 2. Tritium has one proton and two neutrons, so its mass number is 3.

An element's **atomic mass** is an average of the masses of all the element's isotopes. The average is based on the percentage of each isotope in nature. For example, the atomic mass of hydrogen is 1.0079, which is close to the mass number of the most common isotope.

Scientists describe an isotope of an element by giving its symbol, its atomic number, and its mass number. The atomic number appears to the lower left of the symbol. The mass number appears to the upper left. For example, hydrogen is ^1_1H, deuterium is ^2_1H, and tritium is ^3_1H.

Review It

1. What does an element's atomic number represent?
2. How does the mass number distinguish an element's isotopes?

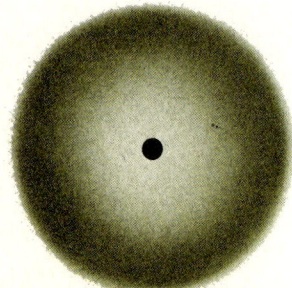

Mass number = 1

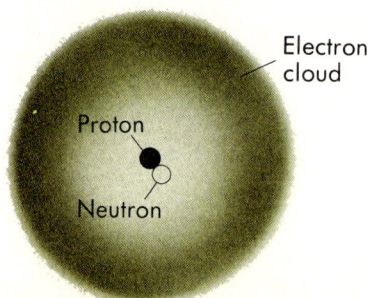

Mass number = 2

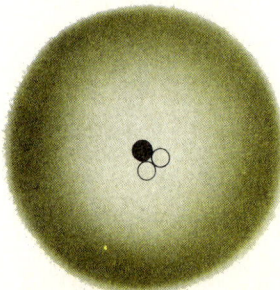

Mass number = 3

Activity

Atomic Models

Purpose
To represent atoms.

Materials
- piece of cardboard
- narrow stick
- clay
- marble or bearing
- blue, red, yellow, and black pencils or markers
- white paper

Procedure

Part A
1. Put the marble in the center of the lump of clay.
2. Shape the clay into a sphere having a diameter of about 5 cm. Be sure that the marble is still in the center.
3. Put a hole in the cardboard with a pencil.
4. Ask your partner to hold the sphere behind the hole in the cardboard.
5. CAUTION: *Do not hit your partner's hand with the stick in this step.* Quickly push the stick through the clay, as shown in *a*. The stick does not have to pass through the center of the sphere.
6. Record what happens to the stick as it passes through the clay.
7. Repeat steps 4–6 nine times. Each time, the partner holding the sphere should move it randomly so that the stick probes different parts of the sphere.

Part B
1. Draw Bohr models for the following elements: $^{11}_{5}B$, $^{20}_{10}Ne$, $^{24}_{12}Mg$, and $^{40}_{20}Ca$. Use one color each for protons, electrons, neutrons, and the energy level.
2. Label the element and the particles, as shown in *b*.

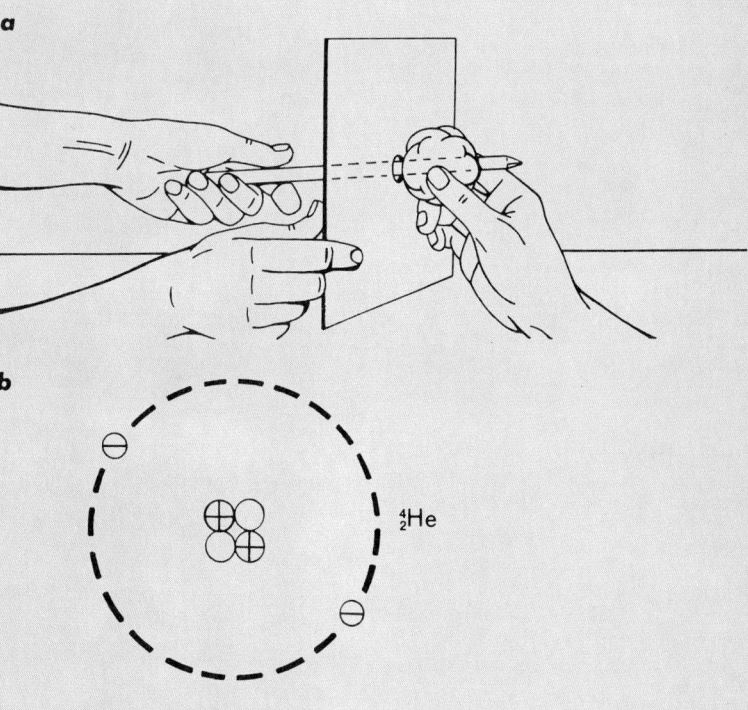

3. Ask your teacher to check your drawings.

Analysis
1. Did the stick pass completely through the sphere more or less often than it struck the marble?
2. What do the marble and clay represent?
3. Compare your Bohr models from Part B with your experimental "evidence" from Part A.

11-3
Classifying Elements

How many ways can you group your classmates? Would you group them by height or by hair color? No matter how you group them, you must look for a common characteristic or pattern among members of each group. In a similar way, we group elements by their characteristics. As you learn to group elements, consider these questions:

a. How are elements classified in the periodic table?
b. What are groups and periods of elements?
c. What properties do elements in certain groups have?

Understanding the Periodic Table

During the 1860s and 1870s, about 60 elements were known. The chemist Dmitri Mendeleev noticed a relationship between the atomic masses and the properties of the known elements. He listed the elements in a table in order of their atomic masses, as shown below. He had to leave spaces in between some elements to make the properties of the known elements fall into groups.

Mendeleev put elements with similar characteristics in columns. He noticed that similar chemical properties appeared at regular intervals—periodically. The table was called the **periodic** (pir/ē od/ik) **table of elements.**

Mendeleev believed that the spaces marked elements not yet known. He hoped that scientists would discover them. He predicted that these elements would have properties like those of other elements in a column.

Mendeleev's table of 1872

GROUP		I	II	III	IV	V	VI	VII	VIII
SERIES	1	H(1)							
	2	Li(7)	Be(9.4)	B(11)	C(12)	N(14)	O(16)	F(19)	
	3	Na(23)	Mg(24)	Al(27.3)	Si(28)	P(31)	S(32)	Cl(35.5)	
	4	K(39)	Ca(40)	-(44)	Ti(48)	V(51)	Cr(52)	Mn(55)	Fe(56), Co(59), Ni(59), Cu(63)
	5	[Cu(63)]	Zn(65)	-(68)	-(72)	As(75)	Se(78)	Br(80)	
	6	Rb(85)	Sr(87)	?Yt(88)	Zr(90)	Nb(94)	Mo(96)	-(100)	Ru(104), Rh(104), Pd(106), Ag(108)
	7	[Ag(108)]	Cd(112)	In(113)	Sn(118)	Sb(122)	Te(125)	I(127)	
	8	Cs(133)	Ba(137)	?Di(138)	?Ce(140)	—	—	—	
	9	—	—	—	—	—	—	—	
	10	—	—	?Er(178)	?La(180)	Ta(182)	W(184)	—	Os(195), Ir(197), Pt(198), Au(199)
	11	[Au(199)]	Hg(200)	Tl(204)	Pb(207)	Bi(208)	—	—	
	12				Th(231)		U(240)		

196

For example, Mendeleev recognized that titanium (Ti) has properties similar to those of carbon (C) and silicon (Si). He placed it in the same column with these elements, which left a blank. Mendeleev predicted that an element with an atomic mass between that of calcium and titanium would be found to fill this gap.

Years later, the element was discovered and named scandium (Sc). It had the properties and the atomic mass that Mendeleev predicted. Scientists discovered other elements by using the periodic table in this way.

In 1913, the physicist Henry Moseley found that an element's properties are more closely related to its atomic number than to its atomic mass. Since then, the periodic table has been based on the elements' atomic numbers.

The present periodic table appears on the following pages. It contains the 106 elements known today. Eighty-nine of these elements occur naturally on earth. The other 17 are made artificially. Some of them exist for only a fraction of a second.

Each box in the periodic table contains an element's symbol and name. The atomic number appears above the symbol. The atomic mass is below the name.

The two rows of elements below the main table are usually shown there to save space. The diagram below shows where these rows fit into the table.

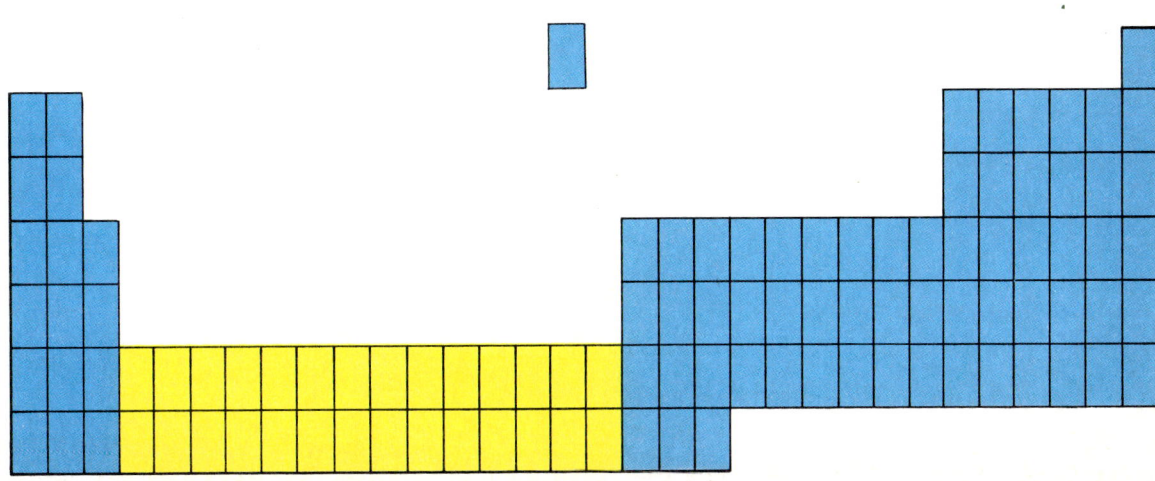

Periodic Table of the Elements
(Based on Carbon 12 = 12.0000)

Group	IA	IIA	IIIB	IVB	VB	VIB	VIIB		VIII								
1	1 H Hydrogen 1.01 (1)																
2	3 Li Lithium 6.94 (2,1)	4 Be Beryllium 9.01 (2,2)															
3	11 Na Sodium 22.99 (2,8,1)	12 Mg Magnesium 24.30 (2,8,2)															
4	19 K Potassium 39.10 (2,8,8,1)	20 Ca Calcium 40.08 (2,8,8,2)	21 Sc Scandium 44.96 (2,8,9,2)	22 Ti Titanium 47.90 (2,8,10,2)	23 V Vanadium 50.94 (2,8,11,2)	24 Cr Chromium 52.00 (2,8,13,1)	25 Mn Manganese 54.94 (2,8,13,2)	26 Fe Iron 55.85 (2,8,14,2)	27 Co Cobalt 58.93 (2,8,15,2)								
5	37 Rb Rubidium 85.47 (2,8,18,8,1)	38 Sr Strontium 87.62 (2,8,18,8,2)	39 Y Yttrium 88.91 (2,8,18,9,2)	40 Zr Zirconium 91.22 (2,8,18,10,2)	41 Nb Niobium 92.91 (2,8,18,12,1)	42 Mo Molybdenum 95.94 (2,8,18,13,1)	43 Tc Technetium 98.91 (2,8,18,13,2)	44 Ru Ruthenium 101.07 (2,8,18,15,1)	45 Rh Rhodium 102.91 (2,8,18,16,...)								
6	55 Cs Cesium 132.91 (2,8,18,18,8,1)	56 Ba Barium 137.33 (2,8,18,18,8,2)	57 La Lanthanum 138.91 * (2,8,18,18,9,2)	72 Hf Hafnium 178.49 (2,8,18,32,10,2)	73 Ta Tantalum 180.95 (2,8,18,32,11,2)	74 W Tungsten 183.85 (2,8,18,32,12,2)	75 Re Rhenium 186.21 (2,8,18,32,13,2)	76 Os Osmium 190.20 (2,8,18,32,14,2)	77 Ir Iridium 192.22								
7	87 Fr Francium (223) (2,8,18,32,18,8,1)	88 Ra Radium 226.02 (2,8,18,32,18,8,2)	89 Ac Actinium (227) ** (2,8,18,32,18,9,2)	104 (260) (2,8,18,32,32,10,2)	105 (260) (2,8,18,32,32,11,2)	106 (263) (2,8,18,32,32,12,2)											

*	58 Ce Cerium 140.12 (2,8,18,19,9,2)	59 Pr Praseodymium 140.91 (2,8,18,21,8,2)	60 Nd Neodymium 144.24 (2,8,18,22,8,2)	61 Pm Promethium (145) (2,8,18,23,8,2)	62 Sm Samarium 150.40 (2,8,18,24,8,2)	63 Eu Europium 151.96
**	90 Th Thorium 232.04 (2,8,18,32,18,10,2)	91 Pa Protoactinium 231.04 (2,8,18,32,20,9,2)	92 U Uranium 238.03 (2,8,18,32,21,9,2)	93 Np Neptunium 237.05 (2,8,18,32,22,9,2)	94 Pu Plutonium (244) (2,8,18,32,24,8,2)	95 Am Americium (243)

Legend:
- ■ Solid
- ■ Liquid
- ■ Gas
- □ Made artificially

NONMETALS →

← METALS

			IIIA	IVA	VA	VIA	VIIA	VIIIA
								2 He Helium 4.00 (2)
			5 B Boron 10.81 (2,3)	6 C Carbon 12.01 (2,4)	7 N Nitrogen 14.01 (2,5)	8 O Oxygen 16.00 (2,6)	9 F Fluorine 19.00 (2,7)	10 Ne Neon 20.17 (2,8)
	IB	IIB	13 Al Aluminum 26.98 (2,8,3)	14 Si Silicon 28.09 (2,8,4)	15 P Phosphorus 30.97 (2,8,5)	16 S Sulfur 32.06 (2,8,6)	17 Cl Chlorine 35.45 (2,8,7)	18 Ar Argon 39.95 (2,8,8)
28 Ni Nickel 58.71 (2,8,16,2)	29 Cu Copper 63.55 (2,8,18,1)	30 Zn Zinc 65.38 (2,8,18,2)	31 Ga Gallium 69.74 (2,8,18,3)	32 Ge Germanium 72.59 (2,8,18,4)	33 As Arsenic 74.92 (2,8,18,5)	34 Se Selenium 78.96 (2,8,18,6)	35 Br Bromine 79.90 (2,8,18,7)	36 Kr Krypton 83.80 (2,8,18,8)
46 Pd Palladium 106.40 (2,8,18,18)	47 Ag Silver 107.87 (2,8,18,18,1)	48 Cd Cadmium 112.41 (2,8,18,18,2)	49 In Indium 114.82 (2,8,18,18,3)	50 Sn Tin 118.69 (2,8,18,18,4)	51 Sb Antimony 121.75 (2,8,18,18,5)	52 Te Tellurium 127.60 (2,8,18,18,6)	53 I Iodine 126.90 (2,8,18,18,7)	54 Xe Xenon 131.30 (2,8,18,18,8)
78 Pt Platinum 195.09 (2,8,18,32,17,1)	79 Au Gold 196.97 (2,8,18,32,18,1)	80 Hg Mercury 200.59 (2,8,18,32,18,2)	81 Tl Thallium 204.37 (2,8,18,32,18,3)	82 Pb Lead 207.20 (2,8,18,32,18,4)	83 Bi Bismuth 208.98 (2,8,18,32,18,5)	84 Po Polonium (209) (2,8,18,32,18,6)	85 At Astatine (210) (2,8,18,32,18,7)	86 Rn Radon (222) (2,8,18,32,18,8)

64 Gd Gadolinium 157.25 (2,8,18,25,9,2)	65 Tb Terbium 158.93 (2,8,18,27,8,2)	66 Dy Dysprosium 162.50 (2,8,18,28,8,2)	67 Ho Holmium 164.93 (2,8,18,29,8,2)	68 Er Erbium 167.26 (2,8,18,30,8,2)	69 Tm Thulium 168.93 (2,8,18,31,8,2)	70 Yb Ytterbium 173.04 (2,8,18,32,8,2)	71 Lu Lutetium 174.97 (2,8,18,32,9,2)
96 Cm Curium (247) (2,8,18,32,25,9,2)	97 Bk Berkelium (247) (2,8,18,32,27,8,2)	98 Cf Californium (251) (2,8,18,32,28,8,2)	99 Es Einsteinium (254) (2,8,18,32,29,8,2)	100 Fm Fermium (257) (2,8,18,32,30,8,2)	101 Md Mendelevium (258) (2,8,18,32,31,8,2)	102 No Nobelium (259) (2,8,18,32,32,8,2)	103 Lr Lawrencium (260) (2,8,18,32,32,9,2)

Copper, Cu

Silver, Ag

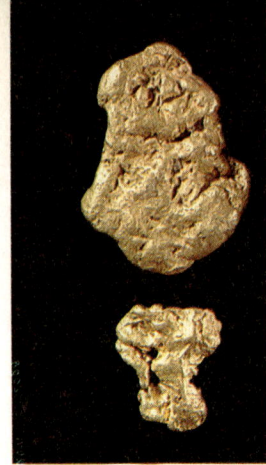
Gold, Au

Using the Periodic Table

Examining an element's position in the periodic table can tell us about the element. Each row is called a **period.** Along a period, a gradual change in chemical properties occurs from one element to another. For example, elements on the left are more metallic than those on the right. Changes occur because the number of protons increases by one. The number of electrons also increases by one. This increase is important because the number of outer electrons determines an element's chemical properties.

Each column is called a **group.** Notice the list of numbers in each box of the periodic table. These numbers tell how many electrons occupy each energy level in that atom. Elements in a group have similar properties because electrons fill their outer energy level in the same way. The elements in a group are like the members of a family. Each is different, but all are related by common characteristics.

The pictures show the Group IB elements. All are metals with similar properties. The atoms of each have one outer electron.

Characteristics of Some Groups

If you know something about one element in a group, you know something about them all. For example, the elements of Group IA are the most metallic elements. Each is a light, soft, shiny, silvery metal that reacts with oxygen and water. These metals undergo chemical changes easily.

Group VIIA contains the most nonmetallic element—fluorine. Chlorine and fluorine are gases. Bromine and iodine easily become gases. These elements also undergo chemical changes easily.

Group VIII elements are all gases. They are the most stable of the elements—they do not naturally undergo chemical changes. This characteristic occurs because their outer energy level is filled with the maximum number of electrons. When excited, these elements give off light as shown. These elements are often used in signs.

The transition elements take up the middle section of the periodic table. They all have one or two electrons in their outer energy level. Most are heavy, hard metals. Chemical changes of the transition elements produce colorful new substances, as the picture below shows.

Group VIII

He

Ne

Ar

Kr

Xe

Rn

Review It

1. How is the table of elements periodic?
2. What information can the periodic table give us?
3. Describe the elements of Group IA.

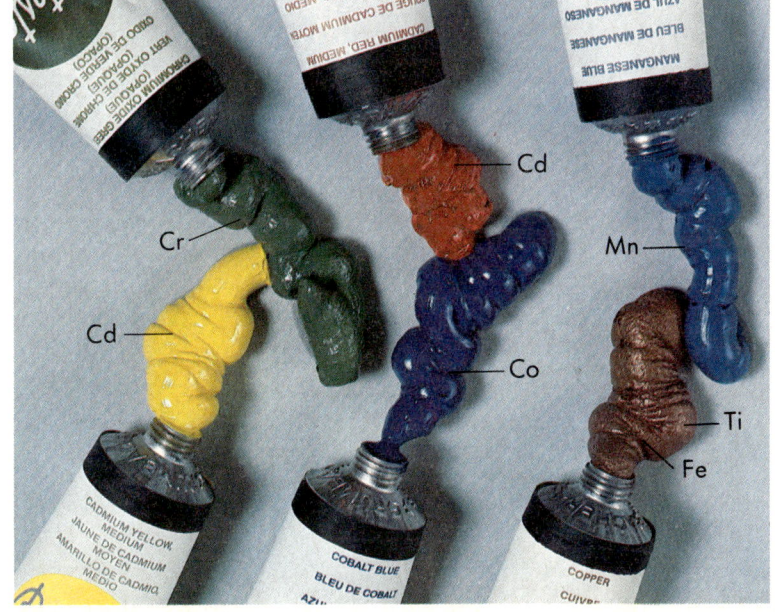
Some substances containing transition metals

Activity

Using the Periodic Table

Purpose
To practice getting information from the periodic table.

Materials
- paper and pencil
- ruler
- periodic table

Procedure
1. Copy the chart shown. Make it large enough to contain information about 20 elements.
2. Using information from the periodic table, fill in the chart for the first 20 elements. One element is done for you as an example to follow.
   You can find uses of the elements in this book or a dictionary. You may add any that you know already. You will need the mass numbers of the most common isotopes of elements 1–20 to complete the chart. These numbers follow.

hydrogen-1
helium-4
lithium-7
beryllium-9
boron-11
carbon-12
nitrogen-14
oxygen-16
fluorine-19
neon-20
sodium-23
magnesium-24
aluminum-27
silicon-28
phosphorus-31
sulfur-32
chlorine-35
argon-40
potassium-39
calcium-40

Analysis
1. How would the periodic table be different if elements were listed according to mass number instead of atomic number? (Hint: Look at the last three elements in the list.)
2. How many outer electrons do metals have? nonmetals?

Data chart

Element	Symbol	Number of Protons	Number of Neutrons	Number of Electrons	Number of Electrons in each Energy Level	Uses
Lithium	$^{7}_{3}$Li	3	4	3	2, 1	

Did You Know?

Alchemy

Can gold be made from lead? Is a frog's stomach the secret to eternal life? Are such ideas the dreams of madmen or the schemes of crooks? Perhaps. But these questions led to the beginnings of modern chemistry.

For nearly a thousand years, alchemists were the closest thing to scientists the world knew. The drawing shows a typical alchemist's study. Two great objectives guided their research. First, alchemists tried to change common metals into gold. Second, they tried to find ways of curing disease and giving people eternal life.

A typical recipe for changing lead to gold might include blood, urine, wine, eggs, mercury, and silver. The whole mixture would then be cooked and distilled. Some of the alchemists' ideas seem strange to us now. However, alchemists had different ideas about matter than we do. Our modern theory of the atom was still hundreds of years in the future.

Alchemists believed that all kinds of matter were very much the same. Everything was made from a combination of three or four basic elements. Substances differed only in the way these basic elements were put together. So, changing a common metal such as lead into a beautiful, valuable metal such as gold seemed possible. Rearrange the basic elements. Presto! A new substance would be created.

Not all alchemists were interested in learning about the nature of matter. Some preferred to find ways to cheat kings and other people. One trick was to cover a gold bar with a layer of iron. The bar was then dropped into a "magic liquid." This liquid was actually a strong acid that dissolved the iron. Lo and behold! A bar of gold appeared.

A way of making gold from silver was never found. The secret of eternal life was never discovered. However, many of the methods invented by alchemists are still used by chemists today. We can truly say that modern chemistry grew out of the work alchemists began.

For Discussion
1. What were the two main goals of alchemy?
2. In what ways were the alchemists' ideas about matter different from ours?

203

Chapter Summary

- An atom consists of negatively charged electrons that whirl around a positively charged nucleus. (11–1)
- An electron gives off energy when it moves from a higher energy level to a lower one. (11–1)
- Each element can be identified by its spectrum. (11–1)
- The number of protons in an atom determines which element it is. (11–2)
- An atom has as many electrons as protons. (11–2)
- Atoms of one element can have different numbers of neutrons. (11–2)
- The chemical properties of an element depend on the number of electrons in the atoms of that element. (11–2)
- An element's atomic mass is an average of the masses of the element's isotopes, based on the percentage of each in nature. (11–2)
- The periodic table of elements arranges the elements in groups that have similar chemical properties. (11–3)
- Scientists used the periodic table to predict the existence and chemical properties of elements. (11–3)

Interesting Reading

Asimov, Isaac. *How Did We Find Out About Atoms?* Walker, 1976. Traces the history of the knowledge of atoms, beginning with the ancient Greeks, and discusses the work of many scientists.

Gallant, Roy A. *Explorers of the Atom.* Doubleday, 1974. Describes our ideas about the atom through the centuries and how we have learned to use the atom.

Questions/Problems

1. Compare Dalton's atomic model with our current model.
2. What experimental evidence convinced Rutherford that the atom is mainly empty space?
3. Draw a Bohr diagram of fluorine (F).
4. What is the mass number of the lead isotope containing 125 neutrons?
5. How many outer electrons does chlorine (Cl) have?
6. How many protons, electrons, and neutrons does an atom of $^{90}_{38}Sr$ contain?
7. What are the atomic number and mass number of an element with twelve protons, twelve electrons, and thirteen neutrons in its atoms?
8. Why would it be easier to find an element if you could predict its properties?
9. In what ways are all the elements in Group VIII alike? How is helium different?
10. What factor determines how easily the atoms of an element undergo chemical changes?

Extra Research

1. Choose an element that interests you. Find out the following about it: a) when, how, and by whom it was discovered; b) its physical and chemical properties; c) some of its uses.
2. Make your own three-dimensional model of an atom. Choose a smaller element, since smaller elements are simpler to model. You might use cotton batting, coat hangers, polystyrene, and small rubber balls to represent the parts of an atom.

Chapter Test

A. Vocabulary Write the numbers 1–10 on a piece of paper.
Match the definition in Column I with the term it defines in Column II.

Column I

1. positively charged particle in the nucleus
2. particle that travels around the nucleus
3. particle that has no charge
4. pattern of energy released by an element's atoms
5. atoms of an element with different mass numbers
6. approximate path of electrons around the nucleus
7. number of protons in an atom of an element
8. occurs when an atom gains energy and electrons can jump to higher energy levels
9. central mass of an atom
10. average based on the percentage of an element's isotopes in nature

Column II

a. atomic mass
b. atomic number
c. electron
d. energy level
e. excited state
f. isotopes
g. neutron
h. nucleus
i. proton
j. spectrum

B. Multiple Choice Write the numbers 1–10 on your paper.
Choose the letter that best completes the statement or answers the question.

1. According to the current atomic theory, an atom is best described as   a) a ball.   b) a central mass surrounded by an electron cloud.   c) the solar system.   d) none of the above.

2. Because the electrons move only in certain paths,   a) they do not fall into the nucleus.   b) they have only certain energies.   c) they have any energy.   d) a and b.

3. The particle with the least mass is the
a) electron.   b) atom.   c) proton.   d) neutron.

4. Isotopes of the same element have the same
a) mass.   b) number of protons.   c) number of neutrons.   d) b and c.

5. An atom is neutral because it has the same number of electrons as   a) energy levels.   b) neutrons.   c) protons.   d) isotopes.

6. In the symbol $^{35}_{17}Cl$, 35 is the   a) proton number.   b) electron number.   c) mass number.   d) atomic number.

7. Mendeleev based his periodic table on the elements'   a) repeating properties.   b) atomic numbers.   c) atomic masses.   d) a and c.

8. Today, elements in periods are arranged according to   a) neutron number.   b) mass number.   c) atomic mass.   d) atomic number.

9. Elements in Group IA have the same number of   a) outer electrons.   b) neutrons.   c) protons.   d) isotopes.

10. The periodic table   a) can be used to find the number of protons and electrons in any element.   b) lists elements in groups with similar chemical properties.   c) still contains many gaps.   d) a and b.

Chapter 12
The Atomic Nucleus

This picture shows tracks that electrons, protons, and other small particles made in liquid helium. Scientists use pictures like these to learn about the particles that make up the nucleus. The study of nuclear particles deals with the basic question, "What is matter made of?"

This chapter is about the atomic nucleus and about particles within the nucleus. The chapter explains how nuclei of one element change into nuclei of other elements. It also discusses nuclear reactions and how mass and energy change into each other.

Chapter Objectives
1. Describe the structure of the nucleus and of the particles within the nucleus.
2. Explain how radioactive decay occurs.
3. Define nuclear fission.
4. Describe nuclear fusion and contrast it with fission.
5. Discuss the law of conservation of mass-energy.

12–1
The Structure of the Nucleus

The nucleus of an atom is so tiny that, if all the atomic nuclei in your body could be packed together, they would fit on the tip of your little finger. Scientists believe that the protons and neutrons in the nucleus are themselves made up of other particles. Answer these questions as you continue reading about the nucleus:

a. What holds the particles of a nucleus together?
b. What do protons and neutrons consist of?
c. Why are more than three kinds of quarks needed?

The Force that Holds the Nucleus Together

Forcing all the atomic nuclei in your body into a tiny space would be very difficult. All nuclei have positive electric charges, so they repel one another. You could not push hard enough to make nuclei touch.

Since all the protons within a nucleus also have positive electric charges, they repel each other. Because the nucleus does not fly apart, some stronger, attracting force must hold it together. The force binding nuclear particles together is the **strong force.** This force holds together nuclear particles that are extremely close to one another. At distances greater than the size of the nucleus, the strong force has no effect.

The strong force, like gravity or the force of electricity and magnetism, is one of nature's major forces. In the first drawing, the electric force repels a high-speed proton that passes near a nucleus. But in the other picture, the strong force attracts and captures a high-speed proton moving close enough to the nucleus.

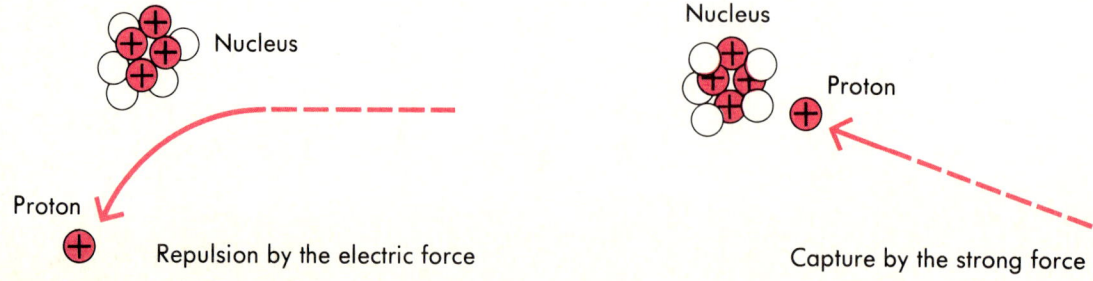

Repulsion by the electric force

Capture by the strong force

The strong force does not affect the electron cloud surrounding a nucleus. Therefore, this force does not affect chemical and physical properties and changes.

Most nuclei have neutrons in addition to protons. Neutrons have no electric charge. However, they add to the strong force holding the nucleus together.

The Building Blocks of Protons and Neutrons

Physicists use very large machines to explore the nuclei of atoms. They shoot high-speed beams of protons or electrons at nuclei in machines such as the one at the right. In the resulting collisions, the nuclei split apart. The parts of the nuclei cannot be seen, but they make tracks in liquid helium. The tracks led physicists to discover many nuclear particles besides protons and neutrons.

In 1963, physicists explained these observations. According to their theory, nuclear particles consist of basic units called **quarks** (kwôrks). At first, physicists believed three kinds of quarks could be enough to form all the known nuclear particles. In a spirit of fun, they named these quarks "up," "down," and "strange." They assigned the up quark an electric charge of +2/3, the down quark −1/3, and the strange quark −1/3.

The chart and the drawing show combinations of different quarks. Protons consist of two up quarks and one down quark. Neutrons are one up quark and two down quarks. The strange quark is found in other, smaller particles.

Quarks in a proton
Quarks in a neutron

Particle	Quark parts	Quark charges	Total charge
Proton	up, up, down	$+\frac{2}{3} + \frac{2}{3} - \frac{1}{3}$	+1
Neutron	up, down, down	$+\frac{2}{3} - \frac{1}{3} - \frac{1}{3}$	0

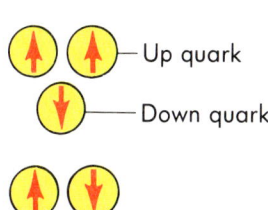

— Up quark
— Down quark

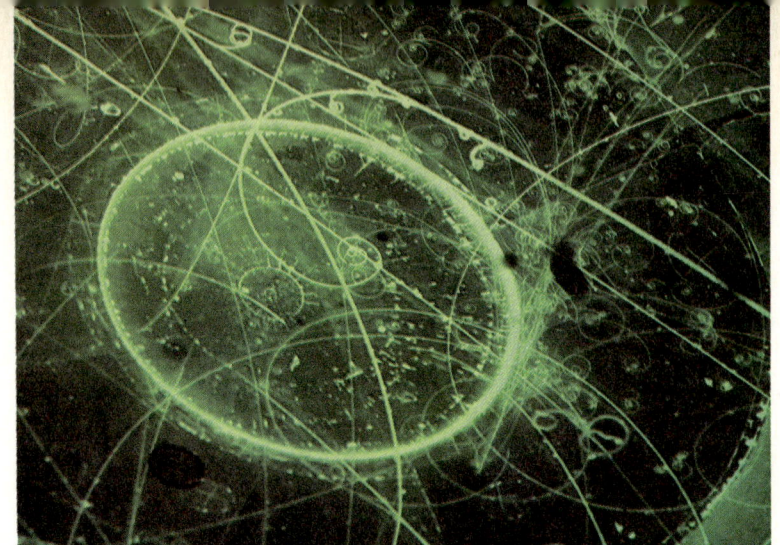

Challenge!

Look up *Wilson cloud chamber* and *bubble chamber* in an encyclopedia. Find out how they are used to study nuclear particles. A bubble chamber is shown below.

More Kinds of Quarks

Since 1963, physicists working in huge laboratories have discovered still more nuclear particles. Their discoveries keep bringing great excitement to the scientific world. Many research groups have been competing to be the first to identify new particles.

However, too many new particles were found. So many particles could not come from only three kinds of quarks. The photograph above shows just a few of these particles. To explain the new particles, physicists needed three more kinds of quarks, named "charm," "truth," and "beauty." Again, the names are just labels. They have nothing to do with the everyday meaning of the words.

We cannot study single quarks. In fact, most physicists think that they may never separate quarks from the particles they form. But the theory that quarks are the basic building blocks of the nucleus does explain the evidence. Physicists continue to study nuclear particles in search of quarks.

Review It

1. What is the force that binds nuclear particles?
2. What are quarks?
3. Name the six kinds of quarks.

Breakthrough

Particle Accelerators

What is an atom made of? One way to find an answer to this question is to break an atom apart. Then we can "pick up the pieces" to see what is inside. Unfortunately, the right "bullet" needed to break apart an atom is hard to find.

Electrons do not make good "bullets." Electrons are repelled by an atom's own electron cloud. A proton could get through the electron cloud, but would then be repelled by the nucleus.

Whatever they use as "bullets," scientists have found a simple way to break atoms apart. The trick is to get the "bullet" going very fast—as fast as 250 million meters per second. Then the "bullet" will smash into the nucleus before it can be pushed away.

Machines designed to make the "bullets" move very quickly are called particle accelerators or atom smashers. Linear accelerators are built in a straight line. They must be very long so that particles can build up speed. One accelerator in California is over three kilometers long.

Fifty years ago Dr. E. O. Lawrence suggested that electrons moving round and round in a circle could be the "bullets." A circular accelerator, called a cyclotron, gets the electrons moving. Then the electrons' paths are bent into a circle by giant magnets. The electrons go around in a circle millions of times per second. Their speed is fast enough to enter a nucleus and smash it apart. You can see the main ring of the Fermilab accelerator in the photograph. Its circumference is about 6.5 kilometers.

Accelerators have taught us a great deal about atoms. At one time, protons, neutrons, and electrons were the only known parts of the atom. Now, more than 200 kinds of particles have been discovered within the atom.

For Discussion
1. What do scientists learn when they smash atoms?
2. What is the advantage of building a cyclotron rather than a linear accelerator?

12–2 Radioactivity

In 1896, the French scientist Henri Becquerel placed a sample of uranium salt near an unopened container of photographic film. To his surprise, the film changed just as if it had been exposed to light. He discovered accidentally that uranium gives off powerful invisible rays. As you read about these rays, think about these questions:

a. What is radioactivity?
b. What are some effects and uses of radioactivity?

Radioactivity Results from the Breakdown of Nuclei

Uranium nuclei are unstable, which means they spontaneously decay (break apart) into lighter nuclei. This decay is **radioactivity** (rā′dē ō ak tiv′ə tē). Isotopes of an element are either radioactive or stable. Notice in the graph that more radioactive than stable isotopes exist.

The nuclei in a given amount of a radioactive element steadily decay until few are left. No one can tell when any particular nucleus will decay. But we can predict how many will decay in a certain time. For example, half the nuclei in a sample of $^{235}_{92}U$ will decay in 713 million years. The time it takes half of the nuclei of a radioactive isotope to decay is that isotope's **half-life**. The half-lives of radioactive isotopes are constant.

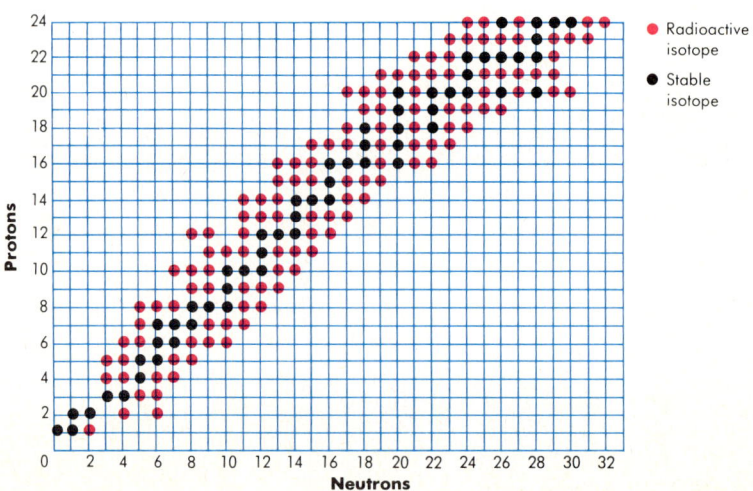

Comparing stable and radioactive isotopes

If a sample of $^{235}_{92}$U contains 1,000 nuclei, 500 of those nuclei would remain after 713 million years. After another 713 million years, 250 nuclei would remain. The graph at the right shows how the number of nuclei decreases over time. The decaying nuclei form lighter elements, some of which are also radioactive. The table below gives the half-lives of some radioactive isotopes.

The Half-lives of Some Radioactive Isotopes

Element	Isotope	Half-life
Polonium	$^{214}_{84}$Po	0.0001 seconds
Iodine	$^{131}_{53}$I	8 days
Strontium	$^{90}_{38}$Sr	28 years
Carbon	$^{14}_{6}$C	5,730 years
Uranium	$^{238}_{92}$U	4.5 billion years

Stages of radioactive decay
Start

| 1,000 $^{235}_{92}$U nuclei |

713 million years

| 500 $^{235}_{92}$U nuclei | Product isotopes |

1426 million years

| 250 $^{235}_{92}$U nuclei | Product isotopes |

2139 million years

| 125 $^{235}_{92}$U nuclei | Product isotopes |

As nuclei decay, they release particles and rays called **radiation** (rā′dē ā′shən). Decaying nuclei give off three kinds of radiation, which were named after the first letters of the Greek alphabet: alpha, beta, and gamma. Later, we learned what these kinds of radiation are. An alpha particle is a helium nucleus. A beta particle is an electron created during the decay of a neutron in a nucleus. (A proton remains in place of the neutron.) A gamma ray is a bundle of energy similar to, but much more powerful than, a bundle of light energy.

Radiation can pass through some substances. The diagram shows how a thin sheet of aluminum foil stops an alpha particle. A five-centimeter sheet of aluminum is needed to stop a beta particle. It takes one meter of concrete or five centimeters of lead to stop a gamma ray.

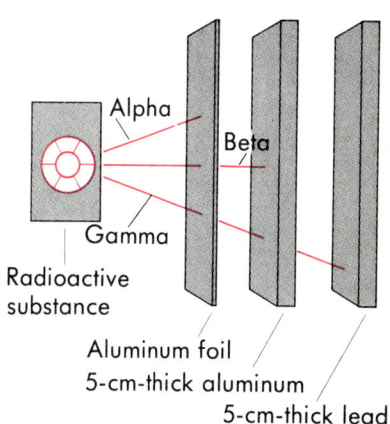

Comparing the strengths of alpha, beta, and gamma radiation

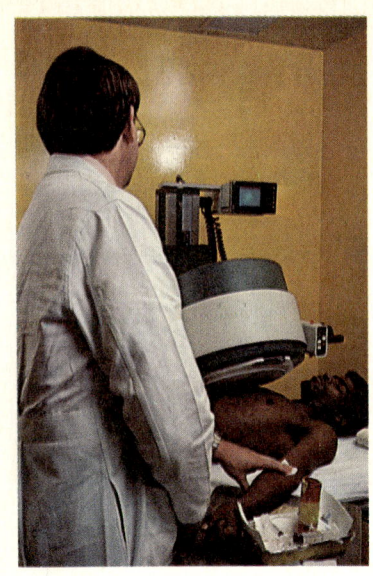

The Effects and Uses of Radioactivity

Radioactivity occurs normally all around us. Some nuclei in all matter are continually decaying. The amount of this natural radiation is generally small.

Exposure to large doses of radiation can cause radiation sickness, which can be fatal. Exposure to even small amounts of radiation over a long time may cause cancer. Radiation can also produce changes in our genes, which determine inherited traits. For this reason, women and girls who may someday become pregnant should try to avoid sources of radiation.

Radiation can be helpful when it is controlled. For example, the cancer patient shown is being treated with radiation because it can kill diseased cells. But radiation can also kill healthy cells. So it must be aimed carefully to hit only the diseased cells.

Radioactive isotopes have helped us learn about earth's history. For example, geologists have estimated the earth's age by measuring the fractions of $^{238}_{92}U$ and $^{235}_{92}U$ in rocks. By making calculations based on the half-lives of the uranium isotopes, geologists have concluded that the earth is about 4.5 billion years old.

Scientists use the radioactive isotope $^{14}_{6}C$ to determine the ages of fossils, which are the preserved remains of dead plants and animals. All living organisms maintain the same amount of $^{14}_{6}C$ in their bodies in relation to other isotopes of carbon. After an organism dies, it stops taking in any kind of carbon. Years later, scientists can measure the amount of $^{14}_{6}C$ that decayed and so determine how long ago the organism died.

Review It

1. What does the term "half-life" mean?
2. How is radiation used in medicine?

Activity

Radioactive Decay

Purpose
To graph data from the decay of a radioactive isotope.

Materials
- graph paper
- pencil

Procedure
1. Copy the grid shown. Use at least one-half of a sheet of graph paper for your grid.
2. An experiment was performed to determine the half-life of an isotope. Some data recorded during the experiment are listed here. Plot the results on your grid.
3. Draw a smooth curve through the data points. Use your graph to answer the questions below.

Analysis
1. What is the half-life of this isotope?
2. The grid ends at 30 days. On what day will the next half-life be completed? How many radioactive nuclei will remain on that day?
3. It would be wrong to continue your calculations after getting the answer to #2. Use your knowledge of the atom and radioactivity to explain why.

Radioactive Nuclei Remaining	Time (Days)
80	0
71	1
59	3
50	5
46	6
31	11
26	13
24	14
22	15
20	16
13	21
12	22
11	23
10	24
8	27

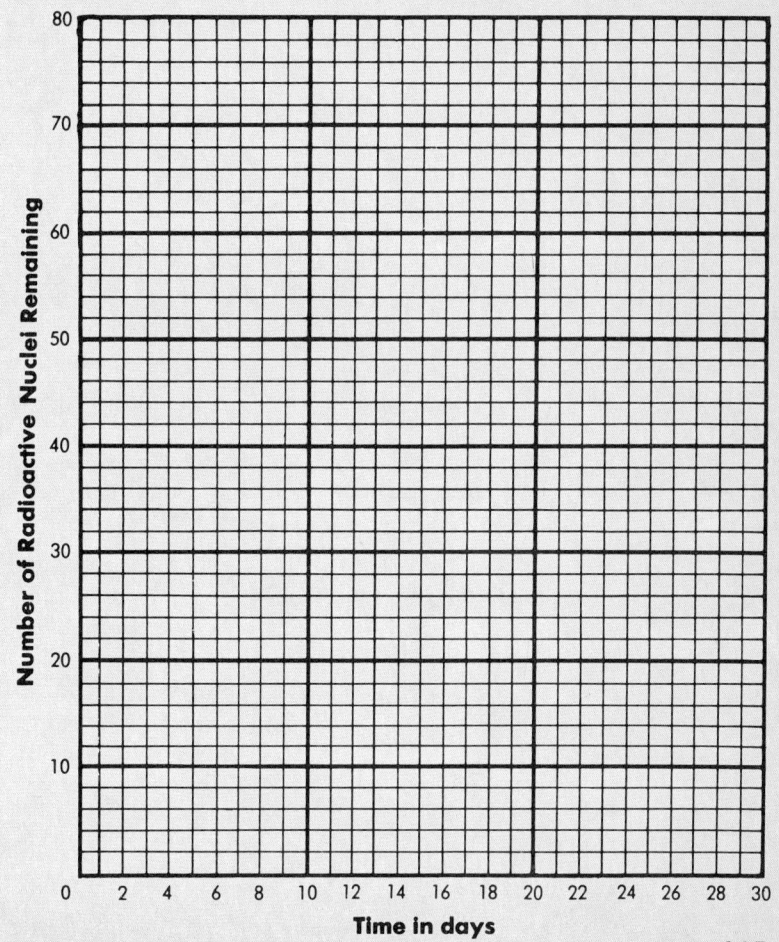

Decay of a radioactive isotope

12-3
Nuclear Fission

Long before people knew about atoms and nuclei, alchemists tried to change one element into another. Little did they know that this change occurred all the time in the process of radioactivity. Physicists now know how to make some heavy nuclei break apart to form lighter elements. As you read about the splitting of nuclei, keep these questions in mind:

a. What is nuclear fission?
b. What are some uses of nuclear fission?

What Happens During Fission

In 1939, scientists found that certain heavy nuclei split when they absorb slowly-moving neutrons. The splitting of a nucleus in this way is nuclear **fission** (fish′ən).

The picture shows a typical reaction when a slow-moving neutron splits a uranium nucleus ($^{235}_{92}U$). In this particular example, three neutrons and two lighter isotopes, barium ($^{141}_{56}Ba$) and krypton ($^{92}_{36}Kr$), are formed.

The number of protons after fission occurs (56 + 36 = 92) is the same as the number of protons before fission. The total number of nuclear particles afterwards is 141 + 92 + (3×1) = 236. This number is the same as the original number of particles, 1 + 235 = 236. The nuclei and particles at the end, though, have slightly less total mass than the uranium nucleus and neutron at the beginning. The energy of this missing mass is still present. It went into making the nuclei speed away from each other. These nuclei hit material around them and heat it.

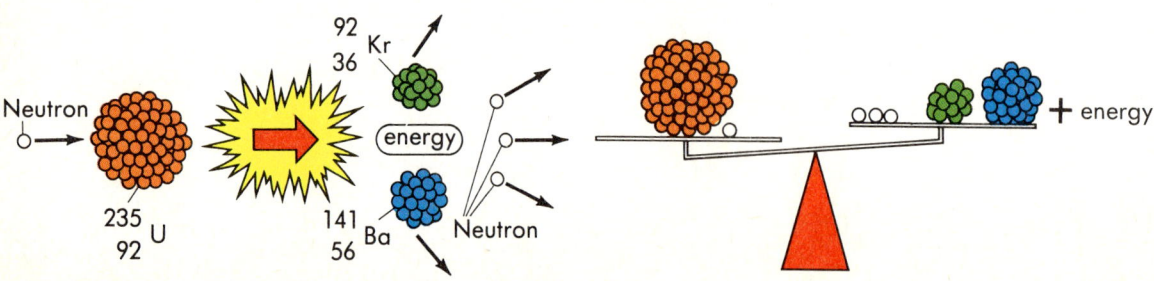

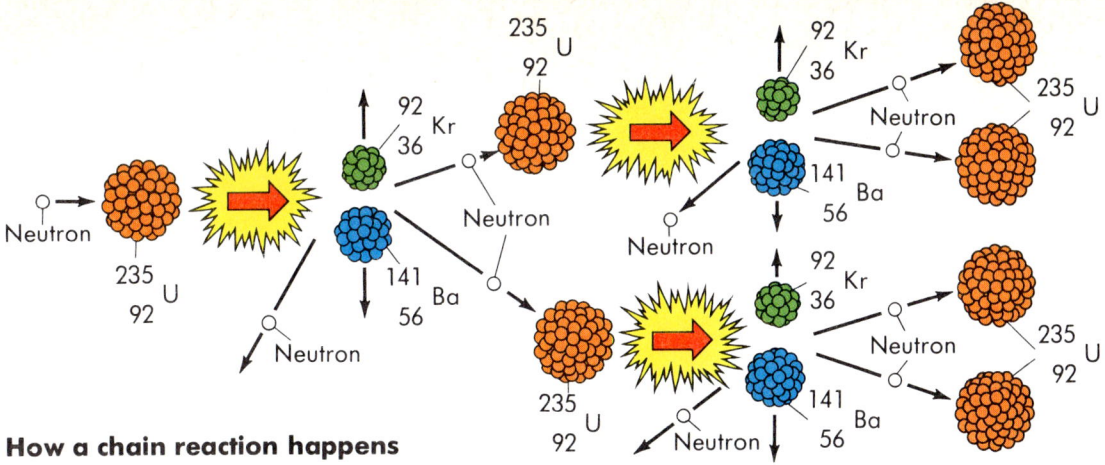

How a chain reaction happens

The amount of energy given up in a fission reaction can be found using Einstein's famous equation, $E = mc^2$. E is the energy released by some quantity of mass, m. The speed of light, c, is 300,000,000 meters per second. So the speed of light squared is a very large number. This equation means that a tiny amount of mass can be changed into a huge amount of energy.

Once nuclear fission starts, it can continue in a **chain reaction** if enough of the right fuel is present. The pictures above demonstrate how a chain reaction works. First, a neutron enters a $^{235}_{92}U$ nucleus. Next, the nucleus splits and releases neutrons. Then one or two of these neutrons enter and break up other nuclei. Still more neutrons are produced. They collide with and split still other nuclei. Each splitting nucleus releases energy.

If a large number of uranium nuclei are available for fission, the chain reaction continues to grow. The radiation and fast-moving nuclear particles produce a huge amount of energy. Whenever more than about four kilograms of uranium nuclei are packed together, so much energy is released so quickly that it acts like a bomb. A nuclear bomb is a runaway chain reaction.

We can control nuclear chain reactions by using such a small amount of uranium that it could never explode. Materials like cadmium or zirconium capture some of the neutrons created when nuclei split. The captured neutrons cannot hit other nuclei, so the reaction is slowed. By capturing some—but not all—of the neutrons, scientists maintain a chain reaction at a steady rate.

Have You Heard?

To understand how atoms interact, scientists accelerated charged particles close to the speed of light. The particles knocked out pieces from bismuth atoms, leaving gold atoms! But we will still have to mine gold. It cost $10,000 to operate the machines used to make the gold. But the gold was worth less than one-billionth of one cent.

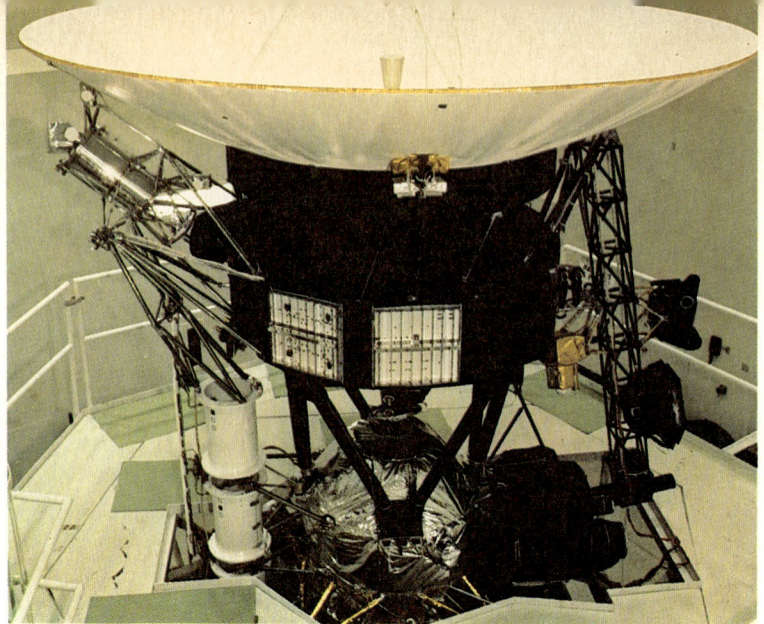

Uses of Nuclear Fission

The energy from nuclear fission can be helpful or harmful. A fission bomb can easily destroy a whole city. But a nuclear power plant supplies energy for many homes and factories. It produces huge amounts of electricity from small amounts of fuel.

Most nuclear power plants produce electric energy by heating water under pressure to a high temperature. The fast-moving particles and radiation produced during fission heat metal pipes. Water passing through the pipes boils into steam. The rest of the process is the same as it is for power plants that run on other fuels. The steam pushes fanlike blades to make electricity.

Nuclear power can also be used on a smaller scale. Submarines with nuclear power plants can travel under water for long periods. Spacecraft traveling away from the sun to distant planets can get their energy from the fission of a radioactive element. The fission power plant in the *Voyager* spacecraft shown above is in the three white cylinders.

Review It

1. How does a chain reaction work?
2. How does nuclear fission make electricity?

Activity

Chain Reactions

Purpose
To show how a chain reaction works.

Materials
- 24 dominoes
- paper and pencil
- small piece of cardboard

Procedure

Part A
1. Set up 10 dominoes about 10 cm from one another, as in *a*.
2. Knock over 2 or 3 of the dominoes. Note what happens to the others.

Part B
1. Set up 24 dominoes in a line, as in *b*. Space the dominoes about 1 cm apart.
2. Knock down the first domino. Note how it affects other dominoes and how many dominoes fall. This situation is like a nuclear chain reaction.
3. Repeat steps 1–2 a few more times. Each time, stand the cardboard between two rows. Note how having the cardboard there affects the time all the dominoes take to fall.

a

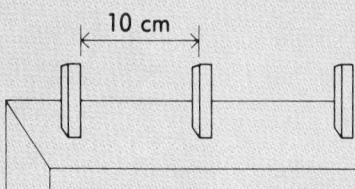

b

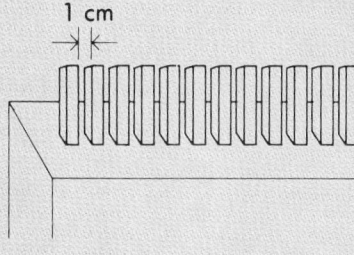

c

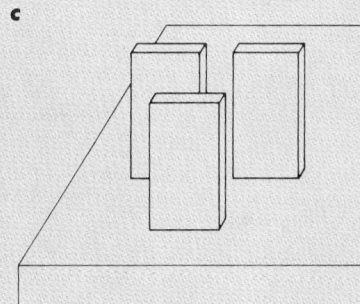

Part C
1. Set up 3 dominoes as in *c*. When the first domino falls, it knocks down the other two. This situation represents the fission of one uranium nucleus causing the fissions of two uranium nuclei.
2. Set up one, two, and four dominoes so that one domino knocks over two, which in turn knock over four.
3. Set up 24 dominoes so that the one in the first row can fall and cause the other 23 to fall. The setup should have as few rows as possible. The fewer the number of rows, the faster the chain reaction.
4. Make a drawing of the setup that used the 24 dominoes in the fewest rows.

Analysis
1. How can you speed up, slow down, or stop a chain reaction of dominoes?
2. How can you make the chain reaction include the most dominoes?
3. How could scientists slow or speed a chain reaction of uranium nuclei?

12–4 Nuclear Fusion

Stars shine by changing mass into energy in a process different from fission. The process that causes stars to shine also makes the elements. As you learn about this process, think about these questions:

a. What is nuclear fusion?
b. How are the elements in the universe formed?

How Nuclear Fusion Happens

In nuclear fission, a heavy nucleus breaks apart into lighter particles. In nuclear **fusion** (fyü′zhən), two or more light nuclei combine to produce a heavier nucleus. The diagram below shows one fusion reaction in the sun.

The fusion of this new hydrogen isotope ($^2_1$H) with another proton makes an unusual isotope of helium, $^3_2$He. Then the fusion of two of these helium isotopes makes ordinary helium, $^4_2$He. The resulting helium nucleus and the particles that went off contain less mass than the total mass of the protons that formed it. The missing mass was changed into energy, according to the equation $E = mc^2$. This energy produces the sun's heat and light.

Large amounts of energy are required to start the fusion process and keep it going. This energy produces temperatures of millions of degrees. At such high temperatures, nuclei move at rapid speeds. Some come close enough for the strong force to capture them.

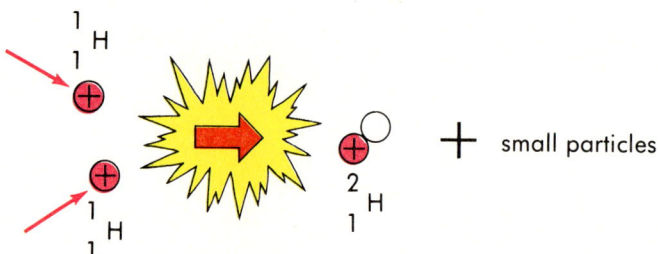

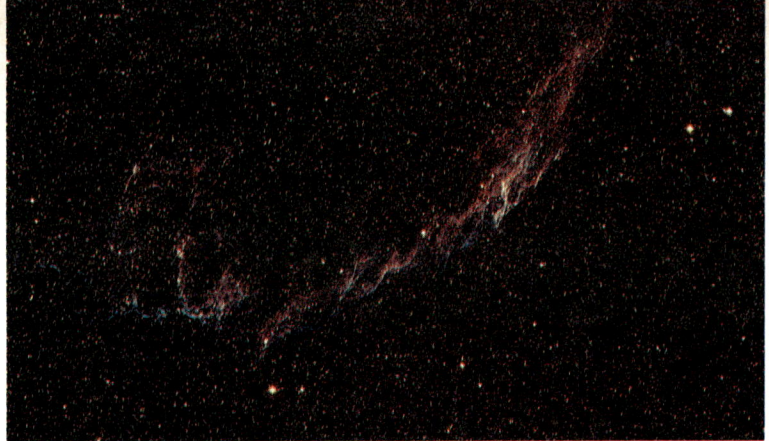
The Veil Nebula

Fusion occurs continuously inside the sun and other stars. On earth, no container can hold matter at fusion-level temperatures for very long. Scientists are experimenting with fusion as an energy source. They are trying to contain the hot material in a laboratory for a time long enough to produce a fusion reaction.

The Elements Were Formed by Fusion

The iron, carbon, and other elements in your body were once parts of shining stars. Fusion in stars produced the nuclei of many elements.

Astronomers believe that about 15 billion years ago, soon after the universe began, only protons and other particles existed. Within a few minutes, fusion began and produced deuterium and then helium nuclei. Similar processes continue today in most stars. All the elements in the periodic table up to iron are made by combining lighter elements to make heavier ones.

After iron forms in a star, the star explodes. The remains of a star that exploded thousands of years ago are shown above. In these explosions, the heaviest elements formed and shot out into space. These elements then became part of stars and planets that formed later.

Review It

1. How does fusion differ from fission?
2. How do elements heavier than hydrogen form?

12–5 Mass, Energy, and the Speed of Light

In fission and fusion, mass is changed into energy. But can energy be changed into mass? Think about this change and the following questions as you read:

a. What did Einstein discover about changes in mass?
b. How are mass and energy related?

Mass Grows as Speed Increases

You have probably felt the excitement of riding a bicycle faster and faster on a smooth, level road. Now imagine your strength is unlimited, your bicycle has hundreds of higher gears, and you can keep increasing your speed. Suppose a friend measures your speed and mass as you pass different spots along a path. Your mass would become greater the closer you came to reaching the speed of light. At nine-tenths the speed of light, your mass would be about twice as large as it was at the start. The extra mass had to come from somewhere.

Einstein was the first person to state that mass grows as speed increases. This growth is shown on the graph. Einstein also stated that the new mass is created from the energy used to make the original mass go faster. As you neared the speed of light, much of the energy of your pedaling would change into mass. The speed of this larger mass would increase slowly.

Einstein's theory predicts that no object that has mass when it is still can go as fast as the speed of light. You can never produce enough energy to make an object go this fast. Instead, the object's mass will keep increasing.

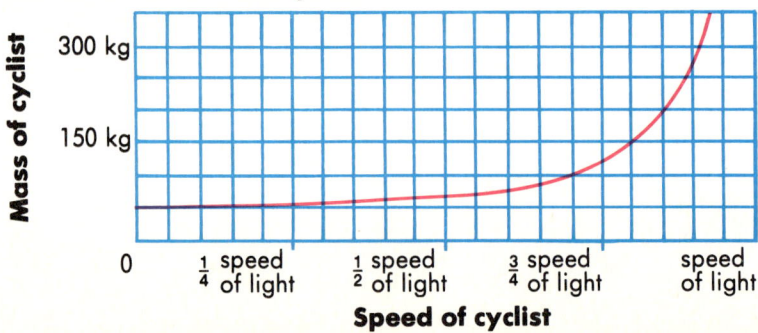

How mass varies with speed

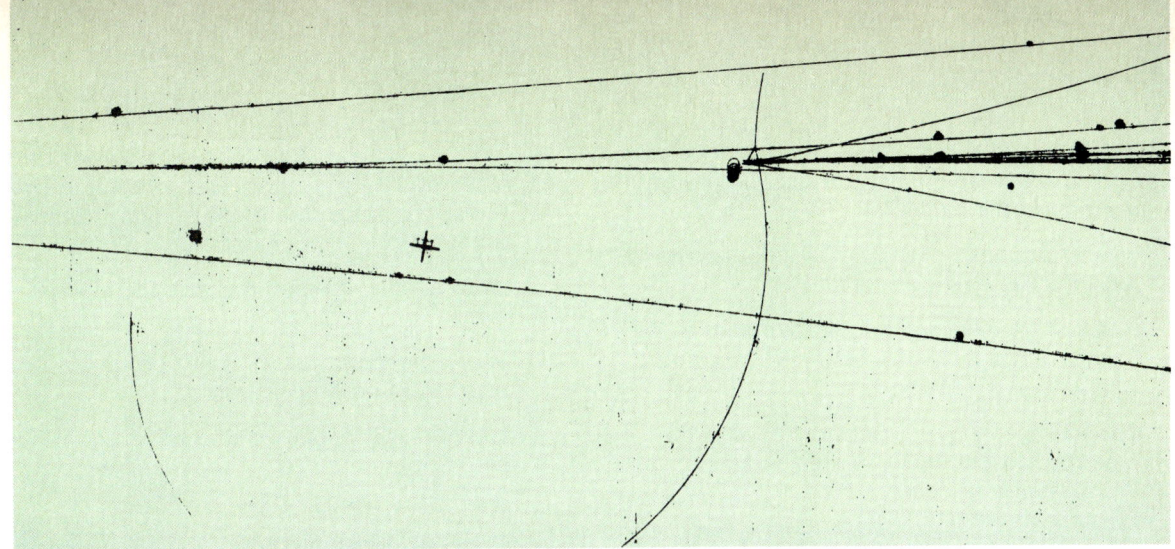

One proof of the law of conservation of mass-energy

Mass and Energy Are the Same

Einstein realized that matter is more than something that has mass and takes up space. He recognized that matter and energy are simply different forms of each other. He pointed out that a small amount of matter can change into a large amount of energy, following the formula $E = mc^2$.

Einstein also realized that the total amount of energy and matter never changes. Energy and matter can change into each other, but nothing is ever lost in the change. This idea is called the **law of conservation of mass-energy.**

Einstein's idea seemed strange when he first suggested it. But the theory has been proved many times in laboratories around the world.

The picture of particle tracks shows one experiment that proves Einstein's theory. Energy entered the chamber on the left in a form that does not make a track. At the V, the energy turned into matter as a pair of particles formed. A similar event occurred on the right side of the chamber. In this chamber, pure energy traveling at the speed of light changed into matter.

Review It

1. What would happen to objects traveling near the speed of light?
2. What is the law of conservation of mass-energy?

Chapter Summary

- The strong force holds neutrons and protons together in the nucleus. (12–1)
- Quarks are basic particles that combine to form nuclear particles. (12–1)
- Radioactive elements release radiation in the form of alpha particles, beta particles, and gamma rays. (12–2)
- The half-life of a radioactive isotope is the time for half the nuclei in a given amount of the substance to decay. (12–2)
- Slowly-moving neutrons can split certain heavy nuclei during fission. (12–3)
- A chain reaction occurs when a nucleus undergoes fission and releases additional neutrons that split other nuclei. (12–3)
- In nuclear fusion, two or more nuclei combine to form a heavier nucleus. (12–4)
- Fusion occurs only at very high temperatures. (12–4)
- Einstein discovered the relationship between mass, energy, and the speed of light. (12–5)
- The law of conservation of mass-energy states that the total amount of mass and energy never changes. (12–5)

Interesting Reading

Asimov, Isaac. *How Did We Find Out About Nuclear Power?* Walker, 1976. Explains how scientists learned about fission and fusion.

Chester, Michael. *Particles: An Introduction to Particle Physics.* Macmillan, 1978. Traces the discovery of both atomic and subatomic particles.

Ellis, R. Hobart, Jr. *Knowing the Atomic Nucleus.* Lothrop, 1973. Discusses the nucleus, including nuclear reactions.

Engdahl, Sylvia, and Robertson, Rick. *The Subnuclear Zoo.* Atheneum, 1977. Describes the nature of nuclear particles.

Questions/Problems

1. Why are neutrons more easily captured by the strong force in a nucleus than protons are?
2. Contrast the strong force with the electric force.
3. What changes in quarks must occur for a neutron to become a proton?
4. How do scientists use radioactive isotopes in determining the ages of rocks and dead organisms?
5. A sample of a radioactive isotope with a half-life of 3 years has 32,000 nuclei. How many nuclei of this isotope will be left at the end of 12 years?
6. Contrast the processes of fission and fusion.
7. In what ways is the control of fission easier than the control of fusion?
8. According to Einstein's theory, what happens when you try to move an object that has mass closer and closer to the speed of light?
9. Explain how changes in mass and energy can occur while mass-energy is conserved.

Extra Research

1. Find out how Marie and Pierre Curie and Frederic and Irene Joliot-Curie contributed to our understanding of radioactivity.
2. At your library, find out what nuclear medicine is and which radioactive isotopes are used to treat people. Then make a report to your class.
3. Some smoke alarms contain a radioactive material. Find out how such an alarm works. You might examine the package of one of these alarms in the store or write to the manufacturer.

Chapter Test

A. Vocabulary Write the numbers 1–10 on a piece of paper.
Match the definition in Column I with the term it defines in Column II.

Column I

1. time for half the nuclei of a radioactive substance to decay
2. splitting of a heavy nucleus
3. combining two or more light nuclei at high temperatures
4. radiation that can be stopped by aluminum foil
5. radiation in the form of electrons
6. radiation that is a powerful bundle of energy
7. idea that the total amount of mass and energy never changes
8. neutrons split nuclei, which release more neutrons, and so on
9. particles believed to make up most nuclear particles
10. equation that related energy, mass, and the speed of light

Column II

a. alpha particle
b. beta particle
c. chain reaction
d. conservation of mass-energy
e. $E = mc^2$
f. fission
g. fusion
h. gamma ray
i. half-life
j. quarks

B. Multiple Choice Write the numbers 1–10 on your paper.
Choose the letter that best completes the statement or answers the question.

1. Which force holds the nucleus together?
a) magnetism b) gravity c) the electric force
d) the strong force

2. The strong force affects a) the electron cloud. b) neutrons. c) chemical changes.
d) physical properties.

3. All radioactive isotopes a) have a specific half-life. b) decay. c) give off radiation.
d) a, b, and c.

4. The half-lives of radioactive isotopes a) do not change. b) depend on temperature.
c) can increase. d) cannot be determined.

5. A nuclear chain reaction can be controlled
a) in a nuclear explosion. b) by providing more nuclei. c) by capturing some of the neutrons created when nuclei split. d) in a nuclear bomb.

6. Nuclear fission can a) power submarines.
b) produce electricity. c) power spacecraft.
d) a, b, and c.

7. Radiation can kill a) healthy cells only.
b) diseased cells only. c) healthy and diseased cells. d) no human cells.

8. Einstein developed a) the equation $E = mc^2$. b) the theory that particles can exceed the speed of light. c) the quark theory.
d) the idea that objects get lighter as they go faster.

9. When mass disappears, it a) becomes motion. b) becomes energy. c) vanishes from the universe. d) is never recovered.

10. The half-life of a radioactive isotope is 6 months. Of 400 of these nuclei, how many will remain after 2 years? a) 25 b) 50 c) 100
d) 200

Careers

Chemist

Not long ago, plastic sandwich bags did not exist. People wrapped their food in waxed or brown paper. Then chemists invented plastic wrap. Food storage is just one area that has not been the same since then.

Chemists work on all sorts of research and development. In the laboratory, a chemist rearranges elements to make chemical compounds with new and different properties. Analytical chemists report on unknown substances. Organic chemists study carbon compounds, while physical chemists examine energy. A chemist uses knowledge of the reactions and structure of matter in order to create new products. Chemists graduate after four years of college study. Some continue their studies for more advanced degrees.

Career Information:
American Chemical Society, 1155 16th St., NW, Washington, DC 20036

Chemical technician

We are more and more careful about the products we buy. Consumers want quality, so businesses pay attention to quality too. They hire chemical technicians to test the things they sell.

Chemical technicians work in laboratories, testing everything from paint to plastics. Technicians check products for strength and durability. They see that products meet health and safety standards. Chemical experiments, X rays, and scale models help technicians control a product's quality.

Technicians might learn to do simple tests on the job. They attend technical school for work that involves complex testing.

Career Information:
American Chemical Society, 1155 16th St., NW, Washington, DC 20036

Assayer

Suppose you are panning for gold and discover a large piece of yellow metal. How can you find out if it is gold? An assayer can tell you the value and composition of your find.

Assayers work in laboratories to analyze metal. They try to discover the type, weight, and purity of metal samples using careful tests and measurements. Often, the assayer looks at many samples from one place to predict the metal content of that area.

Assayers receive science degrees when they graduate from college. They study both chemical and physical properties of metals, and learn methods for doing experimental tests.

Career Information:
American Mining Congress, Library, 1920 N. St., NW, Washington, DC 20036

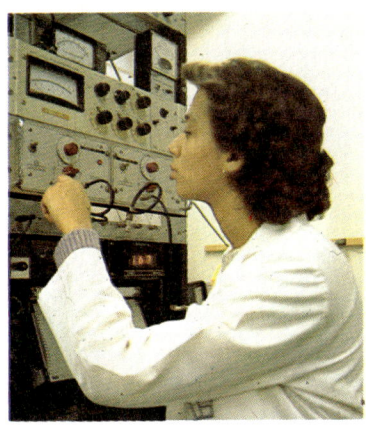 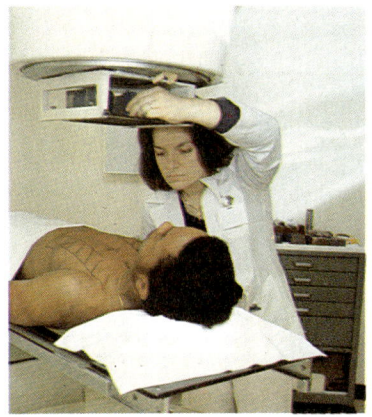 

Elementary particle physicist
When you were younger, you went to elementary school. It was called elementary because you studied the basics of science, math, and reading. The same word describes people who study the basics of matter. These people are elementary particle physicists.

Elementary particle physicists work in laboratories doing research experiments. They use particle accelerators, electron microscopes, and other complex equipment. These physicists study atoms, quarks, and other small particles to understand matter and energy.

The study of elementary particle physics is demanding. Students get a doctoral degree after many years of college. The degree qualifies them to continue the research they began as students.
Career Information:
American Institute of Physics, 335 E. 45th St., New York, NY 10017

Radiation therapist
Scientists still search for a cure for cancer. In the meantime, they use other ways to treat and control the disease. The radiation therapist helps to fight cancer.

Radiation therapy can destroy cancer cells. The radiation therapist concentrates X rays or other radiation on the diseased parts of a cancer patient's body. The radiation might stop the growth of cancer cells. Therapists monitor radiation levels very carefully to prevent any overdose to the patient.

Hospitals and colleges teach radiation therapy. Students work with a radiologist and learn to operate equipment and keep records. Radiation therapy trainees also practice making their patients more comfortable.
Career Information:
The American Society of Radiologic Technologists, 55 E. Jackson Blvd., Suite 1820, Chicago, IL 60604

Scrap handler
A scrap handler collects old cars and junk metal. The scrap is crushed into bales like a farmer's bales of hay.

Scrap handlers operate the equipment for one of our nation's biggest recyclers, the scrap industry. A scrap handler works in yards where old cars, useless machines, and metal parts are collected. The handler sorts, loads, and moves the metal scraps to a baling machine that compresses everything. The bale of metal that comes out of the machine is then melted down, purified, and reused.

Scrap handlers train for work at the scrap yard. They learn safety and operating skills on the job.
Career Information:
Institute of Scrap Iron and Steel, 1627 K St., NW, Suite 700, Washington, DC 20006

UNIT FOUR
CHANGES IN MATTER

What do you think is happening in the photograph? It could be an explosion of paints. Perhaps it is a fire. Maybe it is a swirl of feathers.

Using special filters can change an object's appearance. The photographer used special filters when he took this picture of a crystal of a chemical. The photograph is about 23,000 times larger than the crystal. This chemical resulted from changes in matter, which is the subject of this unit.

Chapter 13 Compounds and Mixtures
All the substances we find around us are combinations or mixtures of the elements.

Chapter 14 Holding Atoms Together
Forces between atoms hold atoms together in the many compounds we find and make.

Chapter 15 Chemical Reactions
The variety of substances we have around us results from reactions among the elements. Some elements react more than others.

Chapter 16 Acids, Bases, and Salts
Certain classes of compounds are important because we use them daily. These compounds have unique properties.

Chapter 13
Compounds and Mixtures

The picture shows a table set for a meal. The food on the table contains various substances combined in different ways. The food you eat, the air you breathe, the house you live in, and even your own body are all made up of various substances mixed together.

This chapter describes various ways in which substances are mixed. The chapter compares different types of mixes. It also discusses many common mixes of substances.

Chapter Objectives

1. Describe compounds and the formulas for compounds.
2. Distinguish between compounds and mixtures.
3. Discuss the properties of solutions, and define solvents and solutes.
4. Identify some kinds of suspensions.

13-1 Identifying Compounds

The bicycle in the picture was left outside in the rain. A rough, brown material appears on the wheel rims. Where did the material come from? Is it part of the metal? Did the rain form it? Did it fall from the air? Think about these and the following questions as you read:

a. What is a compound?
b. How do compounds differ from elements?
c. What are some common compounds?

Elements Combine to Form Compounds

The rust on a bicycle rim is an example of a new substance formed from two other substances. One of the substances is iron, the main ingredient in the steel rim. The other substance is oxygen, a gas in the air. Iron and oxygen join to form rust when water is present.

Iron and oxygen are elements and, therefore, cannot be broken down into simpler substances by ordinary chemical means. A substance formed by the chemical combination of two or more elements is a **compound** (kom′pound). Iron and oxygen join chemically to form a compound commonly known as rust.

The same number of atoms of each element always joins to form a certain compound. For example, the basic combination of atoms in rust is always two iron atoms and three oxygen atoms.

232

A **formula** (fôr′myə lə) states the names and number of atoms in the combination that makes a compound. Fe_2O_3 is the formula for rust. Fe is the symbol for iron. The small number 2 means two atoms. The 3 and the symbol O stand for three oxygen atoms.

The pictures show a new copper utensil and the same utensil after it was heated on a stove several times. A reddish film appears on the copper after heating. This film is a compound of copper and oxygen. The formula for this compound is Cu_2O, which means that two atoms of copper (Cu) have joined with one atom of oxygen. The symbol O without a number stands for one atom of oxygen.

Compounds Can Be Broken Down

Most compounds look different from the elements they contain. For example, the compound rust looks different from iron, which is silvery-white, and from oxygen gas, which is invisible. Also, rust crumbles if you handle it, but you can bend and shape iron.

You can tell a substance is a compound if you can break it down into parts. An element is only one kind of substance and cannot be broken down. A compound, however, contains at least two elements, and these elements can be separated.

Compounds are broken down by various methods. The iron in rust can be separated from the oxygen by heating the compound with the element carbon. In the process, the oxygen combines with the carbon. Liquid iron remains.

The diagram shows how the compound water is broken down by passing an electric current through it. The water separates into its two elements, hydrogen and oxygen.

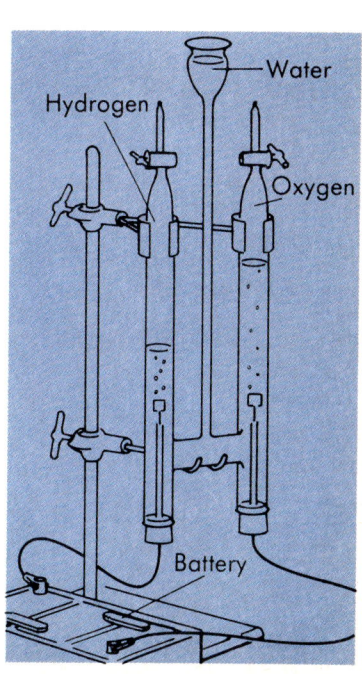

233

Formulas for Common Compounds

Have You Heard?

After diamond, the compound boron nitride (BN) is the hardest substance known. It is about two-thirds as hard as diamond. Boron nitride is made synthetically under high pressure and at high temperature. It is used mainly to grind and polish some kinds of steel.

Millions of compounds exist. You probably know the names of only a few of them. The table lists the formulas of some common compounds, which are also shown in the picture.

Table salt and sugar look similar. But compare their formulas in the chart. Salt is one of the simplest compounds. A salt particle has one atom of sodium for each atom of chlorine. Sugar is much more complex. A sugar particle contains 12 carbon, 22 hydrogen, and 11 oxygen atoms.

Review It

1. How is a compound formed?
2. How are compounds different from elements?
3. Give the formulas of two common compounds.

Name of Compound	Formula
Baking soda	$NaHCO_3$
Cleaning fluid	CCl_2CCl_2
Cream of tartar	$KHC_4H_4O_6$
Sand	SiO_2
Soap	$C_{17}H_{35}COONa$
Sugar	$C_{12}H_{22}O_{11}$
Table salt	$NaCl$
Vinegar	CH_3COOH
Water	H_2O

Common substances that are compounds

Activity

Forming a Compound

Purpose
To observe a compound forming.

Materials
- 2 large test tubes
- 2 wads of coarse steel wool (about 3 cm in diameter)
- baking pan
- 50 mL of vinegar in a paper cup
- ruler
- clock to measure seconds
- water

Procedure

Part A
1. Pour water in the pan to a depth of 0.5 cm.
2. Soak one wad of steel wool in the cup of vinegar for 1 minute. Remove the wad and gently squeeze out the vinegar.
3. Push the moistened steel wool into a test tube. Push the other wad of steel wool into another test tube.

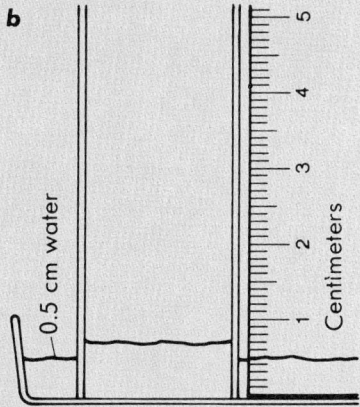

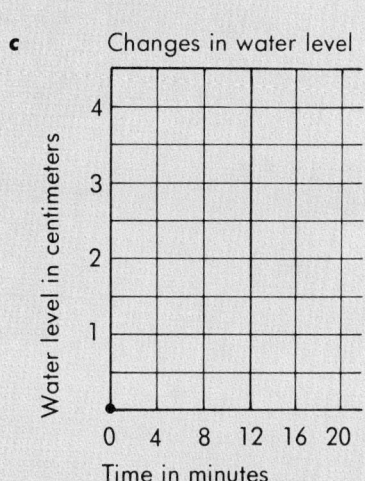

4. Stand both tubes upside down in the water, as in a. The steel wool must stay in place.
5. Measure the height of the water in each tube, as in b. Record the height in each tube at the end of 4, 8, 16, and 20 minutes.
6. Make a copy of c.
7. Plot each height with its time for the wet steel wool. Connect your points with a solid line. On the same grid, do the same for the dry steel wool. Connect these points with a dashed line.

Part B
1. Turn each test tube right side up. Leave the steel wool inside.
2. Fill each tube half full with water.
3. Cover the mouths of the tubes with your thumbs. Shake the tubes.
4. Record any color changes you observe.

Analysis
1. How much did the water level in each tube change during the 20 minutes? What do these results prove?
2. Use your observations to explain what happened in the two tubes during the 20-minute period.

235

13–2
Identifying Mixtures

A popular snack consists of a mix of different kinds of nuts. Each ingredient in this mix keeps its own taste and shape. In a similar way, some substances mix with other substances without joining chemically. Consider the following questions as you read:

a. What is a mixture?
b. How can the substances in a mixture be separated?

Substances in a Mixture Do Not Combine Chemically

If you mix iron dust with powdered sulfur, the two elements do not form a compound. The iron remains a silvery-white metal, and the sulfur remains a yellow powder. Without heat to help the process of chemical combination, iron and sulfur do not form a compound.

Two or more substances mixed together, but not joined chemically, form a **mixture** (miks′chər). The substances in a mixture may be either elements or compounds. Iron dust and sulfur powder form a mixture of two elements.

In some mixtures, you can see each individual substance. In the picture of the ink mixture, you can see the different inks. Another example is soil, which is a mixture of materials such as sand, rock, and clay.

Air is a mixture of gases, but you cannot see each substance in air. Milk is a mixture of water, fats, sugar, and other compounds. Wood, paper, cloth, and glass are all mixtures. In these mixtures, you cannot see the individual compounds.

Unlike a compound, a mixture does not necessarily contain a specific amount of each substance. For example, the compound H_2O must always have two hydrogen atoms for every oxygen atom. But a mixture of hydrogen and oxygen can have any number of each kind of atom.

Separating Substances in a Mixture

You can easily separate the ingredients of some mixtures. For example, you can remove solid particles from a liquid by using a filter. The liquid flows through the filter, but the solid does not. In a similar way, filters in air conditioners remove bits of dust from air.

The substances in some mixtures separate naturally. The pictures show how water and oil separate if they are left standing. The particles of water are heavier than the particles of oil. The water gradually sinks to the bottom of the container.

Review It

1. How can you tell the difference between a mixture and a compound?
2. Why do water and oil separate?

13-3
Solutions—One Kind of Mixture

When you stir sugar, lemon juice, and water in a pitcher, the particles of sugar disappear. The liquid has a yellowish color that is fainter than that of the lemon juice alone. The liquid also has a different taste than the lemon juice or water alone. You have made one type of mixture. Think about these questions as you read:

a. What happens when one material dissolves in another?
b. What are the different types of solutions?

How Substances Dissolve

Lemon juice, water, and sugar form a uniform mixture. Each substance cannot be seen because it has broken down into very small particles. This kind of mixture is a **solution** (sə lü′shən).

The most common kind of solution is a solid in a liquid, such as sugar in water. The diagrams show how some sugar dissolves when individual particles break away from the lump. The particles of the water attract the sugar particles, which spread into the spaces between the water particles. After a while, all the particles are evenly mixed. The different substances do not settle into layers. Instead, they stay mixed. Although the sugar particles do not combine chemically with the water particles, they are too tiny to be seen. The sugar particles are even small enough to pass through a filter.

Some solutions consist of one liquid dissolved in another. Water and alcohol form such a solution, which is used in car radiators to prevent water from freezing.

The dissolving process

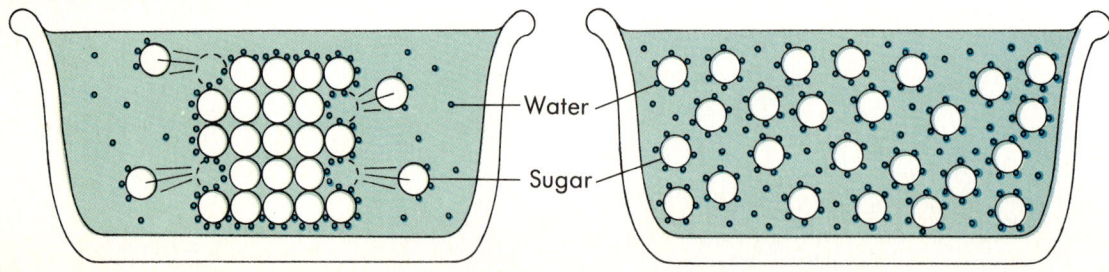

Other solutions contain a gas dissolved in a liquid. Fish breathe air that dissolves in water at the surface of an ocean, lake, or stream. The picture shows the bubbles of gas in a glass of carbonated water. The dissolved gas carbon dioxide gives the water its fizz.

A substance that dissolves other materials is a **solvent** (sol′vənt). When you dissolve sugar in water, water is the solvent. The substance being dissolved is a **solute** (sol′yüt). Sugar is the solute in a solution of water and sugar.

Water is the most commonly used solvent. But not all materials dissolve in water. You use turpentine to remove oil-based paints from brushes because these paints dissolve in turpentine but not in water.

A solution of a gas in a liquid

Describing Solutions

A solution may have a large or small amount of solute. Solutions with a small amount of solute are described as **dilute** (də lüt′). Weak coffee is a dilute solution. A solution with a large amount of solute is **concentrated** (kon′sən trā′tid). Strong coffee is a concentrated solution.

If you stir sugar into cold water and gradually add more sugar, the solution becomes more and more concentrated. Eventually, no more solute can dissolve at that temperature and pressure. The solution is **saturated** (sach′ə rā′tid). If you add any more sugar, it will sink to the bottom of the cup. However, if you heat the solution, still more sugar can dissolve. Usually, more solid solute can dissolve in a liquid solvent as the solution's temperature rises.

Review It

1. How do solutions differ from other mixtures?
2. What are dilute, concentrated, and saturated solutions?

13–4 Suspensions— Another Kind of Mixture

The labels on some shampoos and juices tell you to shake the product before using it. The substances in these products separate after standing awhile. These mixtures are not solutions. They are another kind of mixture. Answer the following questions as you read:

a. How do suspensions differ from solutions?
b. What are some types of suspensions?

Particles in a Suspension

A mixture in which the particles of one substance become scattered through another substance without dissolving is a **suspension** (sə spen′shən). Some shampoos and juices are suspensions. In many suspensions, the substances separate after standing awhile.

The diagrams show a suspension of oil and vinegar. The oil does not break down into particles that stay evenly spread through the mixture. Instead, they rise to the top of the container.

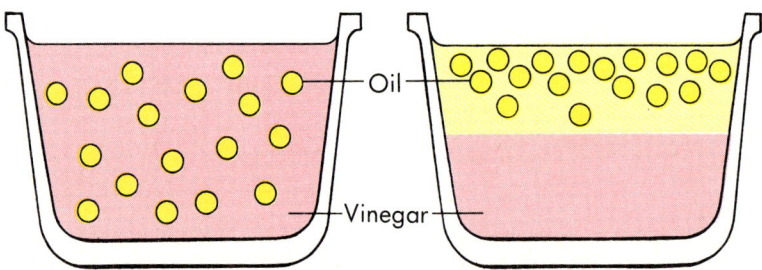

How quickly particles in a suspension will separate depends mainly on their size and weight. In most cases, large or heavy particles separate more quickly than small or light particles.

Kinds of Suspensions

You can make one suspension of a solid in a liquid by adding sand to water and stirring the mixture well. If you let the mixture stand awhile, the sand settles.

Challenge!
Some of the ingredients in foods, cosmetics, and drugs are compounds and suspensions. Check the labels of these products for such ingredients. Then use an encyclopedia or chemistry book to learn more about these substances.

Smoke is a suspension of solid dirt and dust particles in air. After a while, the solid particles fall to the ground.

Shaving cream is a suspension of a gas in a liquid. The gas is air, and the liquid is a creamy fluid. After a short time, shaving cream flattens, because the air escapes.

Not all suspensions contain substances that separate. A suspension of small particles that remain mixed in a gas, liquid, or solid is a **colloid** (kol′oid). A colloid only appears to be a solution. Even though the particles in a colloid are too small to be seen, they are larger than the particles in a solution. As the picture shows, the particles in a colloid are big enough to scatter a beam of light. The particles in a solution are not this big. Milk, blood, and paints are examples of colloids.

A colloid of one liquid in another liquid is an **emulsion** (i mul′shən). For example, mayonnaise is an emulsion made from vinegar and oil plus egg yolk. The egg yolk is an **emulsifier** (i mul′sə fī′er)—that is, it keeps the particles of oil and vinegar mixed. In the diagram, the emulsifier holds the particles of the two liquids in the mixture. One end of the emulsifier attracts vinegar, and the other end attracts oil. Many, but not all, emulsions have emulsifiers.

Comparing a solution and a colloid

Review It

1. How do the particles in a suspension differ from the particles in a solution?
2. What is a colloid? What is an emulsion?

How particles interact in an emulsion

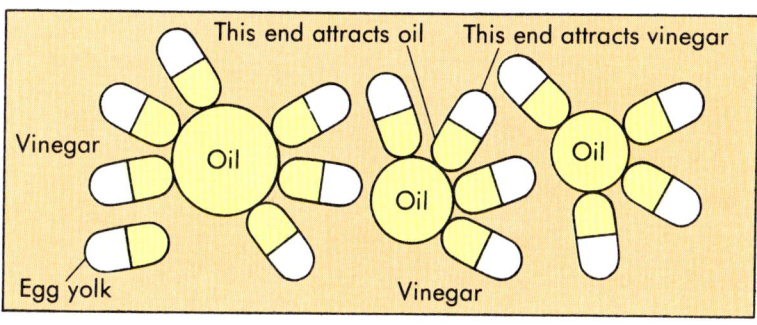

241

Activity

Investigating Suspensions, Solutions, and Emulsions

Purpose
To identify various types of mixtures.

Materials
- 4 test tubes
- medicine dropper
- 2 filter papers
- metal spoon
- candle
- funnel
- matches
- 50 mL of dirt
- 50 mL of sugar
- 50 mL of vegetable oil
- 5 mL of liquid soap
- water

Procedure

Part A
1. Fold the filter paper, as in *a*, and place it in the funnel, as in *b*.
2. Pour dirt in a test tube to a depth of 0.5 cm. Pour sugar in another tube to the same depth. Add water to both tubes until they are about two-thirds full.
3. Shake the tubes to mix the ingredients.
4. Hold the tubes still for 5 minutes. Note what happens to the contents.
5. Place the funnel and filter in an empty test tube.

a
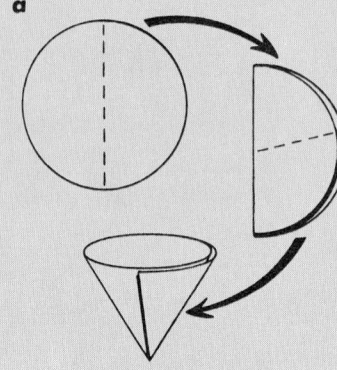

b
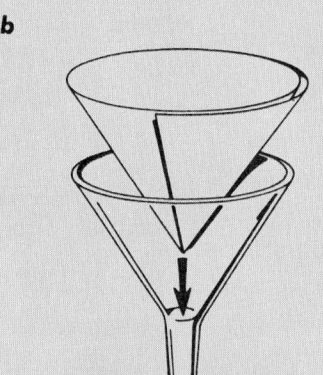

c
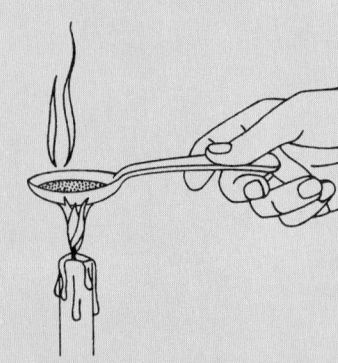

6. Shake the dirt and water mixture well and pour it through the funnel.
7. Record what you observe.
8. Place some filtered dirt and water mixture on a spoon.
9. Using a hot pad, hold the spoon over a lighted candle, as in *c*, until most of the liquid boils away. *CAUTION: Do not touch the hot spoon or any dripping wax.*
10. Record what you observe.
11. Repeat steps 1 and 5–10 for the sugar water.

Part B
1. Fill two test tubes half full with water. Add oil to both tubes until they are three-fourths full.
2. Cover one tube with your thumb and shake it for 10 seconds. Hold the tube still for 5 minutes. Observe the contents.
3. Add a few drops of soap to one tube.
4. Repeat step 2 with both tubes.

Analysis
1. Name and describe the kinds of mixtures you made.
2. What did the soap do to the oil and water?

Breakthrough

Superblanket

The figure in the picture below is not a ghost. It is a firefighter covered with a new kind of fire blanket. The wool blanket has soaked up about thirteen times its weight of a white, goopy colloid. Firefighters use the colloid-coated blanket to smother flames, protect people from heat and smoke, and give first aid to burn victims.

The colloid is a mixture of compounds made from the natural oils and processed natural powders of various kinds of plants. Like catsup, the colloid is thixotropic (thik′sə trop′ik)—thick and jellylike unless it is shaken vigorously. So it will not drip off as the blanket is used. Since the colloid boils only at very high temperatures and does not evaporate easily, it lasts at least long enough to get a person safely out of a burning building.

The colloid-soaked blanket can smother flames. In high heat, the colloid hardens. A thin skin forms that keeps heat out. But the mixture inside is still soupy. It cools hot air that gets through the outer skin of the blanket. Therefore, a person wearing the blanket can move safely through a fire.

The colloid also provides first aid to fire victims. When the blanket is spread over a burn victim, the colloid soaks through clothing and cools the skin. A compound in the colloid prevents the growth of bacteria that cause infections. The colloid dissolves in water, so it washes off the surface of burns easily. It also helps keep burned clothing from sticking to the victim's skin.

In 1981, Alaskan firefighters entered a burning building and found a man with burns over about twenty-five percent of his body. They covered him with a colloid-soaked blanket and took him to the hospital. Emergency room workers left the fire blanket on the burn victim. The colloid was already doing its job—cleaning and sterilizing the burns. When the man arrived at a burn treatment center, doctors found his burns completely clean. No burned clothing stuck to them, and no infection developed afterward.

For Discussion
1. Describe the colloid by listing its properties.
2. How is each of the colloid's properties useful against fires?

Source: "Water-Jel Fire Blanket is First Aid Breakthrough." Trilling Resources, Ltd., Hartsdale, NY.

Chapter Summary

- A compound is formed by the chemical combination of two or more elements. (13–1)

- The number and kind of atoms joined in a compound is always the same. (13–1)

- A mixture consists of two or more substances mixed together, but not joined chemically. (13–2)

- A solution is a mixture in which the substances break down into tiny particles and mix evenly, but do not combine chemically. (13–3)

- Dilute solutions contain a small amount of solute in the solvent. (13–3)

- Saturated solutions contain all the solute that the solvent can dissolve at a certain temperature and pressure. (13–3)

- A suspension is a mixture in which particles of one substance remain scattered through another substance without dissolving. (13–4)

- Colloids are suspensions in which the particles are larger than those in solutions but do not settle out of the mixture. (13–4)

- An emulsion is a colloid of one liquid in another liquid. (13–4)

Interesting Reading

Cobb, Vicki. *More Science Experiments You Can Eat.* Harper and Row, 1979. Describes experiments with food. (Be sure to ask one of your parent's permission before performing any experiments at home.)

Goldin, Augusta. *The Shape of Water.* Doubleday, 1979. Discusses the properties of water that make it unique and important.

Questions/Problems

1. If you stirred sugar and sand together, would you make a mixture or a compound? Explain your answer.

2. Explain the formulas in the table that appears in the first section of this chapter. Tell which elements form the compounds listed, how many atoms of each element are listed in each formula, and the total number of atoms in each formula.

3. You heat water and add as much salt as will dissolve at that temperature. Then you let the solution cool. Later, you notice some salt on the bottom of the pot. Was this solution saturated before or after cooling? Explain.

4. Suppose you were given a bottle that contained a mixture of a liquid and a solid. How could you tell whether the mixture was a solution or a suspension?

5. The water in the oceans contains salt. Explain why a layer of salt does not form on the bottom of the oceans.

6. Why do the beds of streams gradually build up, thus lowering the depth of the water in streams?

Extra Research

1. Visit a water treatment or sewage treatment plant to find out how the plant removes solid substances from water. Make a report to your class.

2. Investigate some common liquids, such as water, vegetable oil, and vinegar, to find out what substances they will dissolve. Use such solutes as salt, sugar, flour, cornstarch, baking soda, and other cooking products.

Chapter Test

A. Vocabulary Write the numbers 1–10 on a piece of paper. Match the definition in Column I with the term it defines in Column II.

Column I

1. containing a small amount of solute
2. a substance that dissolves other substances
3. two or more elements chemically joined in definite amounts
4. two or more substances mixed but not chemically joined
5. a suspension in which the particles stay mixed
6. a substance that is dissolved by another substance
7. containing a large amount of solute
8. a colloid of one liquid in another
9. a mixture in which the substances break into separate particles and become evenly mixed
10. a mixture in which the particles are scattered but not dissolved

Column II

a. colloid
b. compound
c. concentrated
d. dilute
e. emulsion
f. mixture
g. solute
h. solution
i. solvent
j. suspension

B. Multiple Choice Write the numbers 1–10 on your paper. Choose the letter that best completes the statement or answers the question.

1. The atoms in a compound join   a) in definite numbers.   b) chemically.   c) and can be seen individually.   d) a and b.

2. A compound's formula tells   a) which elements are in the compound.   b) the properties of the compound.   c) the compound's size.   d) a, b, and c.

3. The substances in a mixture   a) are always evenly spread.   b) join chemically.   c) do not join chemically.   d) must be in definite amounts.

4. You boil away a clear liquid until a solid remains. The liquid and solid were a(n)   a) solution.   b) suspension.   c) emulsifier.   d) emulsion.

5. The solute in a solution   a) settles.   b) can be filtered out.   c) can be seen.   d) breaks down into tiny particles.

6. Without heating, saturated solutions usually   a) can dissolve more solute.   b) can not dissolve more solute.   c) do not stay mixed.   d) none of the above.

7. Adding more solvent to a solution makes the solution more   a) concentrated.   b) saturated.   c) dilute.   d) emulsified.

8. The particles in a suspension   a) never separate.   b) do not mix.   c) dissolve quickly.   d) none of the above.

9. Emulsifiers cause emulsions to   a) separate.   b) stay mixed.   c) dissolve.   d) become saturated.

10. The particles in a colloid   a) remain mixed.   b) separate.   c) dissolve easily.   d) are very large.

245

Chapter 14
Holding Atoms Together

The glittering diamond in the picture is a remarkable material. It is valued for many qualities besides its beauty. Because carbon in the diamond form is the hardest substance in nature, many industries use diamonds to make cutting tools. A diamond does not melt. It changes into a gas only at the very high temperature of 3500°C. A diamond does not conduct electricity. The connections between carbon atoms account for the distinctive properties of a diamond.

This chapter describes the forces that hold atoms together. It explains how these forces account for many properties of a substance. The chapter also explains the equations used to represent chemical processes.

Chapter Objectives
1. Explain how chemical bonds hold atoms together.
2. Describe how ions form, and explain ionic bonding.
3. Describe how covalent bonds are formed.
4. Explain the use of chemical equations and why they must be balanced.

14–1
Bonding Atoms

Your toothpaste and the nonstick coating on your frying pan might have something in common. Both might contain compounds of the element fluorine. Answer the following questions to learn how the same element can make such different compounds:

a. What are chemical bonds?
b. How do electrons affect bonding?
c. What characteristics do bonds give to a substance?

Chemical Bonds Hold Atoms Together

An electron and a proton attract each other because they have opposite charges. This attraction holds the protons and electrons together in an atom. The attraction, however, does not stop at an atom's outer energy level. The force extends outside the atom, as the drawing shows. Often when two atoms come near each other, the force is strong enough to pull them together. The attractive force holding two atoms together is a **chemical bond.**

Bonds Depend on the Outer Electrons in Atoms

The electrons in an atom are in energy levels. Each energy level can hold only a certain number of electrons. Hydrogen is the only kind of atom in which the innermost energy level is not filled with two electrons. The outer level in most atoms could hold eight electrons. But in most atoms this level does not have all the electrons it can hold.

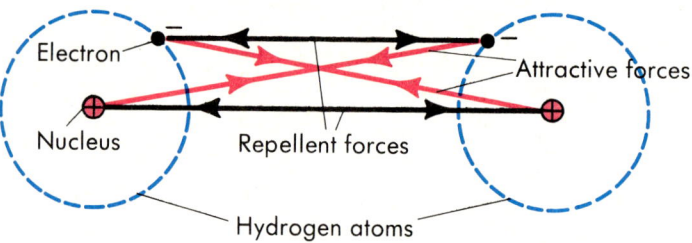

																	VIIIA
						H											He
IA	IIA											IIIA	IVA	VA	VIA	VIIA	
Li	Be											B	C	N	O	F	Ne
Na	Mg	IIIB	IVB	VB	VIB	VIIB	VIIIB			IB	IIB	Al	S	P	S	Cl	Ar
K	Ca	Sc	Ti	V	Cr	Mn	Fe	Co	Ni	Cu	Zn	Ga	Ge	As	Se	Br	Kr
Rb	Sr	Y	Zr	Nb	Mo	Tc	Ru	Rh	Pd	Ag	Cd	In	Sn	Sb	Te	I	Xe
Cs	Ba	La	Hf	Ta	W	Re	Os	Ir	Pt	Au	Hg	Tl	Pb	Bi	Po	At	Rn
Fr	Ra	Ac	(–)	(–)	(–)												

■ gain or share
■ lose or share

How the elements bond

Chemical bonds result from the tendency of an atom to fill its outer energy level. In this condition, the atom is **chemically stable.** Atoms will bond—transfer or share electrons so that each atom's outer energy level is filled.

The periodic table above shows the number of outer electrons in different atoms. Metal atoms, such as sodium and chromium, lose or share their electrons when bonding with other atoms. Nonmetal atoms, such as nitrogen and fluorine, gain or share electrons to complete their outer levels. The helium atom has two electrons in its outer level, and the other members of its group have eight. Because their outer levels are filled, these elements do not usually bond with other elements.

Compounds result from bonds between the atoms of different elements. These atoms can bond in many different combinations. In the drawing at the right, the elements carbon and oxygen bond to make the compound carbon dioxide. Carbon has four outer electrons to offer, while each oxygen atom needs only two electrons to complete its outer energy level. Therefore, one carbon atom bonds with two oxygen atoms to form carbon dioxide.

A CO_2 molecule

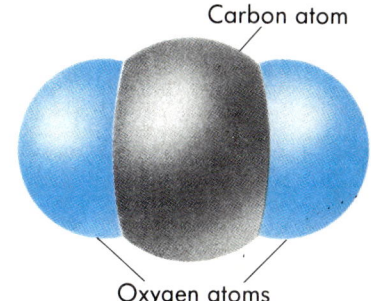

Carbon atom
Oxygen atoms

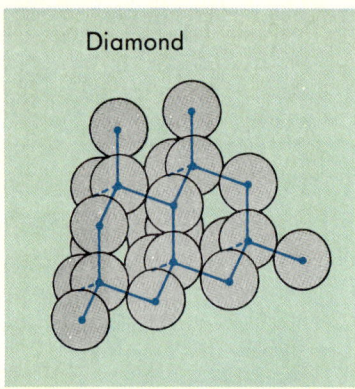

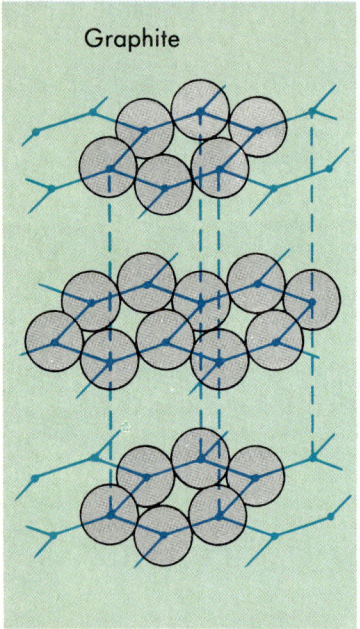

Comparing the structures of carbon

Bonds Determine a Substance's Properties

Chemical bonds cause many of the physical properties of substances. Bonds can exist not only between one atom and another but also throughout a crystal. For example, the graphite form of carbon is a slippery black solid. The bonds in graphite hold the carbon atoms in sheets that slide over one another, as the drawings show. The diamond form, however, is a clear, hard crystal. Bonds in the diamond shown hold each carbon atom to those around it. This tight structure gives the diamond its crystal shape. The hardness of diamonds makes them useful for cutting and grinding softer substances. The slipperiness of graphite makes it useful as a lubricant. Yet both materials are pure carbon.

Another difference in crystal structure occurs when white phosphorus is heated to 250°C in the absence of air. The phosphorus turns red because the bonds between the phosphorus atoms rearrange to form a different crystal structure. Yet both substances are phosphorus.

Melting point is another property that bonds determine. You can heat and melt sugar in a pan on your stove until it turns brown. However, you have to heat quartz to 1600°C before it melts. Bonds within a sugar particle or a quartz particle are strong. But the bonds in quartz extend beyond a single quartz particle. All the particles are linked. The bonds must break before quartz can melt. Therefore, quartz has a higher melting point than sugar.

Review It

1. How are chemical bonds important?
2. How is the number of outer electrons in an atom important in bonding?
3. How do bonds affect a substance?

Activity

Bonds Make the Difference

Purpose
To observe the properties of compounds that have different kinds of bonds.

Materials
- 3 teaspoons
- magnifying glass
- 2 matches
- 2 cm³ of table salt
- 2 cm³ of table sugar
- safety goggles

Procedure
1. Copy the table in *a* for your data. *Put on your safety goggles.*
2. Rub a few grains of salt between your fingers. Do the same with a few grains of sugar. Record which one feels rougher.
3. Place a few grains of salt in one teaspoon and a few grains of sugar in another.
4. Examine the grains with the magnifying glass. In a few words, compare the shapes of the salt and sugar grains. Record your observations.
5. Use the third spoon to crush the grains in the other two spoons, as in *b*. Record which substance is harder to crush.
6. Light a match and hold it above and near the salt in the spoon. Let the match burn for about 10 seconds. Note if any salt melts.
7. Repeat step 6 with the sugar. Record which substance has the higher melting point.

Analysis
1. Summarize in a few sentences the differences in the properties of salt and sugar.
2. Think of each substance as made of atoms locked to one another. Which substance has stronger locks? Explain.

a. Data table

Characteristics	Table salt	Sugar
Roughness		
Grain shape		
Hardness		
Melting point		

b
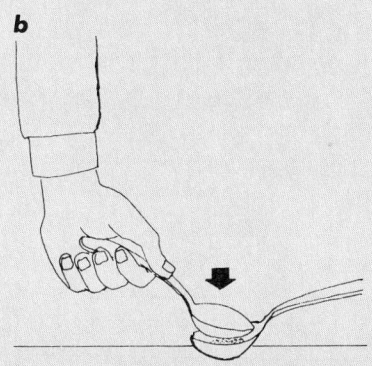

251

14-2
Ionic Bonding

Have you ever been in a tug-of-war? The stronger team usually wins the match. A similar tug-of-war happens between atoms. One kind of chemical bond forms when one atom pulls electrons away from another. Think about the following questions as you read about this kind of bond:

a. What are ions?
b. What are some properties of ionic compounds?

Ions Form Ionic Bonds

An atom is electrically neutral. When an atom loses or gains one or more electrons, a charged particle, called an **ion** (ī′ən), results. When one atom pulls an electron away from another atom, two ions form. The atom that gained the electron becomes a negatively charged ion. The atom that lost the electron becomes a positively charged ion.

Sodium and chlorine atoms become ions easily because of their electron arrangements. The diagram shows how atoms of sodium (Na) and chlorine (Cl) become ions. When a sodium atom loses its outer level electron, a positively charged ion (Na^+) forms. The chlorine atom gains the electron and becomes a negatively charged ion (Cl^-). The oppositely charged ions attract each other, forming a kind of chemical bond called an **ionic** (ī on′ik) **bond**. Sodium chloride (NaCl) is an ionic compound because it consists of sodium ions and chloride ions held together by ionic bonds.

How an ionic bond forms

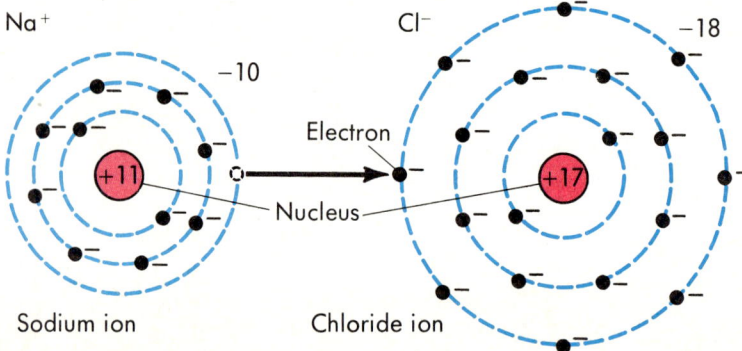

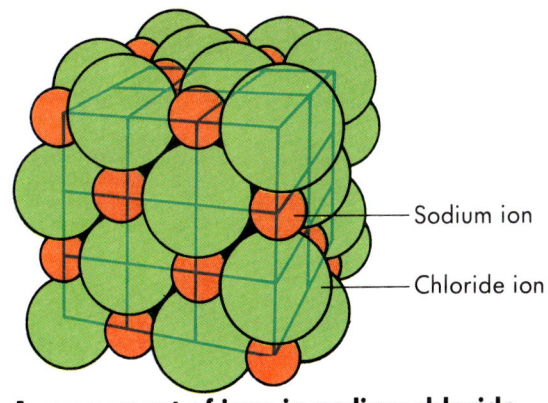

Arrangement of ions in sodium chloride

Properties of Ionic Compounds

The properties of sodium chloride differ from those of either sodium or chlorine. Sodium is a silvery metal that reacts violently with water. Chlorine is a greenish-yellow, poisonous gas. The ionic compound sodium chloride is a white, nonpoisonous solid that dissolves easily in water. You use it as table salt.

The ions in a solid ionic compound are arranged in an orderly way to form a crystal. The diagram shows the structure of the ions in sodium chloride. The orange balls represent sodium ions. The green balls are chloride ions. The photograph shows sodium chloride crystals. Each crystal contains billions of ions.

Ionic bonds are especially strong because they extend throughout an entire crystal. Therefore, most ionic compounds are hard solids at room temperature. They have high melting points. For example, sodium chloride melts at about 800°C.

In the solid form, ionic compounds do not conduct electricity. In the liquid state, or when dissolved in water, the ions separate and move about freely. So in these cases, ionic compounds conduct electricity well.

Review It

1. How do ions form bonds?
2. State three properties of ionic compounds.

14–3 Covalent Bonding

Perhaps you share a room with a sister or brother. If so, you understand how two people can use one thing, even though neither person owns it. (Sometimes it might seem that one of you owns more of the room than the other does!) Another kind of chemical bond happens when atoms share electrons. As you find out about this kind of bond, remember these questions:

a. What is a covalent bond?
b. What are covalent compounds?

How Covalent Bonds Form

When two atoms share a pair of electrons, they are held together by a force called a **covalent** (kō vā′lənt) **bond.** The diagrams show that two hydrogen atoms form a covalent bond. Each atom of hydrogen has one electron. When two hydrogen atoms come together, the electrons pair up. The two atoms share the pair of electrons, which are attracted by both nuclei. This force holds the atoms together. The pair of electrons are in an energy level that both nuclei share. In other words, the electron clouds of the two atoms overlap, as shown.

Molecules Result from Covalent Bonding

Two or more atoms joined together by covalent bonds make a **molecule** (mol′ə kyül). Atoms of the same or different elements can join as molecules. A molecule is the smallest particle of a covalent substance.

Covalently bonded hydrogen is called a **diatomic** (dī′ə tom′ik) **molecule** because it contains two atoms. (*Di-* means "two.") Fluorine, chlorine, oxygen, nitrogen, bromine, and iodine also exist as diatomic molecules.

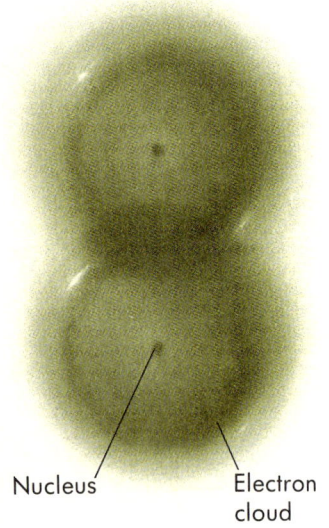

Forming a covalent bond
Hydrogen atoms
Nucleus
Electron cloud

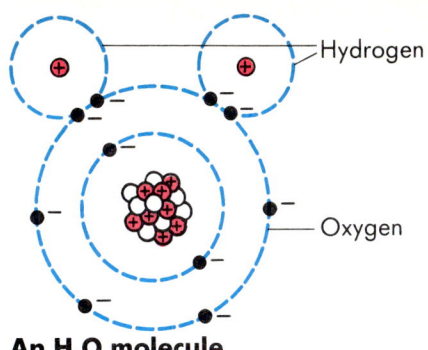
An H₂O molecule

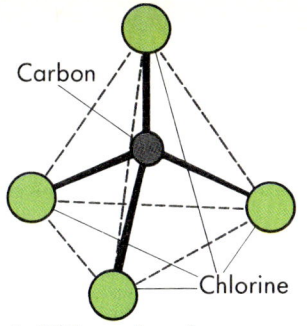
A CCl₄ molecule

Covalent compounds are made of molecules. Water is probably the most important covalent compound. In the water molecule shown above, two hydrogen atoms share their electrons with one oxygen atom. This arrangement completes the outer energy levels of all three atoms: two electrons for each hydrogen atom and eight electrons for the oxygen atom.

Carbon and hydrogen form more covalent compounds than other elements. Carbon forms so many covalent compounds because its atoms have four outer electrons to share with other atoms. Each electron can form one covalent bond.

The other diagram shows one carbon atom bonded with four chlorine atoms. Each chlorine atom needs one electron to complete its outer energy level. The carbon atom completes its outer level by sharing this electron and one of its own with each chlorine atom. This arrangement also fills the outer energy level of each chlorine atom. The result is one molecule of the compound carbon tetrachloride, CCl_4, which was used as a cleaning solvent for many years.

Most covalent substances do not conduct electricity. Some, such as diamond, are very hard because the covalent bonds between atoms are very strong and extend throughout the crystal. Others, such as sugar, are soft and melt at low temperatures because of the weak forces between one sugar molecule and another.

Have You Heard?

Around 1866, the chemist Friedrich Kekulé was struggling to understand the bonds and structure of benzene, C_6H_6. The story goes that he dreamed of a snake chasing its own tail. Kekulé realized that a benzene molecule might look like that snake! Later, experiments proved that benzene is a hexagonally shaped ring. Kekulé's discovery was important because benzene is a basic molecule in many substances.

Review It

1. How do covalent bonds form?
2. How are covalent compounds formed?

14-4 Chemical Equations

A baker uses a recipe to combine the ingredients in making a cake. The scientist shown uses a kind of recipe to join elements and compounds to form new substances. This section describes a kind of "recipe" scientists use. Consider these questions as you read:

a. What does a chemical equation tell you?
b. What are balanced equations?

Chemical Equations Describe Chemical Reactions

Chemical changes result in new substances with different properties. Often a chemical change consists of many different steps. A **chemical reaction** is one of these individual steps. We can describe what happens during a chemical change by writing a sentence for each of its reactions. For example, to describe the formation of sodium chloride, we can say: "Atoms of sodium bond with atoms of chlorine. The compound sodium chloride forms." Chemists, however, use equations to describe chemical reactions in a simpler, shorter way. A **chemical equation** tells how many atoms of each element or molecules of each compound interact to form a different substance or substances. The following equation shows what happens when sodium and chlorine join:

$$2Na + Cl_2 \longrightarrow 2NaCl.$$

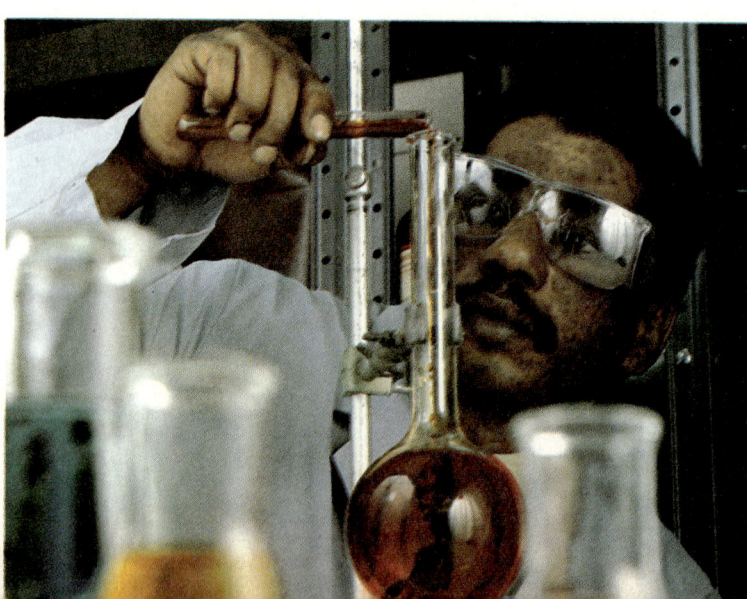

In any chemical equation, the arrow means "produces." The number in front of a symbol or formula tells the number of atoms or molecules.

In the formation of water,

$$2H_2 + O_2 \longrightarrow 2H_2O.$$

Two molecules of hydrogen gas join with one molecule of oxygen gas to make two water molecules.

Chemical Equations Must Be Balanced

During a chemical reaction, atoms are rearranged, but the number of atoms remains the same. This principle follows the **law of conservation of mass,** which states that matter cannot be created or destroyed.

Chemical equations must be balanced because the number of atoms cannot be increased or decreased during a chemical reaction. A balanced equation has equal numbers of each kind of atom on each side of the arrow.

In the photograph, food is being cooked over a propane stove. The equation below describes the combustion of propane as it combines with oxygen:

$$\underset{\text{propane}}{C_3H_8} + \underset{\text{oxygen}}{5O_2} \longrightarrow \underset{\text{carbon dioxide}}{3CO_2} + \underset{\text{water}}{4H_2O}.$$

Count the atoms on both sides of this equation. On the left side, you find 3 atoms of carbon and 8 atoms of hydrogen. To find the number of oxygen atoms, multiply 5 times 2 for a total of 10. On the right side of the equation, you again find 3 atoms of carbon. To find the number of oxygen atoms, multiply 3 times 2 and add 4 (from the water molecules), for a total of 10. To find the number of hydrogen atoms, multiply 4 times 2 for a total of 8. The equation is balanced because each side has 3 carbon atoms, 8 hydrogen atoms, and 10 oxygen atoms.

Challenge!

Find out how baking powder and yeast each react to make bread rise. Look for the answer in an encyclopedia or a book about the chemistry of food.

For Practice

Choose the balanced equation.

- $Na + F_2 \longrightarrow 2NaF.$
- $2Al + 6HCl \longrightarrow 2AlCl_3 + 3H_2.$

Review It

1. What do chemical equations tell us?
2. What is the law of conservation of mass?

Activity

A Chemical Reaction and Its Equation

Purpose
To balance the chemical equation for an observed chemical reaction.

Materials
- 10 cm³ vinegar
- 2 cm³ baking soda
- 200–500 mL beaker or water glass (tumbler)
- 4 matches
- tongs
- safety goggles

Procedure
1. *NOTE: Put on your safety goggles.* The "empty" beaker contains air. Light a match and hold it deep in the beaker with the tongs, as in *a*. Does the match continue to burn, or does it go out? Record what happens to the match.
2. Place the baking soda in the beaker.
3. Pour the vinegar into the beaker. Do not disturb the beaker or its contents. Record your observations about the contents of the beaker.
4. Repeat step 1. Do not let the match touch the contents of the beaker.

a

b

5. Gently tip the beaker, as in *b*. Return it to its normal position. Light another match and slowly move it deeper into the beaker with the tongs. Record your observations.

Analysis
1. Compare the properties of air to the properties of the gas produced during the reaction. Is the new gas heavier or lighter than air? Can objects burn in the new gas? Is the new gas visible, or is it as invisible as air?
2. The chemical formula for baking soda is $NaHCO_3$. Vinegar is a mixture of water and acetic acid. The acetic acid reacts with baking soda. The chemical formula for acetic acid is $HC_2H_3O_2$. The chemical equation for the reaction of the two compounds is:

   __$NaHCO_3$ + __$HC_2H_3O_2$ →
   __$NaC_2H_3O_2$ + __CO_2 + H_2O.

   A solution of a solid in a liquid and carbon dioxide (CO_2), which is the gas you detected, are produced during the reaction. Copy the equation on a piece of paper. Balance the equation by making the number of atoms of each element the same on both sides of the arrow. Place the proper number of atoms in each of the blank spaces.

Did You Know?

Rocket Propellants

"Fill 'er up for a trip to the moon!" What kind of fuel could be used for a voyage like that? Finding just the right propellant for a 70-metric-ton rocket is not easy. Moving a rocket off the earth and into space raises some difficult engineering problems.

All rockets work on the same principle. A heated gas escapes from the back of the rocket at a very high speed. As a result, the rocket moves forward as shown.

Engineers have tried many different kinds of propellants to heat the gas. Each one contains two parts: something to burn (the fuel) and something to make it burn. A fast chemical reaction occurs between the two parts and releases a great deal of energy.

The fuel should give as much push, or thrust, per kilogram as possible. The fuel should also burn a fairly long time and be easy to store. To control flight, the thrust must be turned on and off when needed.

The earliest rockets used solid fuels, such as gunpowder or nitroglycerin. These fuels were easy to carry in the rocket. However, the fuels did not burn very long. Also, thrust could not be turned on and off. Liquid fuels overcame these problems. Liquids are difficult to load and store in the rocket, however.

A mixture of liquid hydrogen and liquid oxygen is one example of a liquid propellant used in modern rockets. Recently, scientists have been trying to improve solid fuels. The perfect rocket propellant may not exist. However, the search for better propellants continues.

For Discussion
1. Why must a propellant undergo a fast chemical reaction in lifting a rocket from earth?
2. What other characteristics should a good rocket propellant have?

Chapter Summary

- The atoms of compounds are held together by chemical bonds. (14–1)
- Chemical bonds are formed by atoms losing, gaining, or sharing electrons from their outermost energy levels. (14–1)
- Bonds determine many of a substance's physical properties. (14–1)
- An ion forms when an atom gains or loses one or more electrons. (14–2)
- Ionic bonds result when atoms transfer electrons, causing oppositely charged ions to form and attract each other. (14–2)
- Ionic compounds are usually solids at room temperature, have high melting points, and have crystal structures. (14–2)
- A covalent bond occurs when two atoms share a pair of electrons. (14–3)
- A molecule is two or more atoms held together by covalent bonds. (14–3)
- The properties of covalent compounds vary. Usually they do not conduct electricity. (14–3)
- Chemical equations use symbols and formulas to describe chemical changes. (14–4)
- A balanced equation shows equal numbers of each kind of atom on either side of the arrow. (14–4)

Interesting Reading

Weiss, Malcolm E. *Why Glass Breaks, Rubber Bends, and Glue Sticks.* Harcourt, 1979. Explains how the structure of molecules determines such properties as electric conductivity and stickiness.

Zubrowski, Bernie. *Messing Around with Baking Chemistry.* Little, Brown, 1981. Explains what happens as you bake bread and cake, especially the effects of baking powder and baking soda.

Questions/Problems

1. What evidence shows that atoms usually combine with other atoms?
2. Why must two atoms of hydrogen bond with one atom of oxygen to form one molecule of water?
3. How does knowing the number of electrons in an atom's energy levels help predict which other atoms it will join with?
4. Explain what happens to two atoms as they form an ionic bond between them.
5. Explain how a covalent bond can form without either atom losing an outer level electron. Include a description of what happens to each atom's electron cloud.
6. What does a chemical equation tell a chemist about making a compound?
7. Explain how the law of conservation of mass governs chemical reactions.
8. Choose the balanced equation(s).

    a) $Zn + H_2SO_4 \longrightarrow ZnSO_4 + H_2$.

    b) $2Li + Br_2 \longrightarrow 2LiBr$.

    c) $2AgNO_3 + NaCl \longrightarrow AgCl + 3NaNO_3$.

    d) $2HgO \longrightarrow Hg + O_2$.

Extra Research

1. Use clay or polystyrene balls and toothpicks to construct models of different crystals. You can find diagrams of the crystal arrangements of atoms in many chemistry books.
2. Visit a chemical factory or a pharmaceutical company in your area. Find out about the work a chemist does and how chemists experiment to produce new substances. Then report on your trip to your class.

Chapter Test

A. Vocabulary Write the numbers 1–10 on a piece of paper. Match the definition in Column I with the term it defines in Column II.

Column I

1. contains two atoms
2. particle made of covalently bonded atoms
3. particle formed when an atom gains or loses an electron
4. formulas and symbols that describe a chemical reaction
5. a bond formed with shared electrons
6. the number and kind of atoms is the same before and after a chemical reaction
7. having a filled outer energy level
8. forces that hold all compounds together
9. a bond formed between electrically charged atoms
10. one step in a chemical change

Column II

a. chemical bond
b. chemical equation
c. chemical reaction
d. chemically stable
e. covalent bond
f. diatomic molecule
g. ion
h. ionic bond
i. law of conservation of mass
j. molecule

B. Multiple Choice Write the numbers 1–10 on your paper. Choose the letter that best completes the statement or answers the question.

1. The number of electrons that can be held in the outer energy level of most atoms is a) two. b) six. c) eight. d) ten.

2. Ionic compounds stay together because a) opposite charges attract. b) like charges attract. c) opposite charges repel. d) electrons are shared.

3. Choose the equation that is *not* balanced.
   a) $Zn + 2HCl \longrightarrow ZnCl_2 + H_2$.
   b) $4Al + 3O_2 \longrightarrow 2Al_2O_3$.
   c) $P_4 + 5O_2 \longrightarrow P_4O_{10}$.
   d) $4Fe + O_2 \longrightarrow 2Fe_2O_3$.

4. The atoms in the smallest particle of water are held together by a) one covalent bond. b) two covalent bonds. c) one ionic bond. d) two ionic bonds.

5. The number of bonds an atom can form depends on the number of a) atoms nearby. b) protons. c) neutrons. d) outer electrons.

6. Fluorine will most probably form a covalent bond with a) fluorine. b) neon. c) helium. d) argon.

7. Lithium will most probably form an ionic bond with a) beryllium. b) helium. c) chlorine. d) sodium.

8. Choose the balanced equation.
   a) $O_2 + H_2 \longrightarrow H_2O$.
   b) $CaCO_3 \longrightarrow CaO + CO_2$.
   c) $C + Cl_2 \longrightarrow CCl_4$.
   d) $H_2 + Cl_2 \longrightarrow HCl$.

9. Choose the element that occurs as a diatomic molecule. a) sodium b) neon c) oxygen d) calcium

10. Which is a property of ionic compounds? a) low melting point b) usually gaseous at room temperature c) poor electrical conductors when dissolved d) usually hard and solid

Chapter 15
Chemical Reactions

In the photo, the campfire supplies energy to cook the food. As the raw eggs cook, their yellow yolks harden, and their thin, watery portions become firm and white. The cooked eggs look, feel, and taste different than raw eggs. Eventually, only a small pile of ashes marks the site of the campfire. The ashes are also very different from the wood that burned in the campfire.

Cooking eggs and burning wood are examples of chemical reactions. This chapter will investigate several kinds of chemical reactions and explore the similarities and differences between them.

Chapter Objectives
1. Describe chemical reactions in terms of changes in substances and energy.
2. Recognize synthesis and decomposition reactions.
3. Recognize replacement reactions.
4. Describe carbon compounds and their reactions.

15-1
Recognizing Chemical Reactions

Warm milk spoils more quickly than cold milk. When you refrigerate food or store medicines in a cool dry place, you control the speed of chemical reactions. As you read about how to recognize and control chemical reactions, answer the following questions:

a. What is a spontaneous chemical reaction?
b. How is energy involved in a chemical reaction?
c. How can you control a chemical reaction?

Many Chemical Reactions Are Spontaneous

A chemical reaction is only one step in any chemical change. In the pictures, chemical reactions take place as the paper burns and the seltzer tablet fizzes. The ashes are not like the original sheet of paper. The gas bubbles from the seltzer tablet are not like the original white tablet. We recognize chemical reactions by the new substances they produce.

In a chemical reaction, the original substances are known as **reactants** (rē ak′tənts). The newly formed substances are known as **products.** In the examples, the original paper, water, and seltzer tablet are reactants. The charred ashes and gas bubbles are products.

Examples of chemical reactions

The result of a chemical reaction on buildings in Washington, D.C.

Some chemical reactions take place at room temperature and without outside help. Dropping the seltzer tablet in water produces gas bubbles immediately. Reactions that occur without outside help are **spontaneous** (spon tā′nē əs) **reactions.** A burning match and a rusting iron chain also show spontaneous reactions. The match burns more quickly than the chain rusts, but both react spontaneously with oxygen in the air.

Color changes often indicate that a chemical reaction has occurred. Toasted bread and tarnished silver are examples of such color changes. The dark surface of the bread and the dull finish of the silver are the products of chemical reactions. The buildings in the picture changed color when the copper in them reacted with the air. The green substance is a compound of copper, oxygen, carbon, and hydrogen. A similar color change happens when you wear copper jewelry. The green stain on your skin shows that a chemical reaction occurred between you and the copper.

Energy in Chemical Reactions

Energy changes accompany chemical reactions. Some chemical reactions give off energy, and other reactions take in energy.

Burning fuels, such as coal, oil, and natural gas, releases large amounts of energy. Chemical reactions that release energy are **exothermic** (ek′sō thėr′mik) **reactions.** Burning coal in a furnace, wood in the fireplace shown, and natural gas in a kitchen stove are examples of exothermic reactions. Rusting is also an exothermic reaction, even though it produces very little heat.

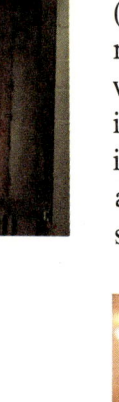

Chemical reactions that absorb energy are **endothermic** (en′dō thėr′mik) **reactions.** Cooking food, such as frying a raw egg, is an example of an endothermic reaction. The egg will not cook without heat. Another endothermic reaction is the cold pack in the picture below. The chemical reaction in the cold pack absorbs heat from the athlete's bruised ankle. This process cools the injured area, which reduces swelling and pain.

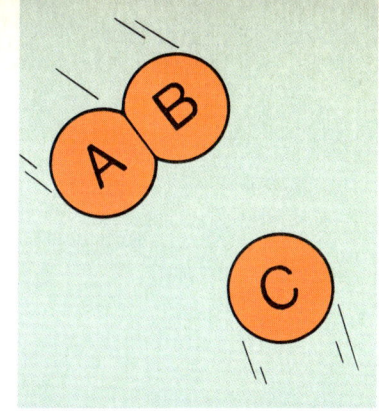  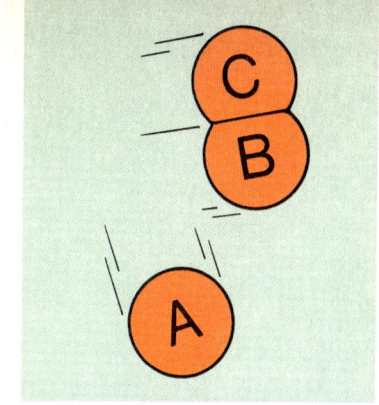

Controlling the Speed of Reactions

Without ways to control chemical reactions, many reactions would take too long, would be very dangerous, or would not occur at all. In controlling reactions, it is often as important to prevent some reactions as it is to speed up others.

Raising the temperature usually speeds reactions. Adding heat to a reactant makes the molecules move faster, as shown above. Because they move faster, they collide more often and form new molecules sooner. The new molecules are the products of the chemical reaction.

You can usually slow a chemical reaction by lowering the temperature. Removing heat lowers the temperature and reduces the number of collisions between reactant molecules. The reaction slows down. Refrigerators slow down the chemical reactions that would spoil food.

Some substances affect the speed of a reaction. A **catalyst** (kat′əl ist) is a substance that changes the speed of a reaction without being affected itself. Some vitamins are catalysts for your body's chemical reactions. A catalyst in your saliva breaks down large starch molecules into smaller ones. The chlorophyll in plants is a catalyst that traps sunlight and splits water during the plants' sugar-making process.

Have You Heard?

Chemicals are added to foods to prevent unwanted chemical reactions. For example, calcium propionate is often added to bread to slow the chemical reactions that help mold grow. Sodium benzoate, which occurs naturally in some foods, is added to other foods because it prevents the growth of yeast and mold.

Challenge!

Look at food labels in your kitchen or at your local food store. Note any chemicals added to these foods to prevent unwanted chemical changes. Such chemicals are *preservatives*.

Review It

1. How do you know when a chemical reaction is a spontaneous reaction?
2. How do exothermic and endothermic reactions differ?
3. How does a catalyst affect a chemical reaction?

Activity

Controlling the Speed of Chemical Reactions

Purpose
To observe how to affect the speed of a chemical reaction.

Materials
- 50 mL vinegar
- baking soda
- thermometer
- hot plate or Bunsen burner, metal tripod, wire gauze, and matches
- two shallow, 1-L aluminum pots
- 50-mL graduated cylinder
- four 100-mL beakers
- two 250-mL beakers
- six ice cubes
- two wood splints
- black paper
- gloves or hot pads
- grease pencil
- safety goggles

Procedure
1. Add the baking soda to 200 mL of tap water in one 250-mL beaker. Stir with a wood splint until the baking soda disappears.
2. Pour half the solution into a small beaker and half into another. Label each "soda."
3. Add 50 mL vinegar to 150 mL of tap water in the other 250-mL beaker. Stir well with a second wood splint.
4. Repeat step 2, but label each small beaker "vinegar."
5. Place one soda beaker and one vinegar beaker in a pot filled 3-cm deep with water.
6. Set up your heating equipment as shown.
7. CAUTION: *Do not touch anything hot. Put on your safety goggles. Handle the thermometer carefully.* Heat the water in the pot until it reaches 65°C. Maintain this temperature for 5 minutes.

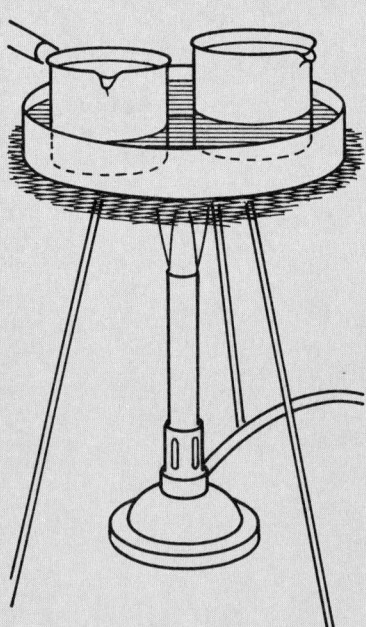

8. While you heat the water, your partner should repeat step 5 with the remaining pot and two beakers of soda and vinegar. Add six ice cubes to the water.
9. Cool these beakers to 20°C.
10. Take the temperature of the soda beakers in each pot. Record each temperature. (Assume the vinegar beakers have the same temperature.)
11. Remove the cold beakers from the pot.
12. Place the cold beakers on black paper and quickly add a small amount from the vinegar beaker to the soda beaker.
13. Record your observations. Explain how rapid any bubbling seems.
14. Keep adding small amounts of vinegar to the soda until you have an accurate idea of the speed of bubbling.
15. CAUTION: *Using the hot pad,* repeat steps 12–14 for the hot beakers.

Analysis
1. At which temperature did the substances react more rapidly? How do you know?
2. Make a statement about the effect of temperature on the speed of this chemical reaction.

Did You Know?

Fireworks

Most of us have seen beautiful fireworks displays similar to the one shown. They are often used on the Fourth of July, Labor Day, or other holidays. In many parts of the world, they are also used as part of religious celebrations. Actually, fireworks have a long history.

Fireworks were probably invented in China in the twelfth century. A mixture of saltpeter, sulfur, and carbon was found to be useful for two quite different purposes: gunpowder and fireworks. Since then, these two uses of explosives have developed at the same time.

Throughout the world, firecrackers, Roman candles, cherry bombs, and rockets are used to celebrate many important events: royal weddings, military victories, important births, peace treaties, and religious holidays. They are also used just for fun. The sound, color, and sparkle of fireworks have fascinated humans for centuries.

However, fireworks are a dangerous kind of fun. They are, after all, explosives. The composition of a Roman candle is not much different than a small bomb. Compounds containing nitrate (NO_3^-) and chlorate (ClO_3^-) ions, sulfur, and charcoal provide the explosion that sends a rocket into the air and then makes it blow apart. Powdered metals —iron, steel, magnesium, aluminum, and titanium—catch fire and produce the sparks and flames. Metallic compounds add color to the sparks. Sodium compounds produce a yellow color; calcium compounds, red; strontium compounds, scarlet; barium compounds, green; and copper compounds, bluish-green. Ammonium compounds change the shades of these colors. Picric acid and sulfur make the colors more brilliant.

Because they can be so dangerous, fireworks are banned in many parts of the United States. Where they are used without qualified supervision, fireworks injure a few children and adults each year.

For Discussion
1. List ten chemicals used in making fireworks.
2. In what ways are fireworks like bombs? In what ways are they different?

15–2 Synthesis and Decomposition Reactions

The tarnish on silverware and the breakdown of limestone in a building are two types of chemical reactions. They represent two opposite processes—forming and breaking down compounds. As you read, consider the following:

a. How do substances combine?
b. What is a decomposition reaction?

Combining Substances

A spoon tarnishes because silver in the spoon joins chemically with sulfur in the air to form a new compound. The chemical joining of two substances to form a compound is a **synthesis** (sin′the sis) **reaction.** During all synthesis reactions, simple substances combine to form more complex substances. Water forms from a synthesis reaction between hydrogen and oxygen.

An important synthesis reaction occurs in the catalytic converter of an automobile. Without a converter, cars release the harmful gas carbon monoxide into the air. Notice the material inside the catalytic converter shown. The material has a large surface area to give the catalyst the greatest effect.

The synthesis reaction in the catalytic converter eliminates carbon monoxide gas. Oxygen joins with the carbon monoxide to form the harmless gas carbon dioxide. Catalysts speed up the process:

$$2CO + O_2 \xrightarrow{Pd,Pt} 2CO_2.$$

The inside of a catalytic converter

Because ammonia compounds are used as fertilizers, the synthesis of ammonia from nitrogen and hydrogen is an important reaction. The following equation shows how ammonia (NH_3) forms:

$$N_2 + 3H_2 \longrightarrow 2NH_3.$$

Breaking Down Substances

The breakdown of a compound into two or more simpler substances is a **decomposition** (dē′kom pə zish′ən) **reaction.** Souring milk is an example. The milk breaks down into simpler sugars and other chemicals. A decomposition reaction is the opposite of a synthesis reaction.

The decomposition of limestone is important to the building industry. The lime produced in the reaction is used to make plaster and cement. The following equation represents the decomposition of limestone:

$$\underset{\substack{\text{calcium carbonate}\\\text{(limestone)}}}{CaCO_3} \longrightarrow \underset{\substack{\text{calcium oxide}\\\text{(lime)}}}{CaO} + \underset{\text{carbon dioxide}}{CO_2.}$$

Another example of a decomposition reaction is the breakdown of hydrogen peroxide, which is used as a hair bleach and a household antiseptic. The equation shows that hydrogen peroxide breaks down into water and oxygen.

$$2H_2O_2 \longrightarrow 2H_2O + O_2.$$

Light speeds up this reaction. To slow the decomposition, manufacturers sell hydrogen peroxide in brown bottles such as the one shown at the right. These bottles block out the light.

Brown bottles block out light

Review It

1. What happens during a synthesis reaction?
2. What happens during a decomposition reaction?

15-3 Replacement Reactions

Long before the days of the Roman Empire, people learned to make iron weapons and tools. In nature, iron is not found as a free element. Iron is bonded with other elements. Ancient peoples discovered how to separate iron from other substances. Keep the following questions in mind as you read about this kind of reaction:

a. What is a replacement reaction?
b. How do we know which elements are more active?

Elements Can Replace Other Elements

The people shown are smelting iron using a very old process. In the smelting process, the element carbon takes the place of the iron in the ore. The following equation represents the removal of iron from one kind of ore:

$$2Fe_2O_3 + 3C \longrightarrow 4Fe + 3CO_2.$$

In this reaction, carbon replaces the iron in the iron ore. In a **replacement reaction,** a free element replaces another element in a compound. Because carbon replaces iron in the reaction above, iron is left behind as a free element.

Smelting iron using a 2,000-year-old method

Making PbCrO$_4$

Two important modern industrial processes are replacement reactions involving bromine and copper. The element bromine is important in making photographic film. Like iron, bromine occurs in compounds and not as a free element. The oceans contain most of the dissolved bromine compounds. Replacing the bromine in these compounds with chlorine releases bromine as a free element. This reaction, which occurs in water, is shown below:

$$Cl_2 + 2NaBr \longrightarrow Br_2 + 2NaCl.$$

Because copper is used in electrical wiring, it is another important element. In nature, some copper exists as a free element, but most of it must be removed from copper and sulfur ores. Oxygen replaces copper as follows:

$$Cu_2S + O_2 \longrightarrow 2Cu + SO_2.$$

In some replacement reactions, two compounds exchange elements. The following reaction between lead and potassium compounds takes place in a water solution:

$$Pb(NO_3)_2 + K_2CrO_4 \longrightarrow 2KNO_3 + PbCrO_4.$$

Both reactants form solutions with water. When the lead (Pb^{+2}) and potassium (K$^+$) ions exchange places, a solid compound of lead ions and chromate ions (CrO$_4^{-2}$) forms. This compound, shown above, is used in paints.

Iron replaces copper in this replacement reaction

The Activity Series

In nature, some elements, such as iron and sodium, exist only in compounds. Other elements, such as silver and gold, exist as free elements. Silver and gold do not react enough to combine with other elements. Chemists recognize that some elements are more active than others.

Some replacement reactions can determine how active a metal is. Above, an iron rod in a copper sulfate solution becomes coated with copper. At the same time, some iron from the rod enters the copper sulfate solution. In this reaction, iron replaces copper. Therefore, iron is more active than copper.

In a similar reaction, magnesium replaces zinc:

$$Mg + ZnSO_4 \longrightarrow MgSO_4 + Zn.$$

In this case, magnesium must be more active than the zinc.

The chart lists metals in the order of their chemical activity. This listing is the activity series. It helps chemists predict how elements will react to make new compounds.

Activity Series

Sodium
Magnesium
Aluminum
Zinc
Chromium
Iron
Cobalt
Nickel
Tin
Lead
Hydrogen
Copper
Mercury
Silver
Platinum
Gold

← more active →

Review It

1. Explain what occurs in a replacement reaction.
2. What is the activity series of metals?

Activity

Determining Chemical Activity

Purpose
To determine a short activity series.

Materials
- eight 100-mL beakers
- 75 mL of dilute $CuSO_4$
- 75 mL of dilute $ZnSO_4$
- 100 mL of dilute HCl
- two 10-cm lengths of bare copper wire, #12 gauge
- three new 10-cm iron nails (10 d)
- two 10-cm strips of zinc
- three 10-cm strips of magnesium ribbon
- 25-mL graduated cylinder
- steel wool
- white paper
- safety goggles
- grease pencil

Procedure
1. Copy the chart shown. *Put on your safety goggles.*
2. Put 25 mL of $CuSO_4$ in each of three beakers. Label each "$CuSO_4$."
3. Clean a nail, a strip of zinc, and a strip of magnesium with steel wool. Note the color of each.
4. Place one metal in each beaker.
5. Set the beakers on white paper. Write the name of each metal near its beaker.
6. Repeat step 2 for the $ZnSO_4$.
7. Repeat step 3 for a piece of copper wire, another nail, and another strip of magnesium.
8. Repeat steps 4–5 for the $ZnSO_4$, nail, copper, and magnesium.
9. Observe the six beakers. Record any change in the color or appearance of the metal strips. Let the beakers stand as you go on to step 10.
10. *CAUTION: HCl can damage skin, eyes, and clothes. Pour carefully.* Put 25 mL of HCl in each of 3 beakers. Label each "HCl."
11. Repeat step 3 for all the metals.
12. Repeat steps 4–5 for the metals and HCl.
13. Observe the 4 beakers for bubbles of H_2 gas. Record your observations.

Analysis
1. A reaction occurs when a more active element replaces a less active element in the solution. Which element was most active? Which was least active?
2. From your observations, write an activity series for the four metals and hydrogen gas.

Data chart

Element	Solution		
	$CuSO_4$	$ZnSO_4$	HCl
Cu	—		
Fe			
Zn		—	
Mg			

275

15–4
Carbon Compounds

Carbon atoms join easily to form useful carbon compounds. Sugars, plastics, and fuels are carbon compounds. Chemically, you are a collection of carbon compounds because every cell in your body contains molecules having carbon atoms. As you explore the chemistry of carbon, consider these questions:

a. How does carbon form different compounds?
b. What are some sources of carbon compounds?
c. What are some reactions that hydrocarbons undergo?

Forming Carbon Compounds

The elements carbon and hydrogen combine to form **hydrocarbon** (hī′drō kär′bən) compounds. Natural gas used for cooking and heating is a mixture of hydrocarbons. The drawing at the left shows the chemical structure of methane, which is the main compound in natural gas. Four of carbon's six electrons are shared in the bonding process. Each of the four electrons pairs with a hydrogen electron. Such covalent bonds are single bonds.

Shown below is a simpler picture of methane. In this version, known as a structural formula, the dash (—) represents a pair of electrons that form a bond. Four dashes around the "C" show that methane has four bonds.

Most natural gas also contains some propane, shown below. Three carbon atoms can bond to each other and eight hydrogen atoms to form propane, C_3H_8.

Carbon occurs in so many compounds for two reasons. One is that carbon forms four strong covalent bonds. The other reason is that carbon can bond with itself. Linking carbon atoms in this way leads to many stable compounds.

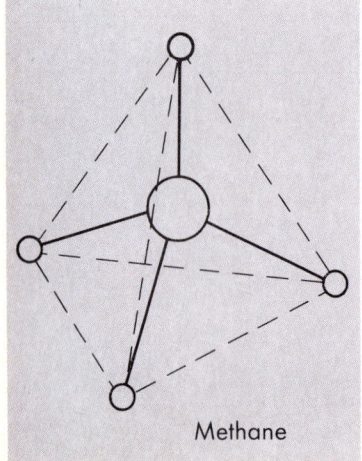
Methane

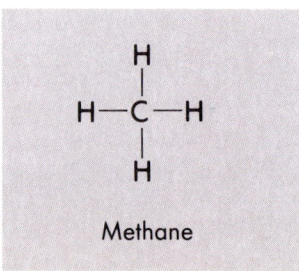
Methane   Propane

276

Hydrocarbon fractions from petroleum

Fraction	No. of C atoms in molecule	Boiling points range (°C)	Uses
Gases	1 to 5	−160 to 30	Heating and torches
Petroleum, ether, and light naptha	5 to 7	20 to 100	Industrial and cleaning solvents
Gasoline	6 to 12	60 to 200	Automobile fuel
Kerosene, fuel oil	12 to 18	180 to 400	Diesel fuel, oil for heating
Lubricating oil	16 and higher	above 350	Motor oil, grease, petroleum jelly
Paraffins	20 and higher	low-melting solids	Candle wax
Asphalt, tar	36 and higher	gummy solid residues	Roofing & road construction

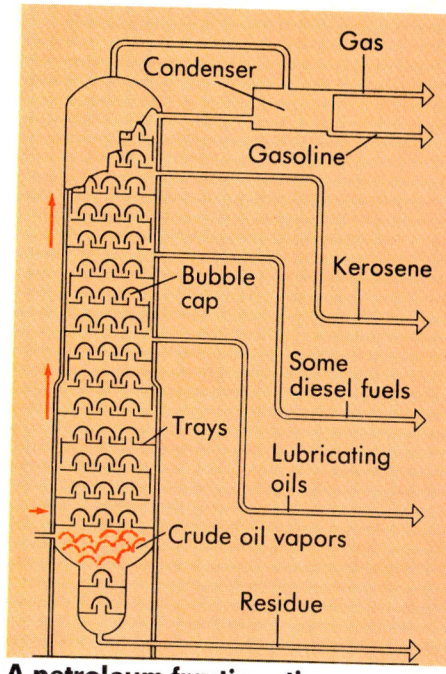

A petroleum fractionating tower

Sources of Carbon Compounds

Most carbon compounds are made from the remains of organisms that lived on earth millions of years ago. These remains became today's coal and oil.

Crude oil, also called petroleum, is mainly a mixture of hydrocarbons. Before crude oil can be used, it must be separated into its components, called **fractions.** The separation process depends on each hydrocarbon's boiling point. The number of carbon atoms determines a hydrocarbon's boiling point. Fractions with more carbon atoms have a higher boiling point. For example, octane (C_8H_{18}) has a higher boiling point than butane (C_4H_{10}).

Crude oil vapors are heated and enter the tower at the bottom, as shown. As the hot vapors rise, hydrocarbons with high boiling points cool, condense, and settle on the lower trays. Hydrocarbons with low boiling points rise higher before they cool and condense. Fractions with the lowest boiling points rise to the top. The process of separating substances by their boiling points is called **fractional distillation** (dis′tl ā′shən). The uses of crude oil fractions are summarized in the table.

Some common objects containing carbon compounds

Reactions of the Hydrocarbons

Hydrocarbons undergo many types of reactions. The products of some of these reactions are shown above.

In many hydrocarbon reactions, the type of bond between some of the carbon atoms changes. For example, when propane is heated above 460°C, it loses some of its hydrogen atoms. The equation for this reaction is:

$$\underset{\substack{H\ H\ H\\|\ \ |\ \ |\\H-C-C-C-H\\|\ \ |\ \ |\\H\ H\ H}}{} \xrightarrow{460°C} \underset{\substack{H\ H\ \ \ \ \ H\\|\ \ |\ \ \diagup\\H-C-C=C\\|\ \ |\ \ \diagdown\\H\ H\ \ \ \ \ H}}{} + H_2$$

Propylene is a product used in carpet fibers, pipes, and unbreakable bottles. Notice that the carbon bonds of propane and propylene differ. Two of the carbon atoms in propylene share two pairs of electrons. This four-electron bond is a double bond. Two dashes (=) represent a double bond. The process in which large, single-bonded hydrocarbons break into smaller, double-bonded hydrocarbons is called **cracking.**

Another important reaction of single-bonded hydrocarbons is combustion. It produces most of our heat.

Both single- and double-bonded hydrocarbons can form compounds having chlorine, oxygen, nitrogen, and other elements in them. In **substitution reactions,** hydrogen atoms of single-bonded hydrocarbons are replaced by other elements. The non-stick material used to line cooking pans is made when fluorine takes the place of hydrogen in a substitution reaction.

In some reactions, long chains of carbon molecules connect to form one giant molecule. The joining of small molecules to form a large molecule is **polymerization** (pol/ə mər ə zā/shən). During polymerization, double-bonded hydrocarbons become single-bonded hydrocarbons.

Rayon and polyester are made through polymerization. Other large molecules make the many plastics and fibers we use. For example, phonograph records and rainwear are made of polyvinyl chloride. Paint and glue are made of polyvinyl acetate.

Nylon is a large molecule that is used in clothes, ropes, and parachutes. Nylon forms by removing atoms from molecules in a **condensation reaction.** Then the remaining molecules are joined during polymerization. In the picture at the right nylon forms at the surface between two chemicals. The nylon is strong enough to be pulled out from the container.

Making nylon in the laboratory

Review It

1. Why do so many compounds contain carbon?
2. How can the many compounds in petroleum be separated from one another?
3. What is polymerization?

Chapter Summary

- Spontaneous reactions are those that occur without outside help. (15–1)
- Reactions can release or absorb energy. (15–1)
- Energy or a catalyst can affect the speed of a chemical reaction. (15–1)
- The chemical combination of substances is a synthesis reaction. (15–2)
- Breaking down compounds into simpler substances is a decomposition reaction. (15–2)
- A free element replaces another element in a compound during a replacement reaction. (15–3)
- The activity series lists elements in order of their chemical reactivity. (15–3)
- Carbon occurs in many compounds because it forms strong covalent bonds with itself or other elements. (15–4)
- Cracking, combustion, substitution, polymerization, and condensation are hydrocarbon reactions. (15–4)

Interesting Reading

Drummond, A. H., Jr. *Molecules in the Service of Man*. Lippincott, 1972. Classic discussion of chemicals derived from petroleum and coal.

Kraft, Betsy Harvey. *Oil and Natural Gas*. Watts, 1978. A discussion of the process of refining fuel.

Williams, Trevor. *Man the Chemist*. Priory Press, 1976. A history of chemistry that includes the making of modern products, such as plastics.

Questions/Problems

1. What is one way of preventing undesirable chemical reactions? How does it work?
2. Why do druggists often keep medicines in refrigerators or dark-colored bottles?
3. Explain how collisions of molecules affect the rate of a chemical reaction.
4. Identify the kind of reaction:
    a) $2KCl + F_2 \longrightarrow 2KF + Cl_2$.
    b) $2HgO \xrightarrow{heat} 2Hg + O_2$.
    c) $Ca + Br_2 \longrightarrow CaBr_2$.
5. Choose the more active metal in the following equation:
    $$Cu + 2AgNO_3 \longrightarrow Cu(NO_3)_2 + 2Ag.$$
6. One kind of gasoline molecule contains 10 carbon atoms. One kind of kerosene molecule contains 16 carbon atoms. Which has the higher boiling point? Which condenses first during fractional distillation?
7. What accounts for the large number and the many kinds of carbon compounds?

Extra Research

1. Write to a petroleum company and ask for information on refining crude oil. Share the information with your science class.
2. Collect products that are made from petroleum. Display these products so they can be studied by your class.
3. Use reference books to find out how plastics are made and what their properties are. Hardness and melting point are two properties you should be sure to read about. Then make a report to your class.

Chapter Test

A. Vocabulary Write the numbers 1–10 on a piece of paper.
Match the definition in Column I with the term it defines in Column II.

Column I

1. reaction that gives off energy
2. original substance in a chemical reaction
3. substance that speeds up a chemical reaction without being changed itself
4. substances that contain only hydrogen and carbon
5. breaking down compounds into simpler substances
6. separating substances by their boiling points
7. combining elements or compounds to make a new substance
8. reaction that absorbs energy
9. newly formed substance in a chemical reaction
10. formation of one giant molecule

Column II

a. catalyst
b. decomposition
c. endothermic
d. exothermic
e. fractional distillation
f. hydrocarbons
g. polymerization
h. product
i. reactant
j. synthesis

B. Multiple Choice Write the numbers 1–10 on your paper.
Choose the letter that best completes the statement or answers the question.

1. In $2H_2O_2 \xrightarrow{MnO_2} 2H_2O + O_2$, which substance is the catalyst? a) H_2O_2 b) O_2 c) MnO_2 d) H_2O

2. $CaCO_3 \longrightarrow CaO + O_2$ is what kind of reaction? a) synthesis b) decomposition c) replacement d) substitution

3. $3Li + AlI_3 \longrightarrow 3LiI + Al$ is what kind of reaction? a) synthesis b) decomposition c) replacement d) substitution

4. A pair of electrons shared between two atoms is a(n) a) ionic bond. b) single bond. c) double bond. d) metallic bond.

5. Plastics result from a) cracking. b) addition. c) substitution. d) polymerization.

6. Large, single-bonded hydrocarbons break into smaller, double-bonded hydrocarbons during a) substitutions. b) cracking. c) combustion. d) condensation.

7. Carbon forms strong bonds with a) itself and other elements. b) itself only. c) other elements only. d) none of the above.

8. Choose the hydrocarbon with the lowest boiling point. a) C_3H_8 b) C_6H_{12} c) C_8H_{18} d) CH_4

9. Raising the temperature during a reaction a) causes more collisions of molecules. b) causes molecules to move faster. c) usually speeds up the reaction. d) a, b, and c.

10. In a series of chemical reactions, sodium replaced zinc, copper replaced mercury, and zinc replaced copper. Which was most active? a) sodium b) zinc c) copper d) mercury

Chapter 16
Acids, Bases, and Salts

The beautiful patterns on this helmet were etched during a chemical reaction. The metal in the helmet reacted with a chemical, leaving the pattern behind. People have known how to use chemicals to etch metal for centuries.

This chapter is about three classes of chemicals. The chapter begins by describing the compounds in each class. Then it explains why they behave as they do. Finally it describes how these chemicals interact.

Chapter Objectives
1. Recognize and list the properties of acids and bases.
2. Describe the properties of acids and bases.
3. Describe the uses of indicators and the pH scale.
4. Describe how a salt forms during neutralization.

16–1
Properties of Acids and Bases

Grapefruit, lemon, and vinegar taste sour. Soap tastes bitter. Once these differences in taste were used to classify chemicals. As you read about identifying chemicals, keep these questions in mind:

a. How can you recognize an acid?
b. How can you recognize a base?

Properties of Acids

Grapefruit, lemons, and vinegar contain chemicals that make them taste sour. Such chemicals are acids. The word *acid* comes from the Latin *acidus,* which means "sour."

Many foods contain acids. Citric acid makes grapefruit and lemons taste sour. Acetic acid makes vinegar taste sour. Milk sours when lactose, which is the sugar in milk, changes to lactic acid.

Since acids can damage the body, tasting chemicals is a dangerous way to identify them. We can safely identify an acid by observing its chemical reactions. For example, lemon juice changes the color of foods, such as tea and red cabbage. Litmus is a dye obtained from mosslike plants. Acids turn blue litmus red.

Acids also react with active metals, such as iron and zinc, in a replacement reaction that produces hydrogen gas. In the picture, notice how battery acid seems to dissolve the metal and form another substance.

Challenge!
Find out how too much stomach acid can cause ulcers. An encyclopedia or a biology text should have an explanation.

Some Common Acids

Name	Formula
Acetic in vinegar	$HC_2H_3O_2$
Carbonic in carbonated water	H_2CO_3
Citric in citrus fruits	$H_3C_6H_5O_7$
Hydrochloric in your stomach	HCl
Sulfuric in batteries	H_2SO_4

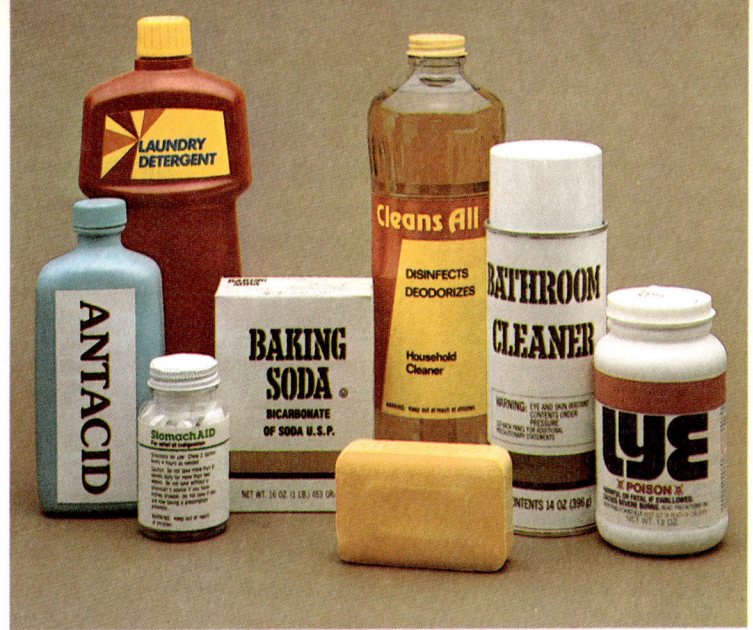

Some Common Bases

Name	Formula
Aluminum hydroxide in water purification	$Al(OH)_3$
Ammonium hydroxide household cleaner	NH_4OH
Calcium hydroxide in mortar	$Ca(OH)_2$
Magnesium hydroxide in milk of magnesia	$Mg(OH)_2$
Sodium hydroxide in drain cleaner; soap making	$NaOH$
Potassium hydroxide soap and glass making	KOH

Properties of Bases

Soap and detergent feel slippery. The slippery feeling is a characteristic of a chemical called a base. Some familiar bases are soapy water, lye, baking soda in water, milk of magnesia, and ammonia water. Many bases, such as those shown, are good household cleaners, but they can damage your body, in particular, your skin.

Bases have the opposite properties of acids. Bases turn red litmus back to blue. Many bases do not react with metals. Bases taste bitter.

Acids and bases destroy each other. Antacids, for example, contain bases. You can take antacids to relieve an upset stomach. The base in the antacid lessens the excess acid that upset your stomach.

You can clean up an acid spill by adding baking soda. The acid and the baking soda eliminate—or neutralize—each other. You add vinegar to a base spill. The acetic acid in vinegar neutralizes the base.

Review It

1. List three properties of acids.
2. What properties do all bases have?

Have You Heard?

The slippery feel of a base on your skin is caused by a very important chemical reaction. It is the same reaction used in the manufacture of soap. Soap is made when a base, such as sodium hydroxide, reacts with a fat. The reaction of the base with the fat of your skin makes a small amount of soap. The soap causes the slippery feeling.

16-2 Explaining the Properties of Acids and Bases

What is similar about the acid in a battery and the acid in your stomach? You learned some answers to this question in the last section. But what makes acids act alike and bases act alike? As you continue reading about acids and bases, answer these questions:

a. How are different acids alike?
b. How are different bases alike?

Acids Donate Protons

Although acids share the same properties, each acid is a different compound. The table shows the formulas of some common acids. Each formula begins with one or more hydrogen atoms. Chemists think that acids behave in similar ways because they all contain hydrogen. This similarity is the key to the behavior of acids.

Hydrochloric (hī′drə klôr′ik) acid, HCl, is the acid that helps digest food in your stomach. Pure HCl is the covalently bonded gas hydrogen chloride. HCl dissolved in water is hydrochloric acid. As a gas, HCl does not conduct electricity. However, hydrochloric acid conducts electricity well, as the bright light of the bulb shown demonstrates.

The following equation represents this reaction:

$$HCl + H_2O \longrightarrow H_3O^+ + Cl^-.$$

In this reaction HCl gives up a positive hydrogen ion, H^+, and forms the negative chloride ion, Cl^-. Because hydrochloric acid produces many ions in solution, it is a good conductor of electricity.

The hydrogen ion is simply a proton. Because it has no electrons, H^+ is strongly attracted to the electrons of the oxygen atom in the water molecule. Each H^+ attaches itself to a water molecule and forms the $H^+ \cdot H_2O$ ion. This ion is usually written H_3O^+ and is called the **hydronium** (hī drō′nē əm) **ion.**

Sulfuric acid (H_2SO_4), which is battery acid, undergoes a similar dissolving reaction. The equation that represents this reaction is

$$H_2SO_4 + H_2O \longrightarrow H_3O^+ + HSO_4^-.$$

Notice that both hydrochloric acid and sulfuric acid dissolve in water to produce H_3O^+ ions. These ions form because both HCl and H_2SO_4 give away protons to water molecules. An **acid** is any substance that donates protons and forms hydronium ions in water. The hydronium ions are responsible for the similar reactions of all acids.

All acids do not have the same strength. For example, the water solution of acetic acid, which is the acid in vinegar, produces only a few H_3O^+ ions. Because acetic acid produces few H_3O^+ ions, it is a weak acid. Hydrochloric and sulfuric acids are strong acids that produce many H_3O^+ ions. Compare the numbers of ions of hydrochloric and acetic acids shown in the drawing.

Some acids are less effective than others. These acids react slightly with active metals and conduct electricity poorly. For example, acetic acid is a poor electrical conductor. Hydrochloric and sulfuric acids are both good electrical conductors.

Have You Heard?

Acetic acid combines with cellulose as the first step in making rayon and other synthetic fibers. Acetic acid also combines with salicylic (sal′ə sil′ik) acid to become acetylsalicylic acid—aspirin. Salicylic acid is related to oil of wintergreen, which is a flavoring.

A strong acid has more ions than molecules

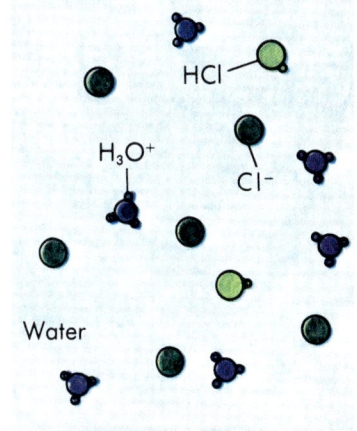

A weak acid has more molecules than ions

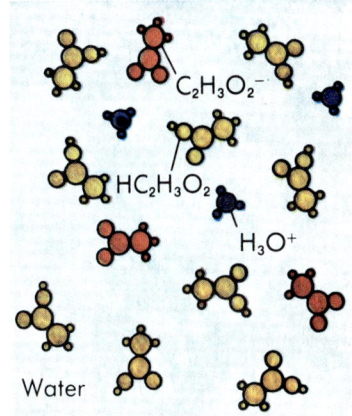

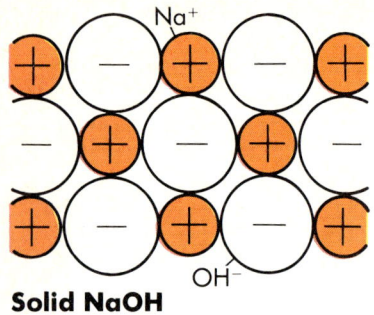

Solid NaOH

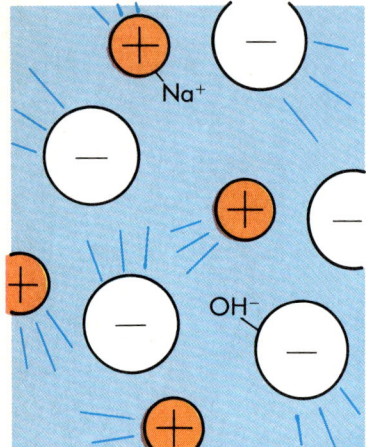

NaOH in solution

Bases Accept Protons

One base that can unclog drains is sodium hydroxide, NaOH. In its pure form, NaOH is a dangerous white solid that damages skin and clothing. Sodium hydroxide contains positive and negative ions. The attraction of the positive sodium for the negative **hydroxide** (OH^-) **ions** holds both tightly together. Because the ions cannot move past each other, solid sodium hydroxide does not conduct electricity. The first drawing shows this structure.

Sodium hydroxide dissolves in water as follows:

$$NaOH + H_2O \longrightarrow Na^+ + OH^- + H_2O.$$

The water separates the solid sodium hydroxide into many Na^+ and OH^- ions. The ions can now move past each other, as shown in the drawing. Therefore, this solution conducts electricity. The many hydroxide ions are responsible for making NaOH a strong base.

Ammonia gas (NH_3) dissolves in water to become the base ammonium hydroxide (NH_4OH):

$$NH_3 + H_2O \longrightarrow NH_4^+ + OH^-.$$

The NH_3 molecule accepts a proton from water. This acceptance produces NH_4^+ and OH^- ions. Because ammonium hydroxide produces few OH^- ions and conducts electricity poorly, it is a weak base. Any substance that accepts protons and forms hydroxide ions in water is a **base.**

Review It

1. Define an acid.
2. How does a strong acid differ from a weak acid?
3. How is a base defined?

Activity

Recognizing Acids and Bases

Purpose
To identify acids and bases by their properties.

Materials
- 10 mL vinegar
- 10 mL lemon juice
- 10 mL HCl
- 10 mL carbonated water
- 10 mL household ammonia
- 8 g washing soda
- 3 g baking soda
- 10 mL liquid drain cleaner
- red and blue litmus paper
- test tube rack
- nine 13 × 100-mm test tubes
- three 100-mL beakers
- one 25-mL graduated cylinder
- aluminum foil
- safety goggles

Procedure
1. Copy the table shown.
2. *CAUTION: Do not spill acids or bases on yourself. They can burn your skin and clothes. Wear your safety goggles.* Dissolve the washing soda in 10 mL of water in a beaker. Label the beaker "washing soda."
3. Repeat step 1 for the baking soda.
4. Measure 7 mL of each of the following into separate test tubes: HCl, vinegar, drain cleaner, washing soda water, household ammonia, lemon juice, baking soda water, and carbonated water. Label the contents of each test tube.
5. Dip one piece of each color of litmus into each of the test tubes. Record any color changes in the table.
6. Drop a small piece of aluminum foil into each test tube. Record any reaction in your table.

Analysis
1. Use your data to classify each substance you tested as an acid or a base.
2. Which kind of substance reacts with metal?

Data Table

Substance	Reaction with Litmus	Reaction with aluminum foil
HCl		
Vinegar		
Drain cleaner		
Washing soda		
Household ammonia		
Lemon juice		
Baking soda		
Carbonated water		

16–3
Indicator Colors and the pH Scale

One reason citrus fruits taste good is that they contain citric acid. Tasting other acids, however, is dangerous. We must have a safe way to identify acids and bases. We can use the color changes acids and bases cause to identify acids and bases. Continue reading to learn how to use such changes, and remember these questions:

a. How do indicators change colors?
b. What is the pH scale?

Indicator Color Changes

Tea, red cabbage, and litmus contain coloring matter known as dyes. When acids are added, some dyes change colors. Purified dyes that show color changes are called **indicators.** The photographs show the colors of some indicators in acid and base solutions.

Acid   Transition   Base

Litmus

Bromthymol blue

Methyl orange

How the H+ ion affects an indicator's color

Indicators can exist in two forms. They are either an acid or a base. For example, the indicator methyl orange is red in strong acid solutions and yellow in base solutions. The pictures show that the red form is a molecule containing a H⁺ ion. It is an acid. It turns yellow as it donates its proton to a base. The yellow form is an ion that lacks a H⁺ ion. This proton difference is responsible for the color change.

The proton difference causes the color changes of other indicators too. For example, the litmus molecule is red. It contains a H⁺ ion. When a base is added to red litmus, the red litmus donates its proton to the base. Blue litmus ions remain.

When an acid is added to blue litmus, acid molecules donate protons to the litmus and leave the litmus red. By donating its proton to the added base, the red form of the litmus molecule acts as an acid. By accepting a proton from the acid, the blue form of the litmus molecule acts as a base. In these reactions, an indicator is both an acid *and* a base. Indicators change color as they donate or accept protons from acids or bases.

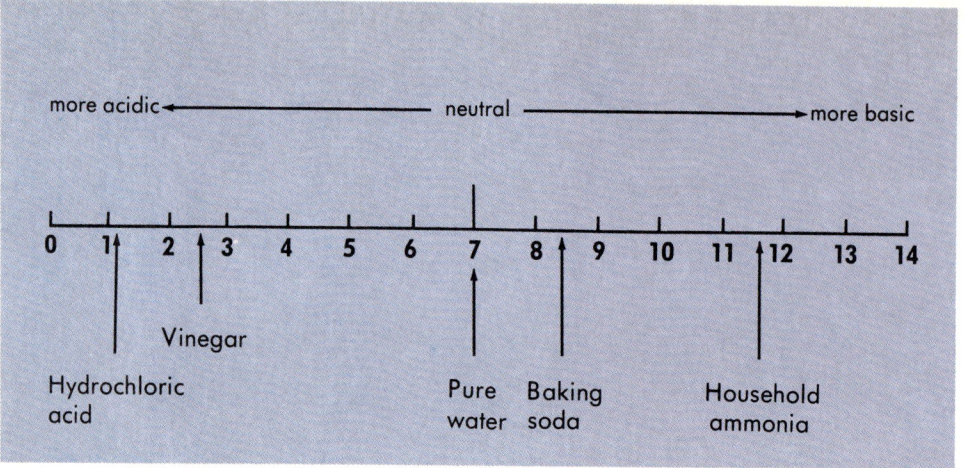

The pH scale

The Meaning of pH

The scale shown above uses the numbers 1–14 to represent hydronium ion concentration. This scale is the **pH scale.** The number seven on the pH scale represents a **neutral** solution, which has as many hydronium as hydroxide ions. Pure water is a neutral solution.

Acid solutions have many H_3O^+ ions, few OH^- ions, and pH values of *less than* 7. Basic solutions have few H_3O^+ ions, many OH^- ions, and pH values *greater than* 7. A strong acid has a low pH, and a strong base has a high pH. As hydronium ion concentration decreases, pH increases.

Each indicator changes color at a certain pH, and is useful only for a certain range of pHs. However, by mixing indicators, we can produce one indicator that operates over the whole pH scale. This mixture is called **universal indicator.** Paper coated with universal indicator turns a different color at each pH. The colors of this paper, known as pH paper, are shown at the left. The paper can be used to test the pH of many acids and bases.

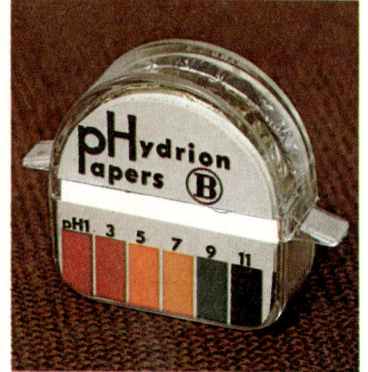

Review It

1. What determines the color of an indicator in an acid or a base?
2. What does the pH tell you about a solution?

Issues in Physical Science

Water Table Pollution

Someone is poisoning our drinking water. We are. We have made great progress in cleaning up our lakes and rivers. However, now the quality of groundwater has become worse.

The reason for this problem is that many activities which make our lives more enjoyable result in harmful waste products. For example, farmers use fertilizers, insecticides, and other chemicals to make crops grow better. But some of these chemicals seep into the earth and pollute groundwater. Salts from mining operations and highway de-icing also add to the pollution of groundwater.

Making plastics, medicines, drugs, cosmetics, and synthetic fibers is also a source of dangerous waste products. Today we dump chemical wastes into landfills, as shown, or bury them underground. Our groundwater is rapidly becoming polluted.

Burning fossil fuels can result in the formation of sulfuric and nitric acids. These acids fall to the earth as acid rain. The acids eventually reach the groundwater. Drainage from industrial dumps and landfills adds many pollutants, including strong bases, to water.

Polluted groundwater is a major concern because most Americans depend on groundwater for drinking, cooking, and other household uses. Over 50 percent of our citizens, including 95 percent of all rural citizens, get their water from groundwater. Nearly half of the water used to irrigate crops also comes from groundwater.

Polluted groundwater is not easy to clean up. Groundwater moves only a few meters a year. Therefore, natural processes can clean only a limited amount of wastes from the water. If too many wastes are dumped into the ground, the pollutants will stay in the groundwater for years.

What can be done about polluted groundwater? We could reduce the amount of waste products we throw out each year. Then we might have to give up some plastics, medicines, synthetic fibers, and other chemical products. Or we could find better ways to eliminate wastes. Then industry would pass these costs onto the cost of their products.

How important is clean groundwater? What price are you willing to pay for it?

For Discussion
1. How is pollution of groundwater a problem?
2. What changes might reduce groundwater pollution in your community?

16-4 Neutralization and Salts

We take antacids (bases) to relieve indigestion (too much stomach acid), because acids and bases are opposites. This section will focus on reactions between acids and bases and the products of these reactions. As you read, think about these questions:

a. What is a neutralization reaction?
b. How do you know when neutralization is finished?

Neutralization Reactions Produce Salts

Perhaps you know someone who has a vegetable garden. If you do, you may have seen this person sprinkle lime, which is a white powder, on the soil around certain plants. Gardeners know that lime makes vegetables less sour because lime in water is a base. Since acids and bases neutralize each other, adding lime to acidic soil raises the pH of the soil by neutralizing some of the acid in it.

A solution of NaOH and HCl is another example of this kind of reaction. When NaOH and HCl mix, each OH^- ion of the base accepts a proton from a H_3O^+ ion of the acid. The following equation represents this exothermic reaction:

$$H_3O^+ + OH^- \longrightarrow H_2O + HOH.$$

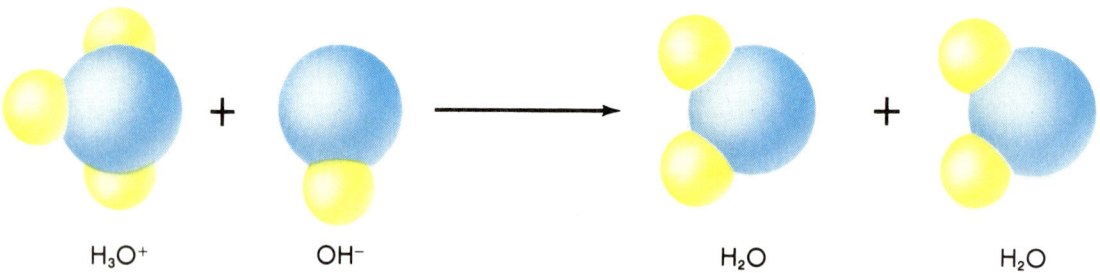

H_3O^+ + OH^- ⟶ H_2O + H_2O

Water is a product of this reaction. Writing the formula for water in two ways (H_2O and HOH) shows how each molecule forms.

Also during this reaction, Na⁺ ions from the NaOH and Cl⁻ ions from the HCl remain dissolved in water. If the water is removed or allowed to evaporate, these ions will attract each other and form table salt, NaCl. The following equation shows this replacement reaction:

$$NaOH + HCl \longrightarrow NaCl + HOH.$$

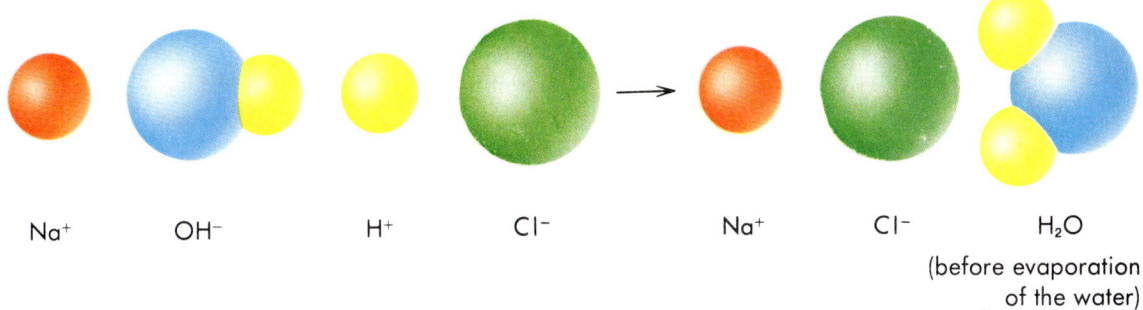

Na⁺   OH⁻   H⁺   Cl⁻   Na⁺   Cl⁻   H₂O
(before evaporation of the water)

A compound, such as NaCl, in which a positive ion from a base joins with a negative ion from an acid, is a **salt.** The positive ion is usually a metal. In this case, sodium is the metal.

The process in which an acid and a base form a salt and water is a **neutralization** (nü′trə lə zā′shən) **reaction.** Table salt is just one of the many salts that might form through neutralization reactions.

Another neutralization reaction happens between the base potassium hydroxide and sulfuric acid:

$$2KOH + H_2SO_4 \longrightarrow K_2SO_4 + 2HOH.$$

Notice that water is still one product of this reaction. However, a different salt, potassium sulfate, is produced.

Salt solutions resulting from neutralization reactions do not always have a pH of 7. A weak acid and a strong base will produce a basic salt solution. A strong acid and a weak base will produce an acidic salt solution. However, a strong acid and a strong base will produce a neutral salt solution.

295

The End-point of Neutralization

An indicator can show the neutral point, called the **end point,** of a neutralization reaction. The indicator phenolphthalein (fē′nōl thal′ēn) is often used for this purpose. In an acid solution, phenolphthalein is colorless. However, when the pH of a solution rises above 8.9, phenolphthalein turns deep pink. The photograph shows this difference.

To completely neutralize an amount of HCl, you add a few drops of phenolphthalein to the acid. You add the base solution drop by drop so the exact amount is known. At the first sign of a pink color, you stop adding the base. All the acid has been neutralized. If the amount of base and acid and the concentration of the base are known, the concentration of the acid can be determined.

Review It

1. What are the two products of a neutralization reaction?
2. How do indicators help determine the end point of a neutralization reaction?

Activity

A Neutralization Reaction

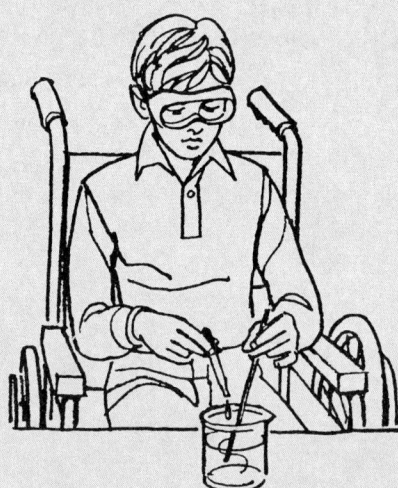

Purpose
To make a salt by neutralizing an acid with a base.

Materials
- 50-mL flask or beaker
- 10-mL graduated cylinder
- 25 mL of HCl
- 25 mL of NaOH
- 1 mL of phenolphthalein
- large medicine dropper
- small medicine dropper
- hot plate
- stirring rod
- porcelain cup
- distilled water
- safety goggles

Procedure
1. CAUTION: Do not spill acids or bases on yourself. They can burn your skin and clothes. Wear your safety goggles. Measure 10 mL of HCl into your graduated cylinder.
2. Add the HCl to 3 drops of phenolphthalein in a small beaker.
3. Rinse out the graduated cylinder with the distilled water. Then dry it. Carefully measure 20 mL of NaOH into the graduated cylinder. Use the large medicine dropper, if necessary.
4. Use the medicine dropper to add 8 mL of NaOH from the graduated cylinder to the HCl sample. Stir well with the stirring rod.
5. Drop by drop, continue adding more NaOH from the graduated cylinder, as shown. Count the number of drops used. Stir with the stirring rod after adding each drop.
6. Stop adding NaOH when a pink color remains, even after you stir.
7. Record the total volume of NaOH added. In a large dropper, 20 drops equal about 1 mL.
8. Rinse the beaker and graduated cylinder. Repeat steps 1–7.
9. Average your two answers to step 7.
10. Heat your second solution in the cup on the hot plate to boil away the remaining liquid.
11. Record your observations about any substance that remains.

Analysis
1. What is the purpose of using phenolphthalein in this experiment?
2. What is the name for the event in the reaction when the pink color remains?
3. Assume that 1 mL of HCl has as many HCl molecules as 1 mL of NaOH. What volume of NaOH would neutralize 10 mL of HCl?
4. If equal volumes of your acid and base did not neutralize each other, what might be true about the concentration of the acid compared to the base?
5. Name the substance that remains after the liquid boils away in step 9.

Chapter Summary

- Acids taste sour, react with active metals, turn blue litmus red, and neutralize bases. (16–1)
- Bases taste bitter, feel slippery, and neutralize acids. (16–1)
- Acids donate protons to form hydronium ions (H_3O^+) in water. (16–2)
- Bases accept protons and supply hydroxide ions (OH^-) in water. (16–2)
- Indicators change color as they donate or accept protons. (16–3)
- The pH scale uses the numbers 1–14 to represent hydronium ion concentration. A low pH indicates an acid. A high pH indicates a base. (16–3)
- Neutral solutions contain equal numbers of H_3O^+ ions and OH^- ions. (16–3)
- Neutralization is a reaction in which an acid and a base produce a salt and water. (16–4)
- An indicator shows the end point of a neutralization reaction. (16–4)

Interesting Reading

Arnov, Boris. *Water: Experiments to Understand It.* Lothrop, 1980. Discover the properties of water through the simple experiments in this book. (Be sure to ask one of your parent's permission before experimenting at home.)

Cherrier, Francois. *Fascinating Experiments in Chemistry.* Sterling, 1979. Explore the chemistry of indicators with the experiments in this book. (Be sure to ask one of your parent's permission first before experimenting at home.)

Questions/Problems

1. What safe tests can you perform to identify a colorless liquid as an acid, a base, or a neutral substance?

2. Suppose your results indicate the liquid in #1 is an acid. How could you tell whether it is a strong or weak acid?

3. Solutions A, B, C, and D have pH values of 6, 8, 1, and 13, respectively. Identify the strongest and weakest acids in this group. Explain your answer.

4. Describe what happens to the molecules of an indicator as the solution containing the indicator changes from an acid to a base.

5. Write the formula of the salt formed when each of the following react: a) $Ba(OH)_2 + H_2SO_4$; b) $Mg(OH)_2 + 2HNO_3$; c) $NH_4OH + HCl$; and d) $KOH + HC_2H_3O_2$.

Extra Research

1. Make your own indicator. Cut some red cabbage into tiny pieces, place them in a pot with enough water to cover them, and heat to almost boiling. Cool and filter the liquid through a coffee or paper filter. The juice is an indicator. Test lemon juice, salt, sugar, tap water, rain, baking soda, detergent, aspirin, and shampoo. Dissolve any solids before testing.

2. Obtain pH paper from your teacher or an aquarium supply store. Test the items in #1 above, and determine the pH of each from the color scale on the pH paper container. Make a similar pH color scale for your cabbage juice. Visit or phone your local water quality agency. Ask them the pH values of rain and local tap water. Compare your values with the official values. (You may find testing shampoos interesting, especially those that claim to be "pH-balanced.")

Chapter Test

A. Vocabulary Write the numbers 1–10 on a piece of paper.
Match the definition in Column I with the term it defines in Column II.

Column I

1. substance that tastes bitter and feels slippery
2. compound that results from an acid-base reaction
3. substance that tastes sour and reacts with active metals
4. dye that changes color in the presence of acids and/or bases
5. reaction between an acid and a base
6. scale representing the amount of hydronium ion in a substance
7. ion formed from the addition of a H^+ ion to a water molecule
8. ion released when a base is added to water
9. point in a reaction when a base has neutralized an acid
10. neither acid nor base

Column II

a. acid
b. base
c. end point
d. hydronium ion
e. hydroxide ion
f. indicator
g. neutral
h. neutralization
i. pH scale
j. salt

B. Multiple Choice Write the numbers 1–10 on your paper.
Choose the letter that best completes the statement or answers the question.

1. Choose the substance that is *not* a salt.
a) $Al(OH)_3$ b) $Ba(NO_3)_2$ c) $MgSO_4$
d) $NH_4C_2H_3O_2$

2. Which combination will form $Mg(NO_3)_2$?
a) $Mg(OH)_2 + 2HC_2H_3O_2$ b) $Mg(OH)_2 + 2HCl$ c) $Mg(OH)_2 + 2HNO_3$ d) $Mg(OH)_2 + H_2SO_4$

3. The pH of solutions A, B, C, and D is 12, 6, 0, and 3, respectively. The most nearly neutral solution is   a) A.   b) B.   c) C.   d) D.

4. Choose the acid.   a) NaOH   b) KOH
c) HCl   d) HOH

5. A solution is a poor electrical conductor, produces few H_3O^+ ions, and reacts slightly with metals. This substance could be   a) soap.
b) drain cleaner.   c) lye.   d) vinegar.

6. Water has a pH of   a) 0.   b) 3.   c) 7.
d) 14.

7. Household ammonia produces few OH^- ions and conducts electricity poorly. It is a
a) strong acid.   b) weak acid.   c) strong base.
d) weak base.

8. To relieve an upset stomach, you might drink   a) saltwater.   b) baking soda in water.
c) water.   d) carbonated water.

9. Which of these chemicals is the best electrical conductor?   a) NaOH in water.
b) NH_4OH c) $HC_2H_3O_2$ d) solid NaCl

10. Indicators can be   a) acids only.   b) bases only.   c) both acids and bases.   d) neither acids nor bases.

Careers

Metallurgist

Can you tell steel and aluminum cans apart? Any metallurgist can tell you that steel cans weigh more. Metallurgists are engineers who study the properties of metals.

A metallurgist searches for the best way to use different metals. First, a metal's strength, melting point, mass, density, and other properties are tested. Metallurgists use these data to find out if a metal is light enough for an airplane, or strong enough for a car. If it resists heat well enough, it could be made into pots and pans.

Metallurgy students learn scientific methods of testing and the chemistry of metals during four years of college, and often during later studies.

Career Information:
The Metallurgical Society of AIME, 420 Commonwealth Dr., Warrendale, PA 15086

Biochemical technician

When you go to the doctor, your blood gets a checkup too. Blood samples go to a laboratory for tests that a biochemical technician runs.

Biochemical technicians use their chemistry backgrounds in their work in hospitals and clinics. A technician might count blood cells, test urine samples, or study foods and drugs. After technicians collect data, they give their results to a doctor, who interprets them and gives a diagnosis.

Technicians need at least two years of training in a hospital or at a junior college. They study chemistry, biology, and lab techniques while in school.

Career Information:
American Society for Biological Chemists, 9650 Rockville Pike, Bethesda, MD 20014

Pharmacist

The Greek word *pharmakon* means "drug." Our word *pharmacist* describes people who work with drugs.

You can find pharmacists in laboratories or stores where prescriptions are filled. They distribute the medicinal drugs that doctors recommend for their patients. Because pharmacists understand the chemical composition of medicines, they can safely mix, measure, and dispense them. Pharmacists also know how different drug doses and combinations affect the human body.

People's health depends on pharmacists doing their jobs well. So pharmacists need a license to practice. Students graduate from pharmacy college after five years and then take their licensing exam.

Career Information:
Student American Pharmaceutical Association, 2215 Constitution Ave., NW, Washington, DC 20037

 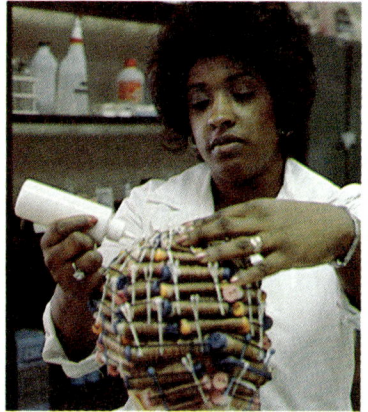 

Welder

Today, most ships are made from large sheets of steel. The problem is how to connect those sheets so they will not spring a leak. The solution calls for a welder to melt metal together.

Welders bond the metal parts of all sorts of ships, cars, planes, and other machines. They use one of three methods. Resistance and arc welding make heat with electricity. Gas welding uses a gas flame to heat the metal. Welders must understand the melting and bonding properties of metals.

Welders start to train by operating automatic welding machines. They do more complex work after they have a few years of experience learning safety and welding techniques.

Career Information:
American Welding Society, 2510 NW 7th St., Miami, FL 33125

Cosmetologist

A permanent wave has nothing to do with water. It is a chemical treatment that curls straight hair. A cosmetologist uses chemistry to change a person's hair, skin, and nails.

Cosmetologists work in beauty or barber shops. They use the chemistry of shampoos, nail polish, and make-up to help people look different. A cosmetologist knows how different treatments will affect your hair, which solutions will dissolve nail color, and how your skin will react to the substances in make-up.

Cosmetologists must have a license. They cut and dye hair, give manicures, or apply cosmetics after they graduate from about one year of training school.

Career Information:
National Beauty Career Center, 3839 White Plains Rd., Bronx, NY 10467

Molder

Perhaps you built sandcastles when you were younger. A molder's job is not that easy. But at a foundry, molders do make things with sand.

A foundry is a place where metal is melted down and poured into the shape of machine parts. Molders pack sand around models of these parts, and then remove the models. What they have left is an imprint of the machine part in sand. Other foundry workers fill this imprint, called a mold, with molten metal. The hot metal cools in the shape of the model that the molder originally used.

Molders gain skill during a four-year apprenticeship. Experienced workers teach new molders the physical and chemical properties of the sand and metals used.

Career Information:
International Molders' and Allied Workers' Union, 1225 E. McMillan St., Cincinnati, OH 45206

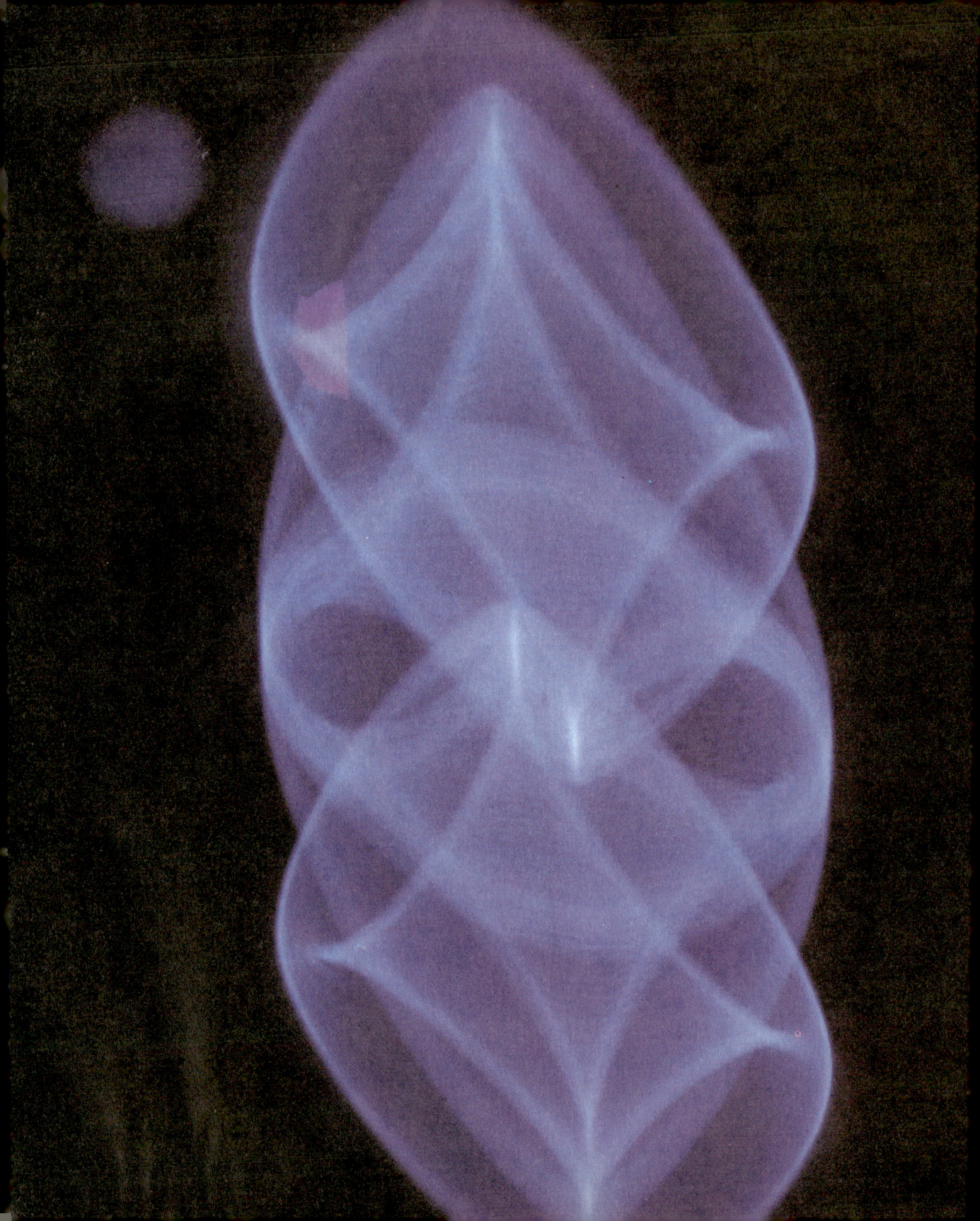

UNIT FIVE
WAVE MOTION

Some people might think this photograph shows a work of modern art. Others might think someone doodled this pattern. What do you think?

If you could see the sound of your voice, it might look like this photograph. Someone spoke into a device that turned sound into a pattern we can see and analyze. Throughout this unit, you will read about the energy and motions of waves, including sound.

Chapter 17 Waves

All waves have basic characteristics. Often a wave's behavior depends on the kind of wave it is and on the material through which the wave passes.

Chapter 18 Light

Light is a special wave. It affects what and how we see.

Chapter 19 Light and Its Uses

We have many devices that use the properties of light. Lenses and mirrors are parts of these devices.

Chapter 20 Sound

Sound waves differ from light waves. Many sounds are noisy. Other sounds are musical.

Chapter 17
Waves

How many kinds of waves do you see in the picture on the left? Surely the ripples are one kind—water waves. Light is another kind of wave present in the photograph. In fact, without light waves, you could not even see the photograph.

Water and light waves are two of many kinds of waves. This chapter discusses the different kinds of waves, their behavior, and their effects on you.

Chapter Objectives
1. Draw and label the characteristics of a wave.
2. Compare the two kinds of traveling waves.
3. Explain how waves act as they pass through different materials and through each other.
4. List the different electromagnetic waves and describe their uses.

17–1
Properties of Waves

You may have watched waves near a beach or in a pool. Did you notice that a ball floating in the water seemed to stay in about the same place? Neither the ball nor the water under it moved very much. Only the wave moved as it passed through the water. As you read more about waves, ask yourself the following questions:

a. How do waves carry energy?
b. How do we describe a wave?
c. How can you find a wave's speed from its wavelength and its frequency?

Waves Carry Energy

Remember how a hard-thrown ball hurt your hand when you caught it? The moving ball carried energy from the person who threw it to you. Waves carry energy from place to place too. But matter does not move with the wave as it moves with a ball. To understand this idea better, imagine that one end of a rope is tied to a fixed object, such as the doorknob shown. You quickly flick the other end of the rope up and down, as shown. The wave passing down the rope carries the energy of the flicking motion. But the rope itself does not move forward.

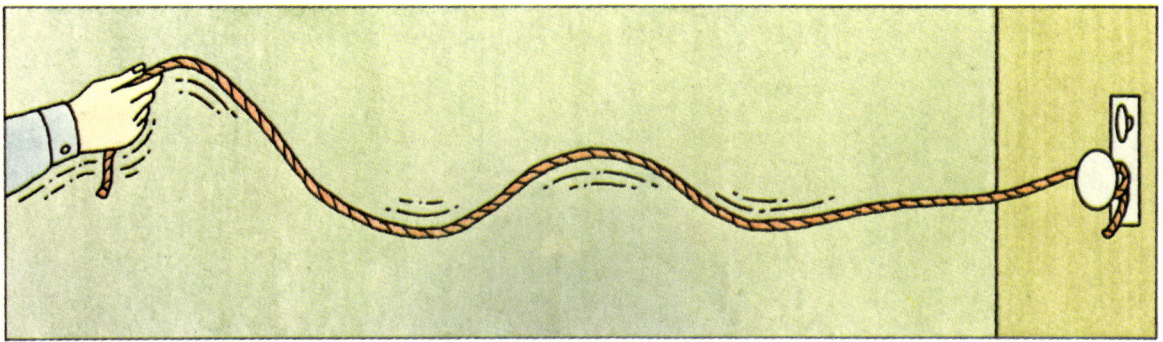

Ocean waves are energy traveling through water. Other waves, such as sound, also require a material to travel through. Some waves, such as light and radio waves, can travel through a **vacuum** (vak′yü əm), which is space empty of matter.

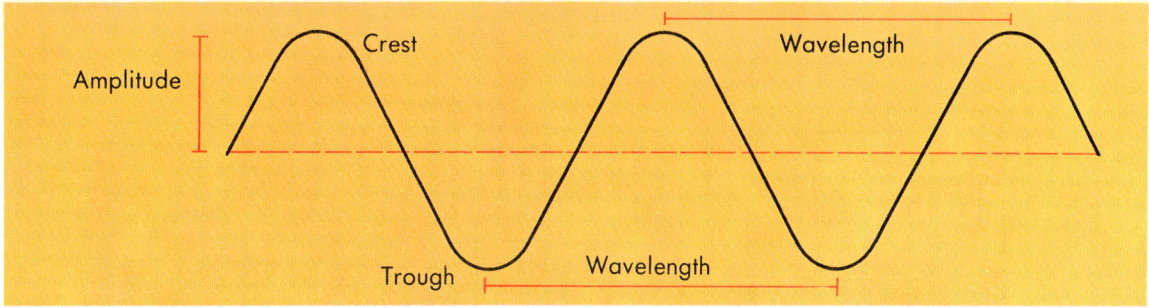
Describing a wave

Describing a Wave

All waves have some properties in common. Think again about a wave on a rope, such as the one above. The waves look like hills and valleys. The **crest** is the highest point of the wave. The **trough** (trôf) is the lowest point of the wave.

Notice the dashed line in the picture. This line marks the rope's rest position before a wave passes along it. The wave's **amplitude** (am′plə tüd) is the distance from the dashed line to the top of a crest or to the bottom of a trough. The amplitude indicates the amount of energy the wave carries. The greater the energy, the greater the amplitude. On a calm day, the amplitude of ocean waves is small. During a storm, the raging wind gives a lot of energy to the waves. A wave's amplitude soon becomes large. The wave at the right has a large amplitude and carries a lot of energy.

The distance from one crest to the next, or from one trough to the next, is the **wavelength.** Some wavelengths are marked in the drawing.

The number of waves that pass a point in a certain time is the **frequency** (frē′kwen sē). The unit used to measure frequency is the **hertz** (hèrts). One wave per second equals one hertz. If two water waves pass the edge of a pier in twenty seconds, the frequency is two waves divided by twenty seconds, or 0.1 hertz.

Speed is another important wave property. A wave's speed depends upon the kind of wave and the material it travels through. For example, light waves travel faster than sound waves. Therefore, you see lightning before you hear thunder, though they start off at the same time.

A large amplitude wave

Have You Heard?

We measure the frequency of radio waves that travel from a station to your AM radio in kilohertz, which is 1,000 waves per second. On a radio dial, 55 is short for 55 kilohertz—55,000 waves per second.

Wavelength and Frequency Affect Speed

Wavelength and frequency are two different ways to measure how close together a wave's crests are. Wavelength and frequency also have a relationship to each other, which is pictured below. Suppose a wave's speed and amplitude remain the same, but the wavelength doubles. The frequency would be halved, as the pictures show. If the frequency is doubled, the wavelength would be halved.

Two waves with equal amplitudes

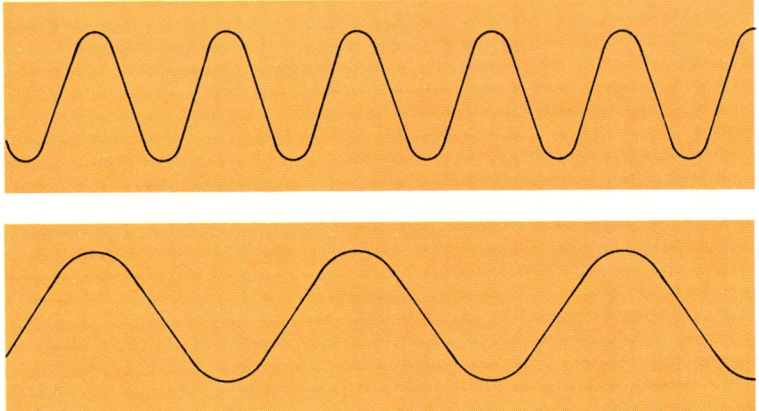

For Practice

Use the wave speed equation to solve these problems.
- What is the speed of a sound wave whose wavelength is 17 m and whose frequency is 20 hertz?
- Calculate the speed of a wave whose wavelength is 1 m and whose frequency is 0.5 hertz.

The equation below shows how wavelength, frequency, and speed are related:

$$\text{wavelength} \times \text{frequency} = \text{speed}.$$

How fast is a wave moving if its wavelength is 10 meters and its frequency is 110 hertz? To find out, multiply 10 meters by 110 hertz. The speed of the wave is 1100 meters per second. If you know any two properties of a wave, you can use the equation to find the third property.

Review It

1. Explain how waves carry energy but not matter.
2. What characteristics can you use to describe a wave?
3. How do you calculate a wave's speed?

Activity

Properties of Waves

Purpose
To observe basic wave properties.

Materials
- shallow pan
- water
- two small rocks of different sizes
- small cork
- coiled spring, at least 10 cm long

Procedure

Part A
1. Fill the pan about 3-cm deep with water.
2. Drop one rock into the middle of the pan. Notice the height of the waves.
3. When the water is calm again, repeat step 2 with the other rock.
4. Compare the heights of the waves. Record what you observe.

Part B
1. Put the cork in the water.
2. Practice making a wave by plunging your finger 1 cm in and out of the water. Do not slosh the water.
3. Repeat step 2 once every second.
4. Observe and record how the cork moves on the water.

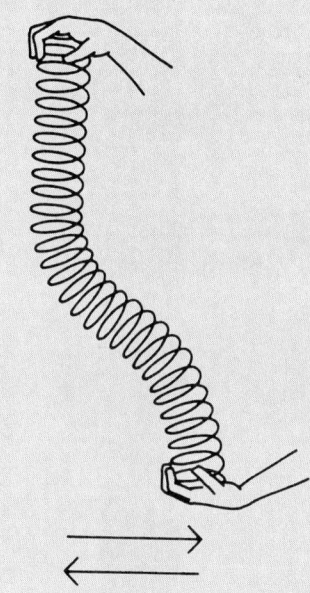
a

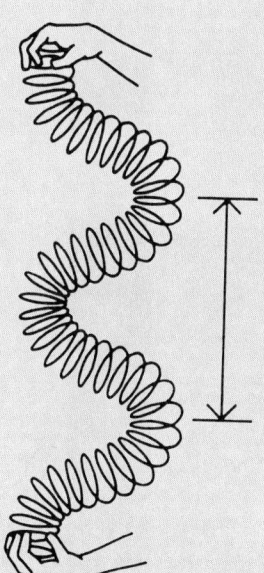
b

Part C
1. Dip your finger 1 cm in and out of the water once each second.
2. Notice the distance between ripples.
3. When the water is calm again, repeat step 1 every 2 seconds.
4. Compare the distance. Record this observation.

Part D
1. Place the coiled spring on your desk. Ask your partner to hold one end steady.
2. Stretch the spring slightly. Jerk your end of the spring quickly about 5 cm to one side and then back again to the center, as in *a*. Practice to make a good wave travel down the spring.
3. Flick your end once each second, as above.
4. Estimate and record the distance between 2 similar points in a row on the wave, as in *b*.
5. Repeat step 3 about twice as often.
6. Repeat step 4.
7. Compare the distances. Record this observation.

Analysis
1. In what ways can you change the shape of a wave?
2. What properties do waves in any material have in common?

309

17–2
Wave Motions

Ships bob up and down on water waves because of one kind of wave motion. Sound moves through air in another kind of wave motion. Consider these questions as you read about the two ways waves move:

a. What are transverse waves?
b. What are compressional waves?

Transverse Waves Have Two Directions

Any moving wave is a traveling wave. Waves in water and light in air are all traveling waves. None of these waves takes matter with it as it moves. However, matter does move somewhat as the wave passes it. Compare the way the beach ball moves with the way the wave moves past it in the pictures. The wave passes the ball from left to right. But the ball moves up on the crest, down into the trough, and back to its original place.

Notice that the wave moves the ball up and down but not away from its original place. Any wave that moves matter in a direction perpendicular to the direction of the wave's motion is a **transverse** (tranz vėrs′) **wave.** Waves on a rope and light waves are transverse waves.

Transverse waves

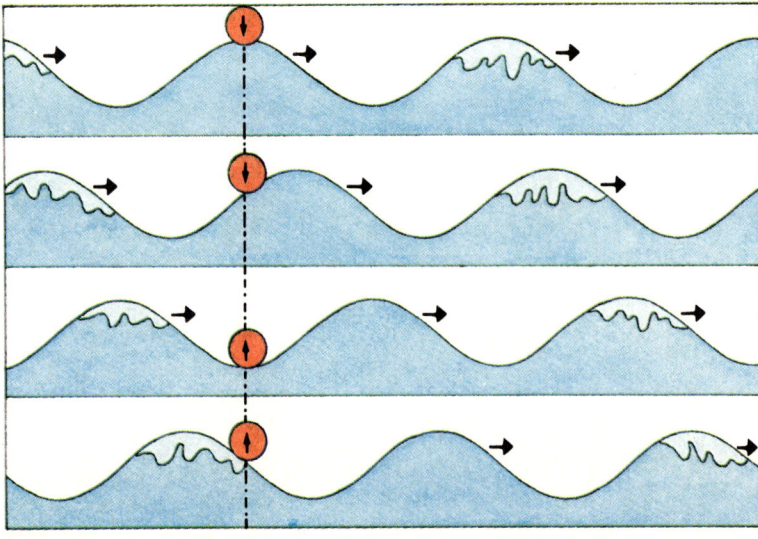

Compressional Waves Pack the Material

The other kind of traveling wave moves the molecules of a material as the coils of a spring move, shown at the right. Here the ribbon is like the beach ball in the previous example. Compare the motions of the wave and the ribbon. The wave passes from left to right. The ribbon moves somewhat to the right as the wave compresses (packs) the coils closer together. This area is a **compression** (kəm presh′ən). Then the ribbon moves somewhat to the left as the wave spreads the coils apart. This area is a **rarefaction** (rer′ə fak′shən). Finally the ribbon returns to its original place.

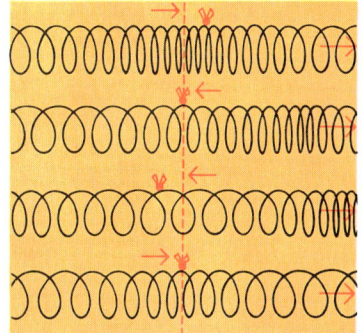
Compressional Waves

Notice that the wave moved the ribbon back and forth, but it did not move the ribbon away from its original place. This kind of wave is a **compressional wave,** because it packs and then spreads the material's molecules. These waves move matter in a direction parallel to the direction of the wave's motion. Sound is a compressional wave.

The picture below compares transverse and compressional waves. Crests are like compressions, and troughs are like rarefactions. A compressional wavelength is the distance from one compression to the next or from one rarefaction to the next.

Review It

1. How does a transverse wave pass through a material?
2. How does a compressional wave move the molecules of a material?

Comparing transverse and compressional waves

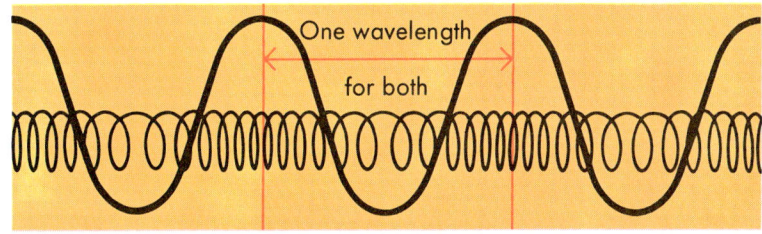

17-3
The Behavior of Waves

Some sunglasses are made to reduce the glare from light waves. This section explains how waves can be affected by different materials and by other waves. Remember these questions as you read:

a. How does a wave refract?
b. How does a wave reflect?
c. Which kind of wave can be polarized?
d. What is interference?

Waves Refract

Sometimes a wave can pass from one material into another. The wave goes straight into the second material in the first picture. It continues moving in a straight line, but its speed changes.

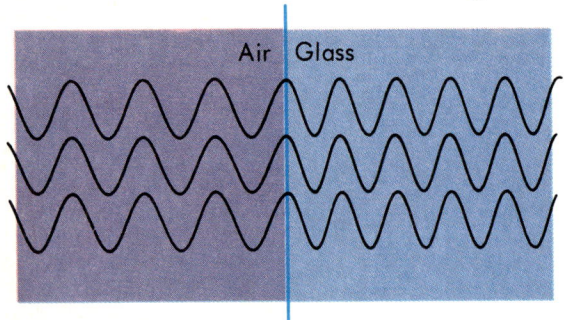
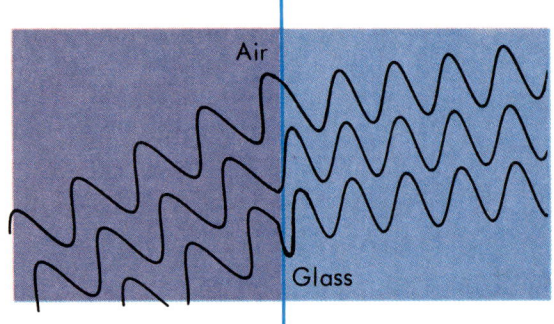

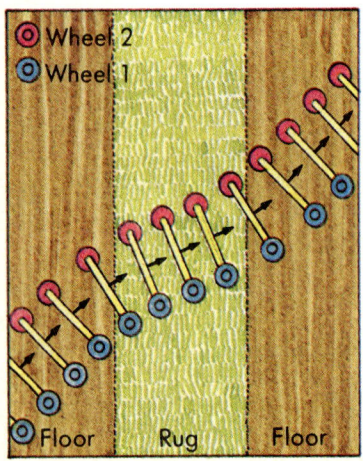

Refraction

Waves can also enter a new material at other angles. Again, speed changes. The second picture shows how the wave also changes direction. **Refraction** (ri frak′shən) is the bending of a wave as it enters a new material.

Refraction is shown on the left. The wheels roll across the smooth floor. Friction slows Wheel 1 as it moves onto a new material, the rug. Wheel 2 keeps moving at its original speed. But since an axle connects the wheels, Wheel 2 turns toward the slower moving wheel. When Wheel 2 reaches the rug, both wheels move at the same slower speed. The result is a change in the wheels' direction and speed. The process reverses as the wheels move from the rug to the floor. Both changes happen to all waves.

Waves Reflect

If you roll a ball straight into a wall, it bounces straight back. If you roll a ball to hit the wall at an angle, it bounces off at an equal angle, but toward the other side.

A wave behaves in the same way. When a wave bounces off the surface of a material, **reflection** (ri flek′shən) occurs. For example, a mirror reflects light to your eyes. A wall reflects sound to your ears.

Notice in the upper picture how the wave reflects from the surface. The angles at which a wave hits and leaves the surface are marked. If you measure these angles from a line drawn at a right angle to the surface, you find that both angles are the same. The angle of a wave hitting the surface always equals the angle of the wave leaving the surface.

Reflection of water waves

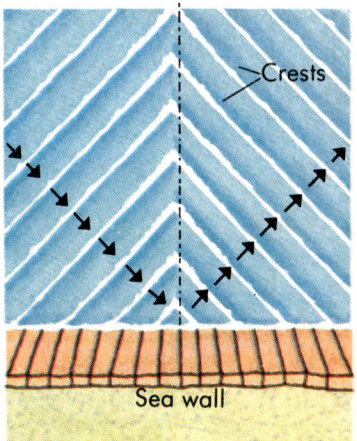

Transverse Waves Can Be Polarized

The crests and troughs of the transverse waves can move in *any* direction perpendicular to the wave direction: side to side, up and down, and so on. Waves with crests and troughs moving in only one perpendicular direction, as shown below, are **polarized** (pō′lə rīzd′).

If any waves are to pass through the grating shown, the crests and troughs must be parallel to the grating's openings. These waves are polarized. In a similar way, light waves that reflect from a road are polarized in one direction. But it is the wrong direction to pass through some sunglasses and camera filters, so glare is reduced.

Challenge!

Find out about earthquake P, S, and L waves. Which kind of traveling wave is each? Where in the earth is each reflected, refracted, or absorbed?

Polarized wave

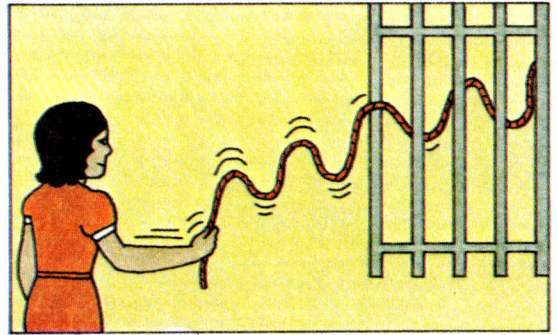

Wave does not pass

Interference

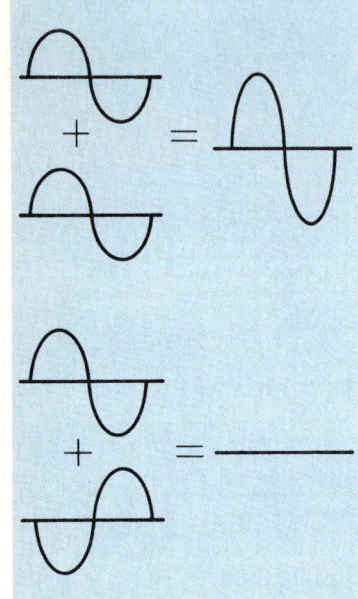

Interference of Waves

A stone thrown in a pond causes waves that move out in circles from a central source. On one side of the photograph, the pattern becomes more complicated as waves from different sources meet.

The drawing shows what happens as waves move through each other. Where two crests meet, the resulting wave is equal to the sum of both crests. The same is true for two troughs. If a crest and a trough meet, the result is no wave or a smaller wave than either of the two that met.

Interference (in′tər fir′əns) is the result of different waves moving through one another. Interference affects all waves. Light waves demonstrate interference in the drawing below. Crests meeting crests or troughs meeting troughs produce the bright bands of light. Crests and troughs that meet cancel each other. They produce the dark bands.

Review It

1. Define and give an example of wave refraction.
2. Define and give an example of wave reflection.
3. How can you polarize transverse waves?
4. How do waves show interference?

The interference of light waves

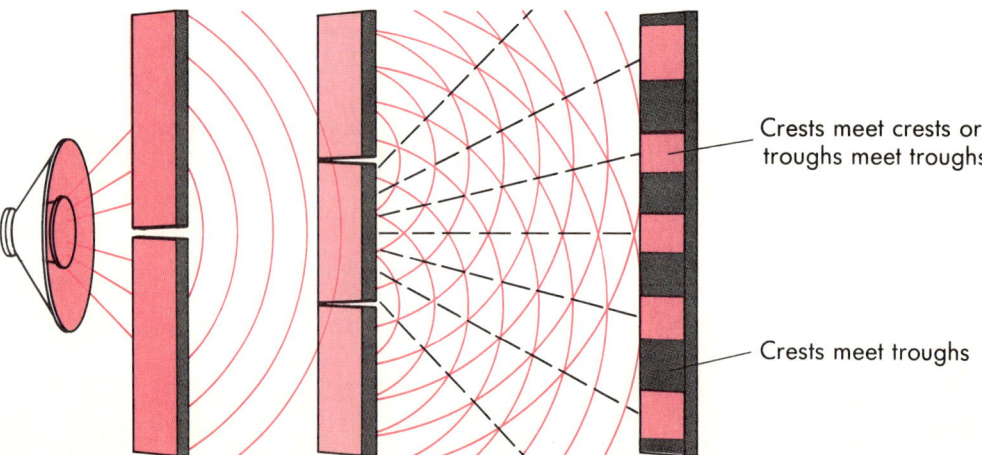

Activity

Reflection

Purpose
To describe and measure the angles of a wave during reflection.

Materials
- sheet of paper
- ruler or straightedge
- pencil
- small ball of clay
- protractor
- flat mirror

Procedure
1. Stand a mirror upright by attaching clay to the corners at the bottom. *NOTE: The mirror must be vertical to the table.*
2. Fold the paper in half lengthwise. Crease it, open it, and draw a dashed line on the crease.
3. Place a dot labeled "X" at one end of the crease.
4. Draw a line from point X, as in *a*. Label the line "Y." The angle between line Y and the dashed line should be less than 90°.
5. Place the paper so that the dashed line is perpendicular to the bottom of the mirror at point X.
6. Position yourself so that you can look into the mirror from the blank half of the paper. Sight along the reflection of line Y.
7. Line up the ruler's edge with line Y's reflection on the blank side of the paper.
8. Make two dots on line Y's reflection, along the ruler, as in *b*.
9. Remove the paper and connect the two dots and point X in a straight line. Label this line "Z."
10. Use a protractor to measure the angles formed with the dashed line, as in *c*.
11. Repeat this experiment two more times. Each time, change the first angle.

Analysis
1. How do the two angles in each set compare with each other?
2. State a general rule about the angles formed during reflection.

a
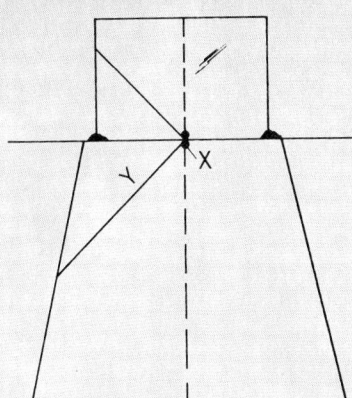

b
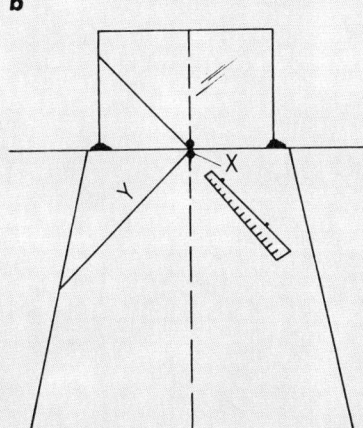

c
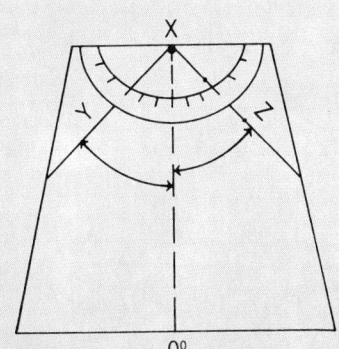

315

17-4 Electromagnetic Waves

At this very moment you are surrounded by thousands of special waves. One kind of wave allows you to see everything around you. Another kind brings radio and TV programs to your home. To find out how these and other waves affect you, consider the following questions:

a. What is the electromagnetic spectrum?
b. Which waves have high energies?
c. How do we use the low energy waves?

The Electromagnetic Spectrum

One kind of transverse wave is a combination of electricity and magnetism. All **electromagnetic** (i lek′trō mag net′ik) **waves** travel through a vacuum at 300,000,000 meters per second. They can travel around the earth seven times in one second! Radio waves, infrared rays, light, ultraviolet waves, X rays, and gamma rays are all electromagnetic waves. An arrangement of these waves by frequency or wavelength is the **electromagnetic spectrum,** shown below. You already know of a small part of this spectrum, the rainbow, which arranges the colors of light as shown.

The electromagnetic spectrum

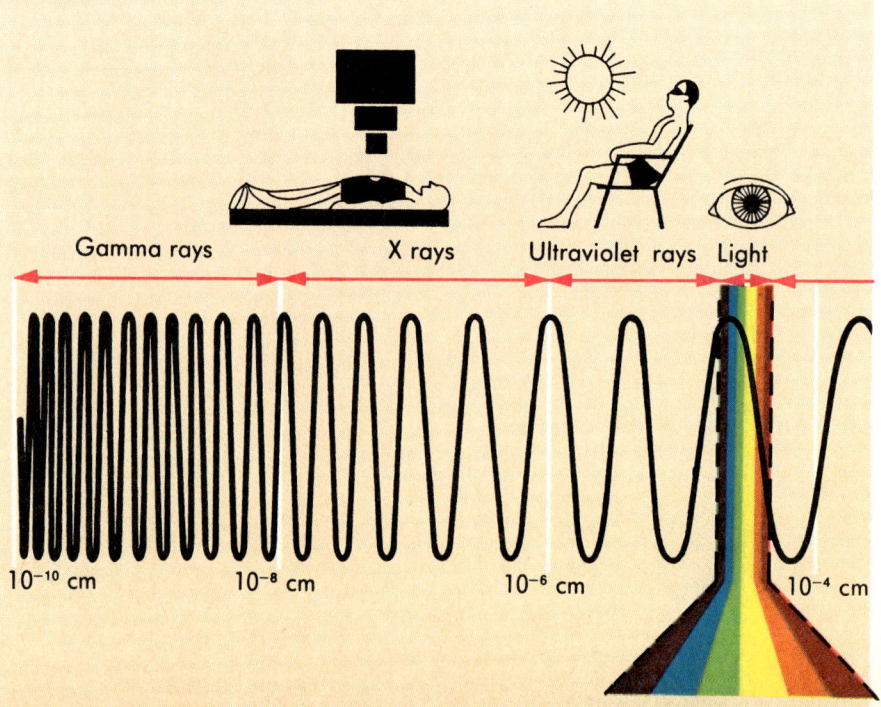

Notice below that radio waves have the longest wavelengths and lowest frequencies. Gamma rays have the shortest wavelengths and highest frequencies. Waves with high frequencies have the greatest energy. How a wave affects us, and how we use it, depends on its energy.

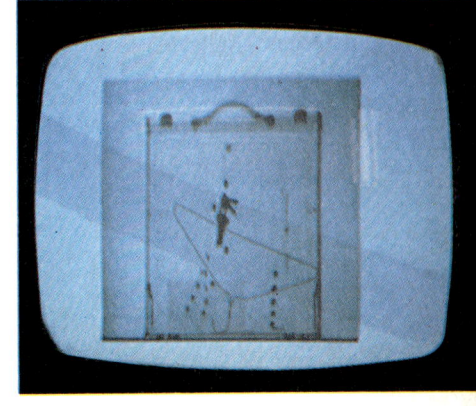

How the High-Energy Waves Affect Us

The high-energy waves are most harmful to people. Yet we can use them effectively if we limit our exposure to them. *Gamma rays* from radioactive cobalt are used to kill cancer cells. *X rays* pass right through the soft parts of your body, but are stopped by bones and teeth. Therefore, doctors can use X rays to "photograph" cavities in your teeth, breaks in your bones, and deep, interior parts of your body. The machines in airports that check the contents of luggage without opening it use X rays, as shown at the right. *Ultraviolet rays* sterilize objects by killing bacteria. They also burn or tan your skin.

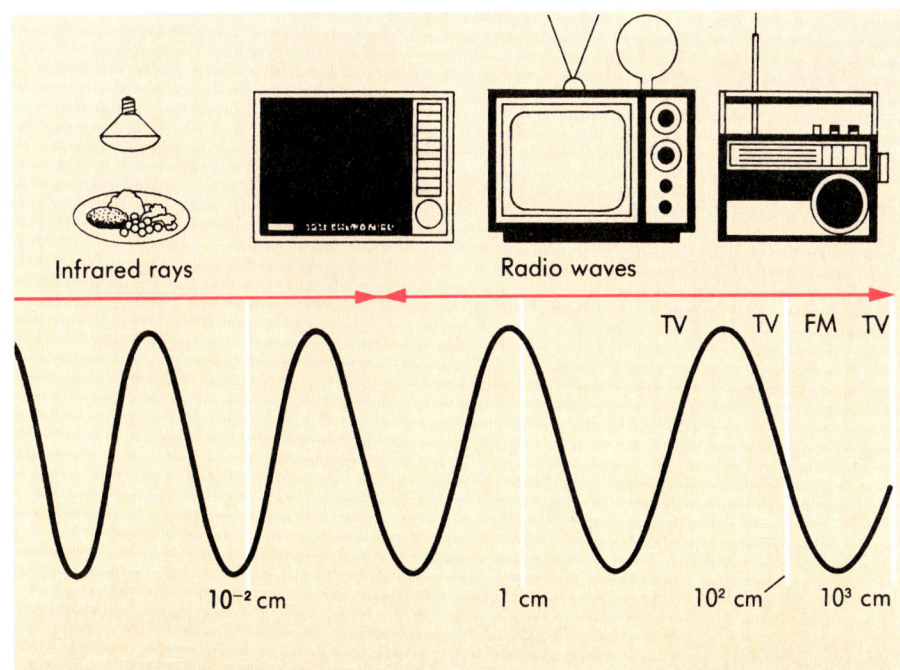

Using the Low-Energy Waves

You can see because objects reflect *light waves* to your eyes. Coils on electric heaters give off *infrared rays*. These rays and *microwaves* in ovens make cooking in homes and restaurants convenient. Radar uses *radio waves* to help pilots navigate and to detect flying objects. A radar device sends out radio waves to an object, such as a plane. The object reflects the waves back to the device. Since the waves' speed is known, the object's distance from the device is determined from the time between sending and receiving the wave.

Weather satellites, such as the one shown at the left, detect the low-energy waves. As a storm approaches, meteorologists work hard to interpret the satellite information about it. The storm's air mass sends out infrared rays and microwaves, which indicate its temperature. The air mass also reflects light waves (during the day), infrared rays, and radio waves. Certain radio wavelengths detect fine particles in clouds. Others can detect raindrops. All this information goes into the weather report you see on television and weather maps, such as the one above.

Review It

1. By what characteristic are waves arranged in the electromagnetic spectrum?
2. State one use for each of the high-energy waves.
3. State one use for each of the low-energy waves.

Issues in Physical Science

Microwave Pollution

We may have a new kind of pollution to worry about. It is microwave pollution. You cannot see, smell, taste, hear, or feel this form of pollution. But it is all around you.

Microwaves are used in many ways.
- They carry television signals.
- They are used in automatic garage door openers and burglar alarms.
- They help send messages through space from telescopes. The telescopes shown receive microwaves from space.
- They are used to heat and cook food in microwave ovens.

Some kinds of radiation can be a health problem. For example, X rays can cause radiation burns. Ultraviolet rays can burn your skin.

Are microwaves a health problem too? Scientists do not have an answer to this question. They do know that very large amounts of microwaves can overheat cells in your body. A microwave oven cooks meat and vegetables in this way. Some people worry that smaller amounts of microwaves may also harm living cells.

Some experiments have been done on small animals. The tests show that microwaves can cause damage to the nervous system and the reproductive system. Even if no harm is done to the body, an animal's behavior might change. The animal becomes nervous and confused when placed near microwaves. An animal exposed to microwaves in a microwave oven would die.

These tests do not tell us enough information to be useful. The level of microwave radiation used in them was fairly large—about ten times the amount you get standing near a microwave oven. You probably do not have to worry about microwaves from ovens. Still, many people are worrying more about microwave pollution. More and more gadgets that produce microwaves are found in many homes and industries. How concerned should we be, when we are not sure what damage they can do? The more we use microwaves, the more we must think about this issue.

For Discussion
1. Why is microwave pollution likely to become a greater problem in the future?
2. How is microwave pollution an issue in your life?

Chapter Summary

- Waves carry energy but not matter. (17–1)
- Wavelength, frequency, amplitude, and speed are four wave properties. (17–1)
- A wave's speed can be determined by using speed = wavelength × frequency. (17–1)
- Transverse and compressional waves are the two kinds of traveling waves. (17–2)
- When a wave moves into a new material, it refracts because its speed changes. (17–3)
- Waves moving from one material to another may be reflected, refracted, or both. (17–3)
- The crests and troughs of polarized waves move only in one direction perpendicular to the wave's direction. (17–3)
- Interference occurs as crests and troughs of one wave move through those of another and either add together or cancel each other. (17–3)
- The electromagnetic spectrum arranges electromagnetic waves in order of their wavelengths or frequencies. (17–4)

Interesting Reading

Bascom, Willard. *Waves and Beaches: The Dynamics of the Ocean Surface.* Doubleday, 1980. An enthusiastic description of water waves on the ocean surface and how they affect the land.

Branley, Franklin M. *The Electromagnetic Spectrum.* Crowell, 1979. An overall treatment of the discovery and uses of all the electromagnetic waves.

Questions/Problems

1. Suppose you make waves in a pond by dipping your hand in the water with a frequency of 1 hertz. What would you do to make waves of a longer wavelength?
2. Use speed = frequency × wavelength to find the wavelength, in meters, of a radio wave traveling at 300,000,000 m/s with a frequency of 540,000 hertz.
3. Draw a series of five transverse waves with a wavelength of 5 cm and an amplitude of 2 cm.
4. Imagine that you are sledding down a snowy hill on a double-runner sled. One runner hits muddy ground. What will happen to the sled? In what direction will the sled move?
5. You stand in the back right corner of a racquetball court and hit the ball to the middle of the front wall. Where should the other player stand to return it?
6. Two sound waves interfere so that the sum of their crests and troughs is 0 cm. What do you hear? Why?

Extra Research

1. Tie two coiled springs of different masses end to end with string. Ask a friend to hold one end. Make transverse or compressional waves at the other end. What happens to the waves at the boundary (the connection) between the waves? What happens to your friend's coil?
2. Cover the bottom of one end of a pan with 1-cm thick plastic or wood. Cover the pan bottom and the plastic or wood with water. Dip a ruler lengthwise into the deep end to make waves. What happens to the waves in the shallow end? Why might ocean waves act the same when they reach a shore?

Chapter Test

A. Vocabulary Write the numbers 1–10 on a piece of paper.
Match the definition in Column I with the term it defines in Column II.

Column I

1. the low point of a transverse wave
2. half the perpendicular distance from the top of a crest to the bottom of a trough
3. transverse waves that include X rays and visible light
4. unit for measuring frequency
5. a traveling wave that causes matter to move perpendicular to the direction of a wave's motion
6. the distance between two crests next to each other
7. the number of waves that pass a certain point in one second
8. a traveling wave that causes matter to move back and forth parallel to the direction of the wave's motion
9. the bouncing of a wave from a surface
10. the result of waves moving through each other

Column II

a. amplitude
b. compressional wave
c. electromagnetic waves
d. frequency
e. hertz
f. interference
g. reflection
h. transverse wave
i. trough
j. wavelength

B. Multiple Choice Write the numbers 1–10 on your paper.
Choose the letter that best completes the statement or answers the question.

1. A wave transfers   a) matter.   b) speed.   c) energy.   d) frequency.

2. A wave with a large amount of energy has a large   a) amplitude.   b) wavelength.   c) speed.   d) a, b, and c.

3. Two waves have the same speed and amplitude. The wavelength of the first is four times greater than the second. The frequency of the first is   a) one-fourth the second.   b) twice the second.   c) one-half the second.   d) four times the second.

4. Choose the compressional wave.   a) radio   b) ultraviolet   c) water   d) sound

5. Refraction occurs because the new material changes the wave's   a) hertz.   b) amplitude.   c) frequency.   d) speed.

6. When a wave moves from one material to another, it   a) changes color.   b) changes from transverse to compressional.   c) may refract.   d) stops.

7. The angles at which a wave hits and reflects from a surface are   a) equal.   b) different.   c) unrelated.   d) unknowable.

8. A wave on a rope approaches two gratings in a row. The wave is polarized perpendicular to Grating 1 and parallel to Grating 2. The wave passes through   a) 1.   b) 2.   c) both.   d) neither.

9. All the following are electromagnetic waves *except*   a) light.   b) sound.   c) X rays.   d) radio waves.

10. The waves with the greatest energy are   a) light waves.   b) gamma rays.   c) microwaves.   d) ultraviolet waves.

Chapter 18
Light

What do you notice first about the picture on the left? Perhaps you notice how colorful it is. Maybe you see the many ways light is used. No matter what catches your eye, light makes it possible for you to see this picture. We depend on light for information about the world.

In this chapter, you can apply your knowledge about waves to waves of light. You will begin with a description of the nature of light and move to an explanation of how colors make up white light. Your reading ends with a section on a special kind of light—laser light—and its many uses.

Chapter Objectives
1. Describe the nature of light.
2. Describe the visible spectrum and how colors of light interact.
3. Contrast ordinary light with laser light, and list some uses of lasers.

18-1
The Nature of Light

What is light? You already know part of the answer to this question, because you know about waves. The story is not over, however, because light is not only a wave. As you learn more about the nature of light, keep these questions in mind:

a. How does light travel?
b. How does a source produce light?
c. In what different ways does light act?

How Light Travels

Light spreads out from its source in straight lines, called rays, much like paint sprayed from a can. Light cannot go around objects. When an object blocks light, a **shadow** forms.

In the drawings below, compare how much less light reaches the same size square as the distance from the square to the source increases. The greater the distance from the source, the less light reaches that square. The amount of light reaching an area is its **intensity** (in ten′sə tē).

Because light waves travel so quickly, you see nearby events almost at the very instant they occur. For example, when you raise a window shade, light enters in a fraction of a second.

How distance from a source affects intensity

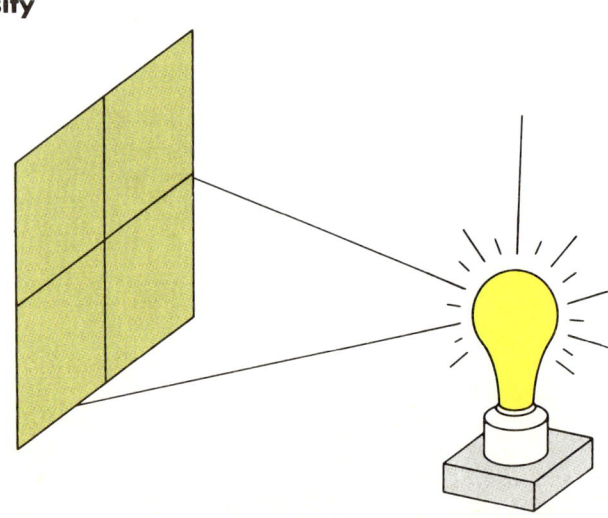

If a light source is very far away, however, the light's travel time becomes noticeable. For example, sunlight takes more than eight minutes to reach earth. Other stars are so far away that their light travels for years before it reaches earth. In fact, some stars may have burned out by the time we see their light.

How Light Is Produced

Atoms are tiny particles that make up all matter. Each atom has a certain amount of energy. An atom can gain energy in different ways. Heating or passing electricity through atoms provides extra energy. If an atom gains energy, the atom becomes excited. The atom must give off this extra energy to return to its usual energy state. Often, we see this released energy as light.

Atoms gain and give off energy in very specific amounts. To help yourself think about this idea, imagine standing on a staircase. You can stand on one step or on the next, but not in between steps. An atom's energy levels are much like the stairs. An atom can gain or lose only certain amounts of energy. Energy levels in between these "steps" are not allowed. An atom releases exactly enough energy to change from a higher energy level to a lower one, as in the drawing. The bundle of energy the atom releases is a **photon** (fō′ton).

Electricity supplies the extra energy for the neon atoms in the sign shown in the photograph. The excited neon atoms release their energy as photons of light.

No matter where extra energy comes from, any object that gives off its own light is **luminous** (lü′mə nəs). All sources of light, such as the sun and light bulbs, are luminous. The moon, however, is not luminous, since it merely reflects the sun's light. A table is not luminous either. It simply reflects light to us.

Have You Heard?

The light year is the unit used to measure the huge distances in space. One light year is 9.6 trillion kilometers, the distance light travels in a year. Light from the closest star to our solar system travels 4.3 light years—or about 41 trillion kilometers—to reach us.

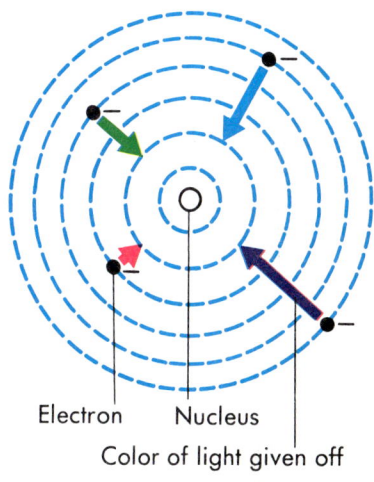

Electron   Nucleus
Color of light given off

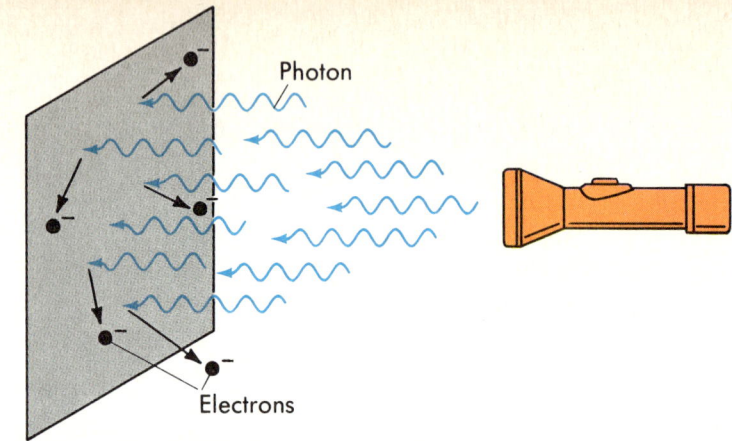

Evidence for the particle nature of light

How Light Acts

For centuries, people disagreed about the nature of light. Some thought that light was a particle that traveled through space in straight lines. Others believed that light was a wave. It seemed reasonable that light was either a particle or a wave, but not both.

When scientists discovered the interference of light, they thought they had proved that light was a wave. Particles did not act this way. Yet, at that time, scientists believed that waves must travel through something. They could not explain how waves of sunlight traveled to earth through the vacuum of space. Later, it was proved that an electromagnetic wave, such as light, can travel through a vacuum.

Still later, evidence for the particle idea appeared. Light shining on certain metals makes electrons jump out of the metal, as shown. Scientists could explain their observations only if light is made of particles of energy (photons). Each photon knocks one electron out of place just as one billiard ball knocks another out of place.

Finally, scientists had to accept that light has properties of both particles and waves. Sometimes it acts as a particle. At other times it acts as a wave.

Review It

1. Describe the path of light from a source.
2. How do luminous objects shine?
3. Why did scientists disagree about the nature of light?

Activity

The Path of Light Rays

Purpose
To demonstrate how light travels.

Materials
- flashlight
- meter stick
- shoe box
- scissors
- 4–6 flat mirrors

Procedure

Part A
1. Cut the box into three sections, as in *a*.
2. Stand the sections as in *b*.
3. Put a small hole in each section, also shown in *b*. The hole must be in the same place on each section. (Check with the meter stick to be sure the holes are in the same place.)
4. Ask your partner to shine the flashlight through one hole.
5. Line up the other sections so that you can still see the light through all the holes.
6. Hold the meter stick over the boxes to check whether or not the holes are in as straight a line as the edge of the meter stick. Record your results.
7. Move one of the sections to the side. Record whether you can still see the light.

Part B
1. If the day is sunny, have someone near a window on the sunny side reflect the sun into the room with one mirror. *CAUTION: Do not look straight at the sun or at its reflection with your eyes. Do not shine the light into anyone else's eyes.*
2. Hand out the other mirrors to persons in your group. Ask each person in turn to hold a mirror so as to reflect the sunbeam toward the next person's mirror.

a

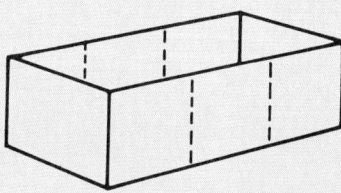

b

3. After the sunbeam is reflected from the last mirror, ask another person to hold up a notebook to see the sunbeam projected on it.
4. Diagram your setup. Using arrows, trace the path of light between the mirrors.

Analysis
1. Explain why you can see the light only when the cardboard sections are lined up.
2. What can you say about the path of light between one mirror and the next in Part B?
3. Use your results to explain how a shadow forms.

18-2
The Visible Spectrum

How dull everything looks to us without color! Color is an important part of our lives, though we often take it for granted. But why do objects appear to have color? Continue reading to find out, and ask yourself:

a. Which colors make up white light?
b. What causes an object to appear colored?

Colors of Light

Light is the **visible spectrum**—the small part of the electromagnetic spectrum we can see. White light, such as sunlight, is really a mixture of all the colors of the visible spectrum. You can prove that white light has this property with a **prism** (priz′əm), which is a wedge-shaped piece of glass or plastic. You can see below how the prism splits light into colors.

Sometimes after a rain shower, water drops in the sky act like prisms. They reflect and refract sunlight, spreading it into its spectrum. A rainbow results.

White light splits into this order: red, orange, yellow, green, blue, indigo (violet-blue), and violet. This order results because each color is a different range of wavelengths that refract by a different amount.

Red light has the longest wavelengths we can see. Violet has the shortest wavelengths. If we think of light as particles instead of waves, the photons of light of a color have a certain range of energies. Photons of red light have less energy than photons of blue light.

  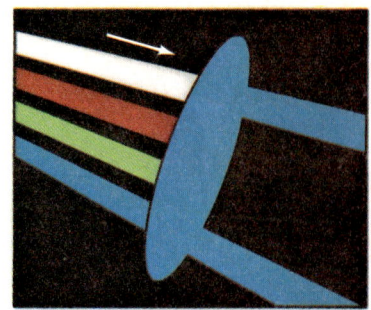

How Objects Appear Colored

The objects we see around us absorb certain colors of light and reflect others. An object's color is really the colored light the object reflects. For example, green leaves absorb most of the colors in white light that shines on them. They reflect mainly green light. Your eye sees the reflected green light.

Only light of the same color as a filter passes through the filter, as shown. If light without green wavelengths strikes a green leaf, the leaf absorbs most of the light. The leaf then appears black, since it reflects almost no light to us. Black is the absence of all light.

Different mixtures of the three primary colors of light—red, green, and blue—make your eye and brain think you are seeing all the colors. The color wheels at the right show a few of the possible combinations. Color televisions work by adding these three primary colors. A close look at the screen shows that each picture is made of only red, blue, and green dots.

While primary colors of light add together, the three primary colors of paints subtract from each other. These primary colors are red, yellow, and blue. Combining them makes black, not white, because the solid particles in paint act as filters. Since the colors cannot pass through each other, they subtract from each other.

The three primary colors of light

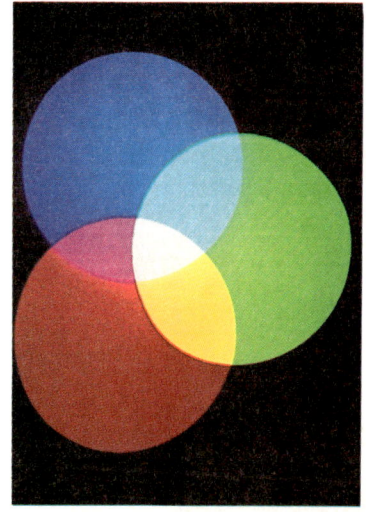

Review It

1. What determines the color of light?
2. Name the three primary colors of light.

Activity

Colors in White Light

Purpose
To investigate how the colors of light make up white light.

Materials
- 2 white cardboard disks, each 5 cm in diameter
- 2 buttons at least 3 cm in diameter
- red, orange, yellow, green, blue, and violet colored papers, felt pens, or paints
- glue or double-sided tape
- 1-m-long piece of string or thread
- pencil
- straightedge
- scissors

Procedure

Part A
1. Use the straightedge to mark off six roughly equal pie-shaped sections on one side of one disk.
2. Put a different color on each section of the disk. Do this by cutting and pasting the colored paper sections or using the markers or paints on the disk.
3. Paste one button flat in the center of the disk on the blank side.
4. Punch two holes in the center of the disk by pushing a pencil point through two button holes.
5. Thread the string or thread through the two holes. Tie the open ends together with a knot.
6. Slip one index finger through each end of the string. With the disk in the middle of the string, twist the string at least 20 times.
7. Pull on the string in opposite directions, as shown. Loosen your pull as the string twists in one direction and then the other.
8. Watch the disk when it twirls the fastest. Record the color you see.
9. Repeat steps 7–8 several times.

Part B
1. Draw some black lines in patterns on the disk.
2. Repeat steps 6–9.

Part C
1. Repeat Part A, except divide the disk into halves, colored red and blue.
2. Record the color you see.

Analysis
1. Explain why you saw different colors in Part A and Part C.
2. What effect did the black markings have on the color you saw in Part B?

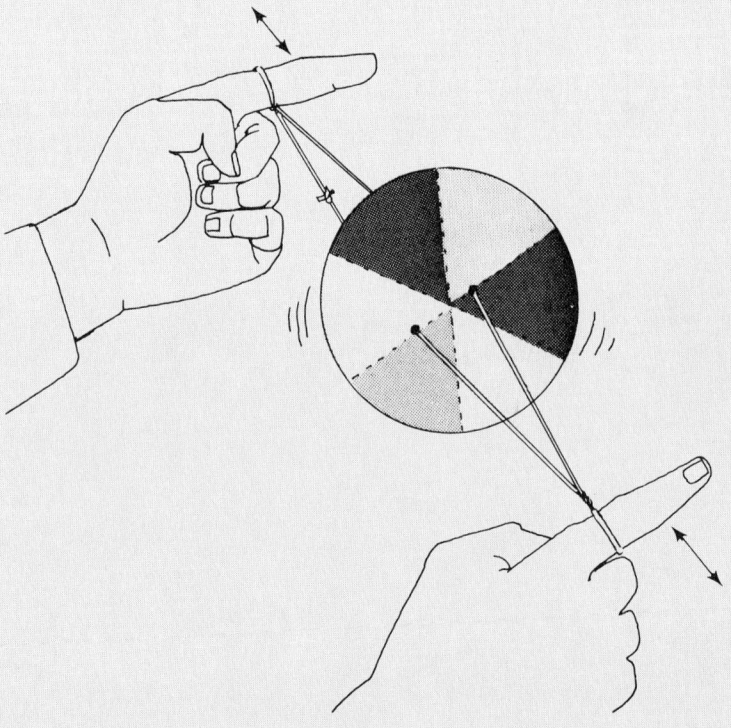

Did You Know?

The Colors in the Sky

At the end of the nineteenth century, the English scientist Lord Rayleigh (rā′lē) figured out how particles in the air reflect light. He found that small particles reflect light differently than large particles. Particles that are much too small to be seen scatter light in many directions. This effect is known as Rayleigh scattering in his honor.

Molecules of air are very small. Rayleigh's rules show that they scatter blue light much more than they scatter red light. During the day, the sun reaches its highest position in the sky. At that time, its rays travel through less air. You are more likely to see blue light than red light. The sky appears blue.

At sunset the sun is low in the sky. Its light passes through more air than it did during the day, as the drawing shows. The air has scattered most of the blue wavelengths away from us. People on parts of the earth where the sun's rays are more direct are seeing blue skies. This scattering leaves more red light in our skies for us to see at sunset.

Water droplets often form around bits of dust in the air, making a cloud. The droplets are too large to scatter light according to Rayleigh's rules. They scatter all colors equally. Since white is a mixture of all the colors of light, the scattered light makes clouds look white.

Since Mars has an atmosphere, scientists thought that it would have blue skies like Earth's. Yet in 1976, the Mars lander pictures showed that the Martian sky is pink during the day! Rayleigh scattering is not the reason for this color, however. Strong wind storms keep a lot of the pink surface dust in the air on Mars. As we look through this dust, everything looks pinkish.

For Discussion
1. Describe how the earth's atmosphere causes the sky to have a certain color.
2. If pink dust can color the Martian atmosphere, how do you think smoke and particles from a factory affect the air over a city?

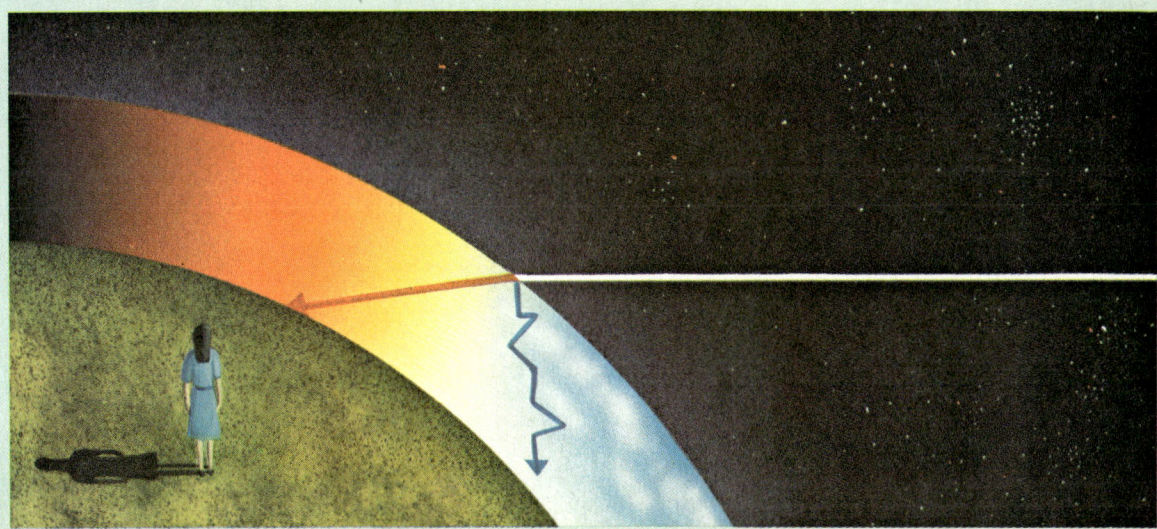

18-3 Lasers

Would you like a beam of light powerful enough to cut through steel? Science fiction writers have imagined beams like this, but now such powerful light beams are really possible. They are a special kind of light. Ask yourself these questions as you read:

a. What is laser light?
b. How is laser light produced?

How Laser Light Differs from Ordinary Light

One kind of light source, called a **laser** (lā′zər), produces a special kind of beam of light. To understand the difference between laser light and ordinary light, think about the water in a crowded swimming pool. The waves move around in every direction, with no definite pattern. The sun, light bulbs, and other ordinary light sources also give off waves that move around in all directions with no definite pattern.

Notice below how the crests and troughs of ordinary light waves do not match. They are out of step with each other. Even light waves of one wavelength can be out of step. But the laser light waves in the second drawing are in step. All the crests and troughs line up. All laser light waves have exactly the same wavelength.

Because the waves have the same wavelength, a laser's light is exactly one color. Ordinary red light has many wavelengths of slightly different reds. Red laser light has only one wavelength and so only one shade of red.

Comparing ordinary and laser light

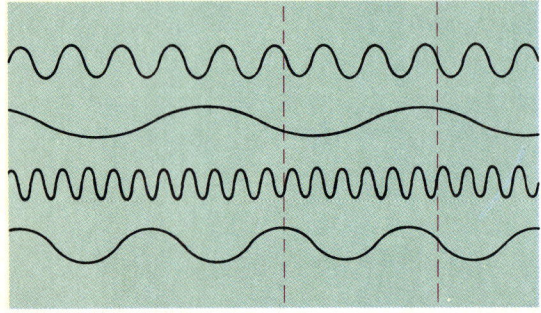
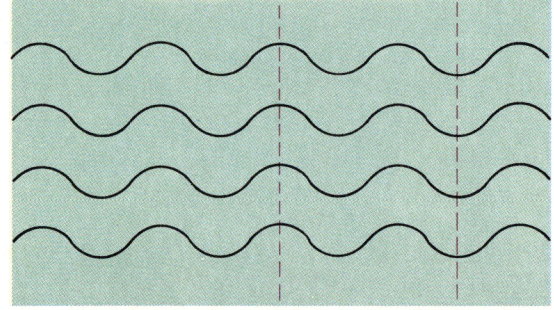

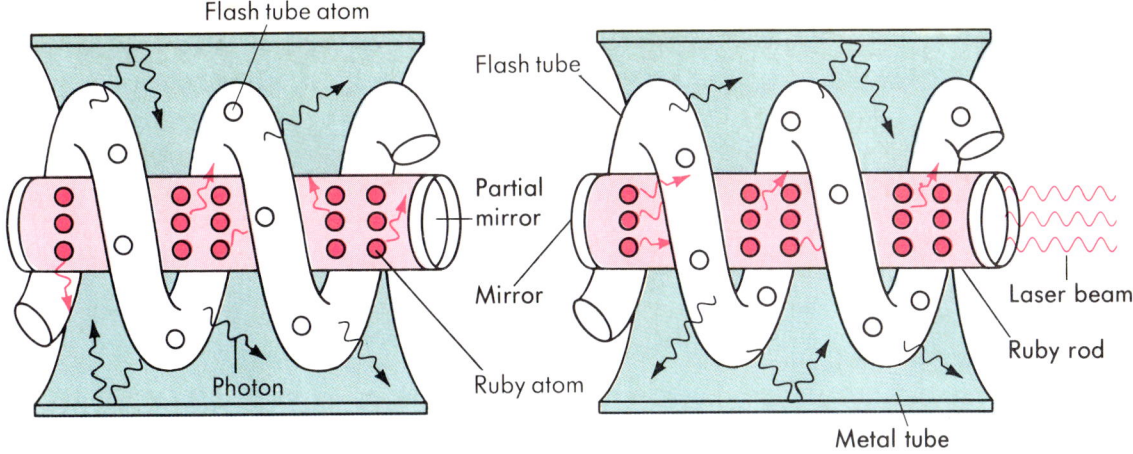

Unlike ordinary light, laser light travels in one direction with little spreading. Therefore, laser beams can be brighter and narrower than ordinary light, as you see at the right.

How Laser Light Is Produced

The picture above shows a basic laser. A flashing lamp gives extra energy to the atoms in the ruby rod. Many of these atoms are excited to a high energy level. When a photon of the right energy hits these atoms, they all drop to the same lower energy level. As they drop, they give off their extra energy as photons of red light. This light bounces back and forth between the mirrors. It makes other atoms release energy as photons of the same red light. As more photons are released, the light becomes more intense. The light that shoots out through the partial mirror is a laser beam.

Because laser light can be so intense, it can seriously damage your eyes. Never look directly into a laser beam.

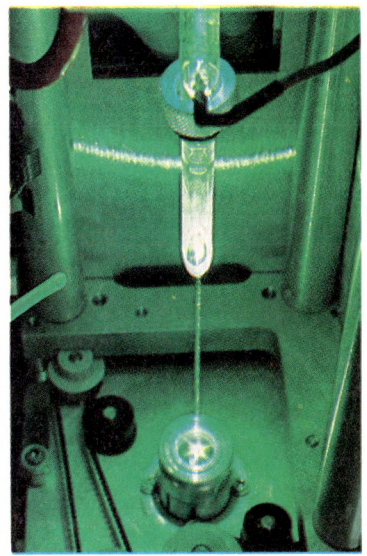

Review It

1. How do laser and ordinary lights differ?
2. How does energy build up in a laser?

Challenge!

Look up *maser* in the encyclopedia. How did the maser lead to the laser?

18–4
Using Lasers

Have you ever sent a message along a piece of string with a tin-can telephone? Would you like to send a message along a light beam instead? As you read to see how lasers have made this possible, ask yourself:

a. How do we use laser beams in optical fibers?
b. What is a hologram?
c. What are some other uses of lasers?

Optical Fibers Carry Messages

Laser beams sent through narrow glass tubes can carry messages. The picture shows these tubes, called **optical fibers** (op′tə kəl fī′bərz), which are as thin as human hairs. The insides of the fibers reflect light—even around curves in the tube. The drawing shows how this reflection moves the beam along the tube.

The telephone company uses laser beams and optical fibers to send telephone calls from city to city. Optical fibers are better than the present copper wiring for several reasons. First, fibers are much thinner and take up much less space under city streets.

Second, we send messages through optical fibers by changing the wave's frequency. The frequencies of the light waves in an optical fiber are so high that we can change them often to send many messages. The signals we send through copper wires change much more slowly. So an optical fiber can carry more messages than a copper wire. Third, static that can affect messages in copper wires does not affect the light in optical fibers.

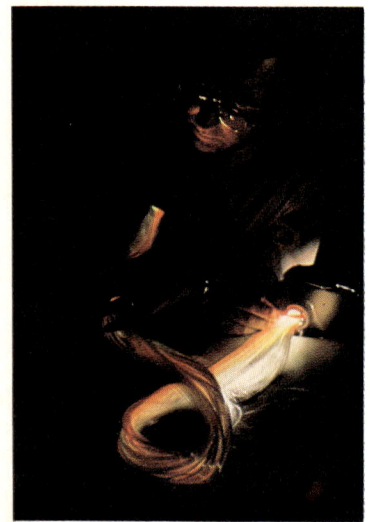
Optical fibers

How light moves through an optical fiber

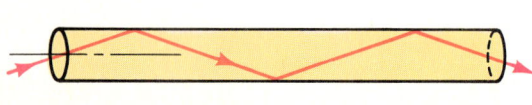
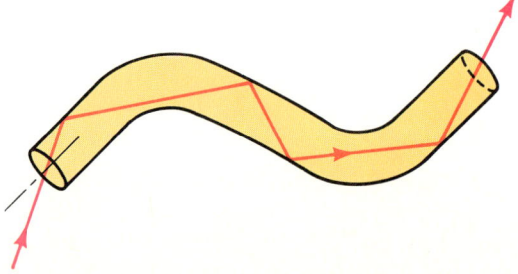

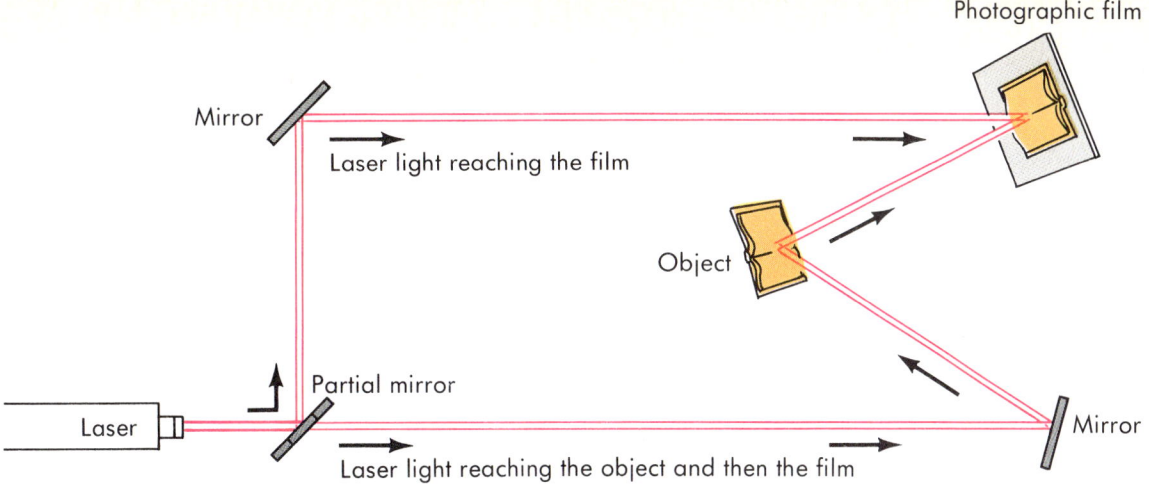

Holograms Give Three-Dimensional Views

A photograph shows you only one flat view of a scene from one angle. With a laser, however, you can produce a three-dimensional image showing the scene from several angles. This photograph is a **hologram** (hol′ə gram).

The diagram shows how splitting one laser beam in two makes a hologram. One beam reflects from the object to a light-sensitive surface, usually a kind of photographic film mounted on glass. The other beam strikes only the film. Because the first beam struck an object before it reached the film, these waves are now out of step with the second beam. The beams interfere, and the interference pattern is recorded on the film.

When you develop the film, you have a hologram. To see the image, you shine the same wavelength of laser light on the hologram. The hologram's interference pattern scatters the laser beam. The scattered light forms an image you can see. The photographs show one hologram seen from different angles.

Ordinary light cannot make a hologram because all its waves are out of step. So many different waves strike the film that no definite image appears. Laser light, with one wavelength in step, is best for making a hologram.

Holograms are used now to test how objects bend under pressure. A hologram of the bent object is compared with a hologram of the unbent object. Sometimes the two holograms appear on the same film as a double exposure.

A universal product code

More Uses of Lasers

Many stores use lasers to speed up checkout lines. Each product carries a Universal Product Code like the one shown. A clerk bounces a laser beam off the bands of the code. A computer senses the beam and identifies the code. It then figures out the bill and keeps track of what the store sells. This information helps storekeepers save money. The laser at the left below "reads" printed words. Its signals are sent by satellite to another city to be printed.

The heat from powerful laser beams welds sheet metal parts and joins glass or plastic parts. Lasers can be used to cut fabric for fifty suits at once or to cut metal, as shown.

Movies, lessons, and other information can now be recorded with a laser on a special disk. This disk is then read in your house with a laser beam instead of with a phonograph needle. The signals can be played back through your home television. Each disk can hold 100,000 separate pictures. A whole encyclopedia can fit on one disk.

Since laser beams travel in very straight lines, companies use them to line up lenses in cameras that they make. Construction companies even use the straight line of laser light to show them the direction in which to dig a tunnel.

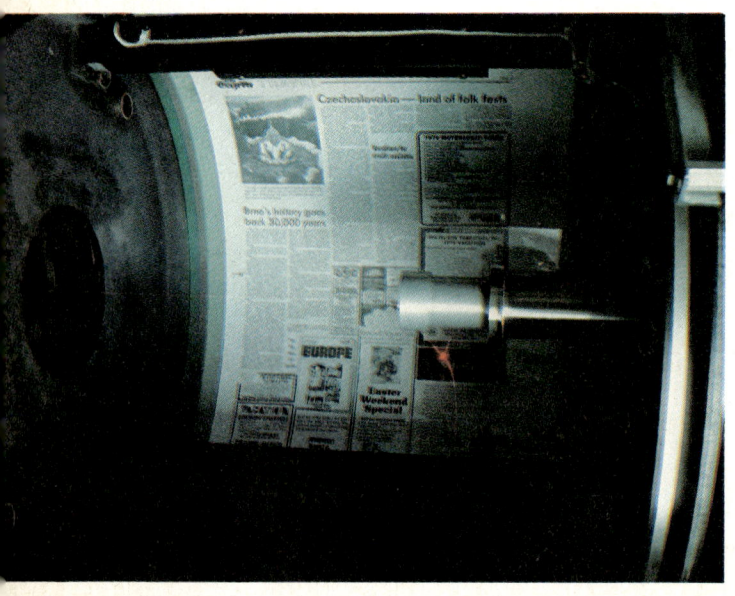

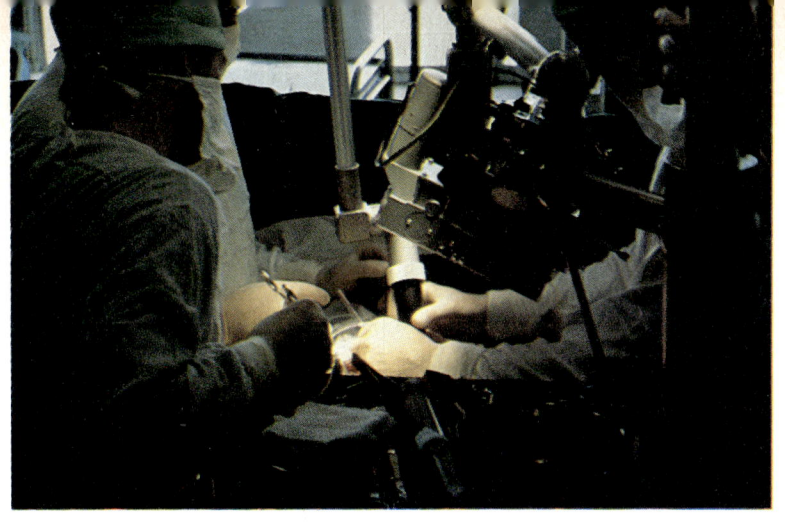
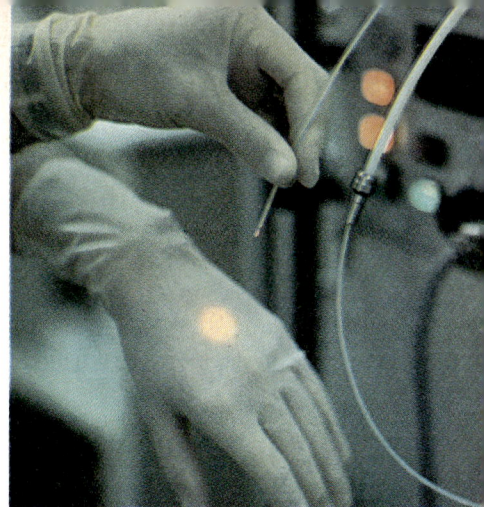

Doctors often use lasers to cut one part of the body without hurting others. The laser's heat cleans and seals the area by boiling away blood, tissue, and germs. A surgical team uses a laser in the photographs above.

The **retina** (ret′n a) covers the back, inside part of the eyeball. When the retina pulls away from the rest of the eye, blindness may result. Now a laser beam can be focused onto the edge of the retina. It welds the retina back in place in a fraction of a second. Operations to do this used to take hours and did not do the job as well.

The astronauts brought mirrors to the moon to reflect laser light. Scientists on earth then bounced laser beams off these mirrors and timed the beam's round trip. They found the distance to the moon with an accuracy of a few centimeters. Sending laser beams to the moon from two different places on earth also tells us about shifts of the earth's surface. This information may help us predict earthquakes.

Scientists bounce laser beams off particles in the atmosphere. How the laser beam bounces tells what kinds of particles are in the air.

Every year, we find many new uses for lasers.

Review It

1. What are optical fibers?
2. What is the advantage of a hologram over a photograph?
3. State four ways we use lasers.

Chapter Summary

- Light travels in straight lines called rays. (18–1)
- The intensity of light from a source depends on the source's distance from the area receiving the light. (18–1)
- A shadow forms when an object blocks light rays. (18–1)
- An excited atom can release energy as light. (18–1)
- Luminous objects give off their own light. (18–1)
- Light has properties of both particles and waves. (18–1)
- The colors of light form the visible part of the spectrum. The colors combine to form white light. (18–2)
- Laser light waves all have the same wavelength in step. (18–3)
- Lasers have many uses in industry, medicine, and research. (18–3)

Interesting Reading

Boraiko, Allen A., and Ward, Fred. "Miracles of Fiber Optics." *National Geographic,* October 1979, pages 516–535. Beautifully illustrated discussion of uses of optical fibers and lasers in medicine and communications.

Branley, Franklyn M. *Color, From Rainbows to Lasers.* Crowell, 1978. Well-illustrated introduction to color.

Lewis, Bruce. *What Is a Laser?* Dodd, 1979. Explanation of lasers using very simple, everyday terms and analogies.

Schneider, Herman. *Laser Light.* McGraw-Hill, 1978. Description of lasers and survey of their uses.

Questions/Problems

1. The astronauts on the moon saw "earthlight" much as we see moonlight. Explain why.
2. When the upper atmosphere contains ice crystals, a halo often appears around the moon. How is the halo similar to a rainbow in the sky?
3. Describe what we see when white light strikes a yellow object.
4. Do lasers use the wave nature or the particle nature of light to produce light? Explain your choice.
5. Explain how laser light moves through an optical fiber.
6. How is the interference of light important in making a hologram? Why can laser light best be used?

Extra Research

1. Cut a 7-cm square out of red paper. Paste it on a piece of white typing paper. Stare hard at the red piece while you count slowly to ninety. Look quickly at another white sheet. What color do you see? Try this with other colors. Then use reference books to find out how this effect happens.
2. Contact a greenhouse or florist in your neighborhood. Ask whether they use special lights to grow plants. Find out why.
3. Partially fill a baking pan with water. Lean a flat mirror inside the pan from the bottom to an upper edge. Shine light from the sun or a flashlight directly on the mirror. Look at a nearby wall and explain what you see.
4. Read about the medical uses of optical fibers in reference books. Then make a report to your class.

Chapter Test

A. Vocabulary Write the numbers 1–10 on a piece of paper. Match the definition in Column I with the term it defines in Column II.

Column I

1. giving off its own light
2. area of darkness formed when an object blocks light
3. bundle of light energy from an atom
4. produces light waves of one wavelength which are all in step
5. thin glass or plastic tube in which light can move
6. band of colors that make up white light
7. back, inside part of the eyeball
8. three-dimensional image produced by laser light
9. wedge-shaped piece of glass or plastic that separates white light into colors
10. the amount of light reaching an area

Column II

a. hologram
b. intensity
c. laser
d. luminous
e. optical fiber
f. photon
g. prism
h. retina
i. shadow
j. visible spectrum

B. Multiple Choice Write the numbers 1–10 on your paper. Choose the letter that best completes the statement or answers the question.

1. Laser light is   a) dangerous to look into.   b) out of step.   c) three shades of one color.   d) made of many wavelengths.

2. We see light from a distant star   a) as soon as the star produces it.   b) only while the star is shining.   c) years after the star produces its light.   d) sooner than light from a nearby star.

3. An atom   a) has only one energy level.   b) can release energy as light.   c) becomes excited when it loses energy.   d) releases energy in waves called photons.

4. Light behaves   a) only as a particle.   b) only as a wave.   c) as both a particle and a wave.   d) as neither a particle nor a wave.

5. Choose the color of light with the longest wavelength.   a) violet   b) orange   c) blue   d) green

6. Choose the distance from the source where its light has the greatest intensity.   a) 8 m   b) 10 m   c) 5 m   d) 1 m

7. We see black when   a) no light reaches us.   b) all the colors of light combine.   c) an object reflects all colors of light.   d) an object reflects black light.

8. White light is   a) an even mixture of all colors.   b) none of the colors.   c) red, blue, and yellow light mixed evenly.   d) yellow, green, and orange light mixed evenly.

9. Laser beams are   a) always red.   b) dimmer than ordinary light.   c) narrower than ordinary light beams.   d) always blue.

10. A hologram produces an image when laser light passes through the hologram's
a) refracting surface.   b) interference pattern.   c) reflection.   d) polarized filter.

339

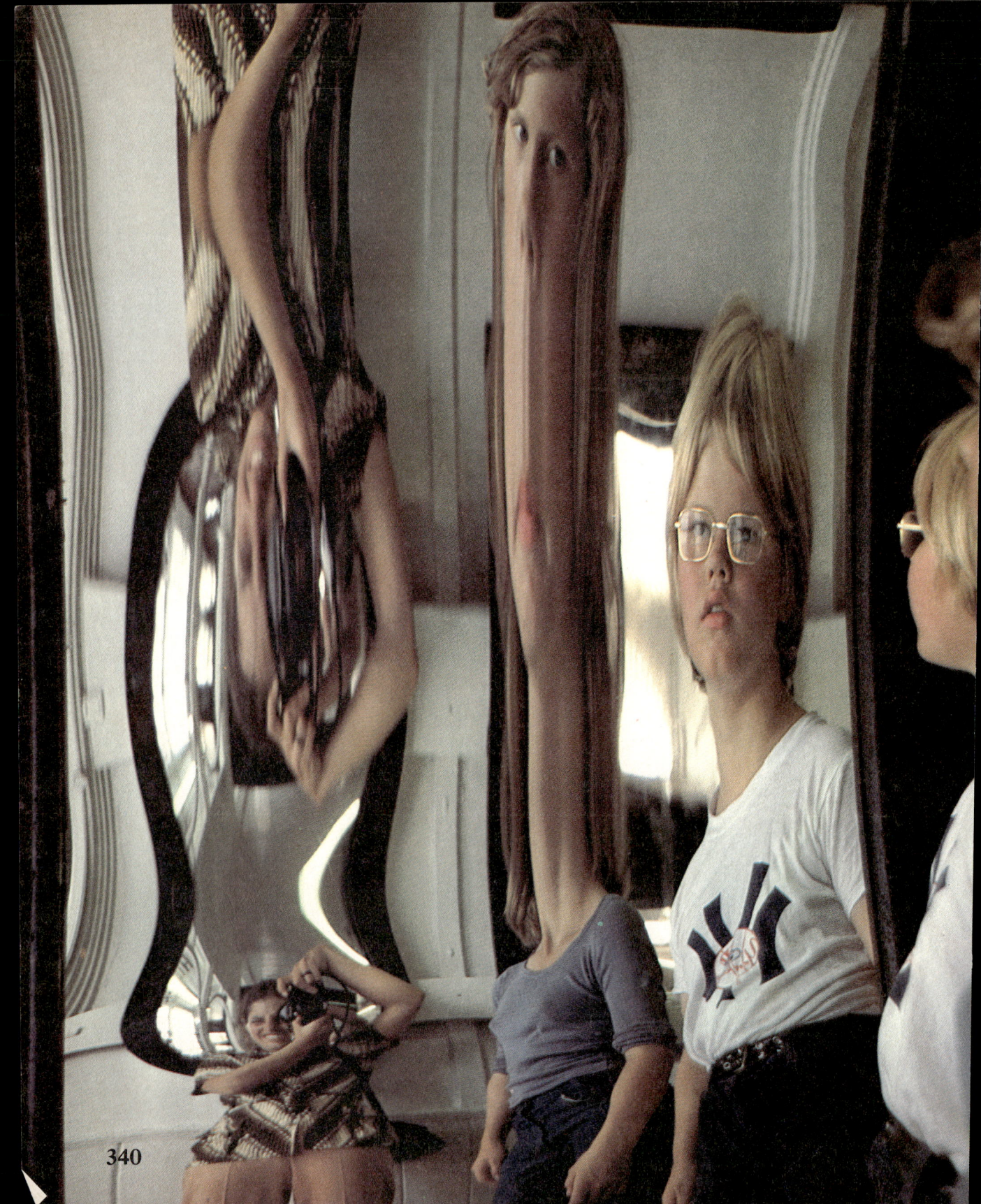

Chapter 19
Light and Its Uses

These people have found a strange mirror that changes their appearance. They are taking a picture of this unusual sight. The mirror and the camera both affect light rays, but in different ways.

In this chapter you will study the paths of light rays. Reflection and refraction of light will be an important part of your study. The chapter explains the function of mirrors and lenses, and describes some devices that use them. The chapter also explains how our eyes work.

Chapter Objectives

1. Explain how a plane mirror makes an image.
2. Compare concave and convex mirrors.
3. Compare convex and concave lenses.
4. Explain how the lens in your eye works.
5. Explain how cameras, telescopes, and microscopes work.

19–1
Plane Mirrors

Folk tales tell us that mirrors have always fascinated people. The witch in *Snow White* had a magic mirror. Narcissus in the Greek legend fell in love with his own image. What you see in a mirror may fascinate you, but it is not magic. You see reflected light. As you read about mirrors, keep the following in mind:

a. How does a plane mirror reflect light?
b. In what ways do we use plane mirrors?

Plane Mirrors Form Images by Reflecting Light

Any smooth surface that reflects light can be called a mirror. **Plane mirrors** have flat surfaces. Usually they are smooth pieces of glass backed by a dark surface.

The rules of reflection apply for light reflected from a plane mirror. In the picture, a beam of light strikes and leaves the mirror's surface at equal angles.

The scene or object you see in a mirror is the image of that object. To us, the image seems to be behind the mirror, because we know light travels in a straight line. The image appears to be as far behind the mirror as the object is in front of the mirror. Yet we know that neither the object nor the image is really behind the mirror. This image is a **virtual** (ver′chü əl) **image,** which means the image seems to be in one place, but it is not really there. If you put a sheet of paper where the image seems to be, no image will appear on the paper.

Forming a plane mirror image

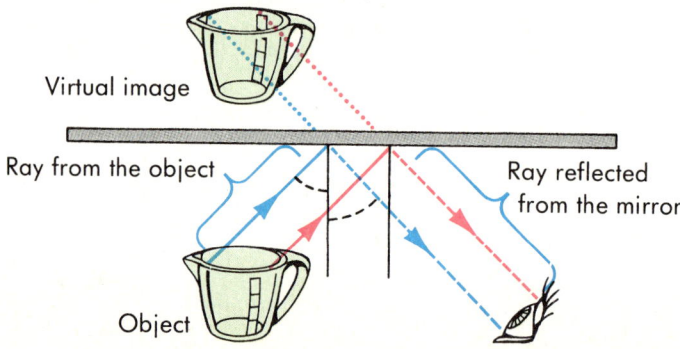

When you look in a plane mirror, the image is right side up and the same size as you are. But the right-hand side of the image looking out is the same as your left-hand side. The image is flipped from left to right.

Uses of Plane Mirrors

Most mirrors in your home are plane mirrors. Bedroom doors and medicine cabinets often have plane mirrors attached to them. We use a plane mirror because the image it produces has the same size and proportions as the object.

Rearview mirrors in automobiles are plane mirrors. The images you see in them show the true size and distance of an object. If you see the image of a car in a rearview mirror, you will know just how far away that car is.

The ambulance at the right has "EDNAJUBMA" written on it. A driver moving through traffic in front of the ambulance will see this word flipped in the rearview mirror, as shown below. Because the mirror flips an image that was backwards, the driver sees "AMBULANCE" written as it should be.

Review It

1. What kind of image does a plane mirror form?
2. State two uses of plane mirrors.

19-2
Curved Mirrors

How does a fun house mirror distort images? Some mirrors make you appear shorter and rounder. Some seem to turn you upside down. The curve of the mirror causes these changes. As you read about curved mirrors, ask yourself:

a. How does a concave mirror affect light rays?
b. How does a convex mirror affect light rays?
c. How do we use curved mirrors?

Concave Mirrors Focus Light Rays

Imagine placing several plane mirrors on a curve, as shown below. The light rays that hit the mirrors at an angle bounce off at the same angle. You can arrange the curve so the mirrors reflect all the parallel light rays to one spot. You focus light rays when you bring them together at one point in this way.

Now consider one curved mirror that follows the curve the plane mirrors were on. In the second picture, light rays are coming toward the mirror from the left. The rays are parallel to each other and to a line through the center of the mirror. This curved mirror focuses the parallel rays to a point, which is called the **focal point.**

Have You Heard?

The story is told that, around 150 B.C., the Greek scientist Archimedes used several shields as plane mirrors. He focused sunlight on an enemy ship with the shields. The sunlight was so intense that it started a fire that destroyed the ship.

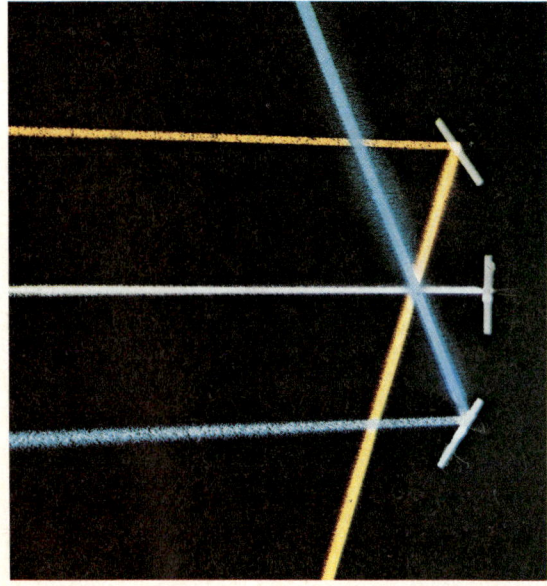
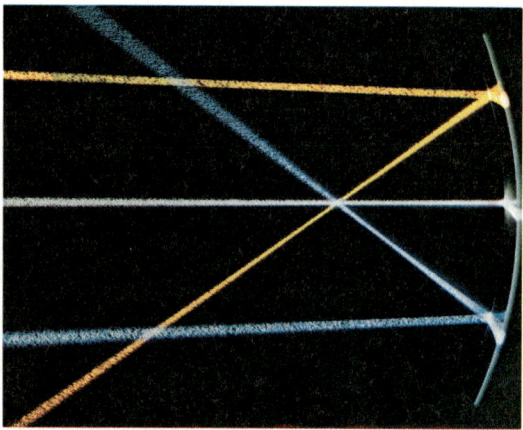

One kind of curved mirror is shaped like part of the inside of a sphere. This mirror's center curves away from the light source. It is called a **concave mirror.** The inside of the spoon shown is a concave mirror.

The image that a concave mirror forms depends on how far the object is from the mirror. Follow the light rays from the distant object shown in the first drawing. Each ray hits the mirror at an angle and bounces off that spot at the same angle. You can imagine that a tiny flat mirror is there. To find out where the top of the pencil is focused, we choose the two simplest rays to follow. An image is formed where these two rays meet again.

Look at the ray from the top of the pencil parallel to the center line. All rays parallel to the center line pass through the focal point. Next, look at the ray directly through the focal point. Rays passing through the focal point bounce off the mirror parallel to the center line. The image appears where the two dotted lines meet.

Tracing similar rays, drawn in blue, shows where the bottom of the pencil appears. Notice that the image of the pencil is upside down and smaller than the pencil itself. The image appears in front of the mirror. Light rays meet at this image, called a **real image.** If you put a sheet of paper there, the image would appear on it.

In the right-hand diagram, the pencil is closer to the mirror than the focal point. Again we choose two simple rays to follow. The image is right side up and larger than the pencil. The image is virtual. No rays actually meet there. The image would not show up on paper.

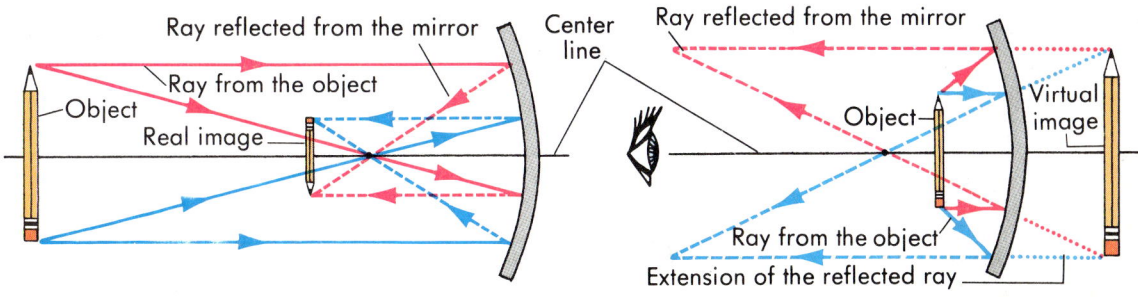

How a concave mirror forms an image

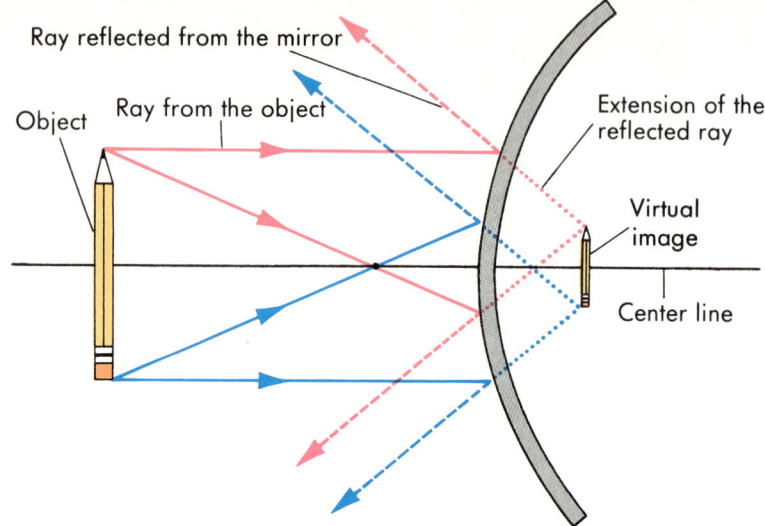

How a convex mirror forms an image

Using a concave mirror to send out light rays

Convex Mirrors Spread Light Rays Apart

If you look at your reflection on the back of a spoon, you will see a small image of yourself right side up. You can see how such an image appears in the photograph at the left. The spoon's center curves toward the light source. This kind of curved mirror is called a **convex mirror.** The simplest convex mirror is shaped like the outside of a sphere.

The drawing shows how a convex mirror forms an image. A convex mirror makes light rays spread out. However, the spreading rays seem to come from the point shown. The image appears at this point. This image is virtual, because light rays only seem to come from this point. The image will not show up on paper.

Uses of Curved Mirrors

Flashlights, searchlights, and headlights have concave mirrors in them. The mirrors are shaped so that light rays from the focal point are sent out in parallel rays. Directing light rays in this way is the opposite of focusing them to a point, as the picture at the left shows.

Radio telescope at Arecibo, Puerto Rico

A radar antenna uses a concave mirror to focus incoming parallel radio waves to a point. The giant radio telescope shown above is the largest in the world. Its concave mirror is wider than the length of three football fields. The mirror focuses the radio waves that hit it. The telescope detects radio waves from many objects in the universe.

Cars and trucks sometimes have convex rearview mirrors in addition to plane rearview mirrors. The convex mirror lets the driver see objects that are farther off to the side than a plane mirror does. Convex mirrors are also sometimes used in banks and stores. These mirrors give clerks and store detectives a wide view of an area, so they can see almost everything in a room. This use is shown at the right.

Using convex mirrors

Review It

1. How does a concave mirror focus light?
2. How does a convex mirror make light rays spread?
3. State three uses of curved mirrors.

19-3 Lenses

In the Sherlock Holmes stories, the famous detective often used a magnifying glass to examine the evidence from a crime. How does a magnifying glass make objects appear larger? As you read about lenses, ask yourself:

a. How do convex lenses affect light rays?
b. How do concave lenses affect light rays?

Convex Lenses Focus Light Rays

Refraction is the bending of any wave as it passes from one material to another. Light waves refract when they pass from any transparent material into another. How much the light bends depends on what the two materials are made of and on the angle at which light hits the surface.

You can shape a piece of glass or other transparent material so that light hitting it always bends to make an image. Then you have made a lens, as shown below. Several shapes of lenses are pictured at the left.

Shapes of lenses

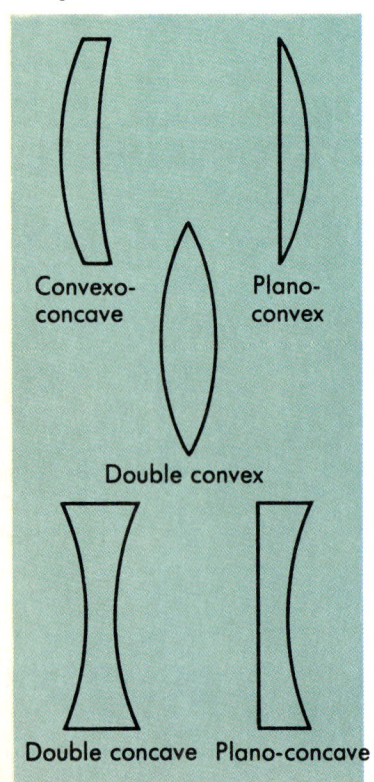

Refraction of light through a lens

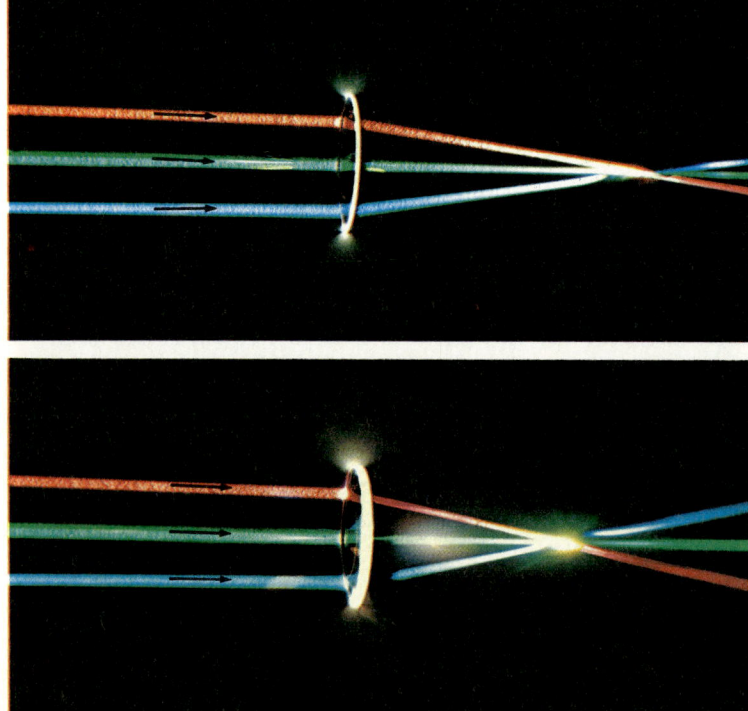

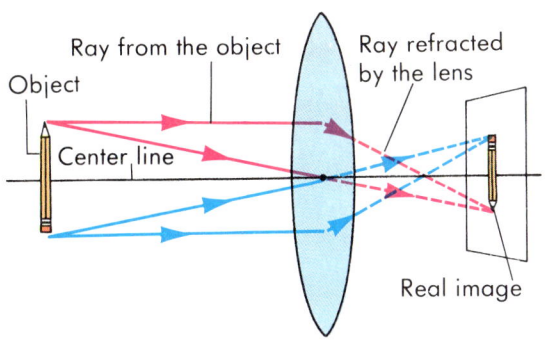
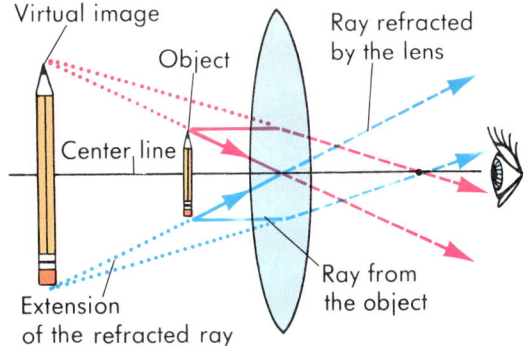

How a convex lens forms an image

The lens at the upper left focuses parallel light rays to a point. The middle of a **convex lens** is thicker than its edges. A convex lens bends light toward the center line. The kind of image it forms depends on the object's distance from the lens.

For the distant object shown, rays come together at the focal point. Notice the ray that hits the lens parallel to the center line. It is simple to follow because it bends to go through the focal point. Another ray that is simple to follow passes through the center of the lens. The opposite sides of a lens are parallel to each other at the center of the lens. So rays enter and leave the center without bending. An image appears on paper placed near the focal point, so this image is real. Notice how rays that pass through the lens form the image. The image is upside down and smaller than the object.

The image of an object between the focal point and the lens forms as shown at the upper right. The image appears right side up, larger than the object, and on the same side of the lens as the object. However, no image shows up on paper placed where the image appears to be. The image is virtual.

A magnifying glass contains a convex lens to make objects appear larger. These lenses are used to examine small objects, such as the small print at the right.

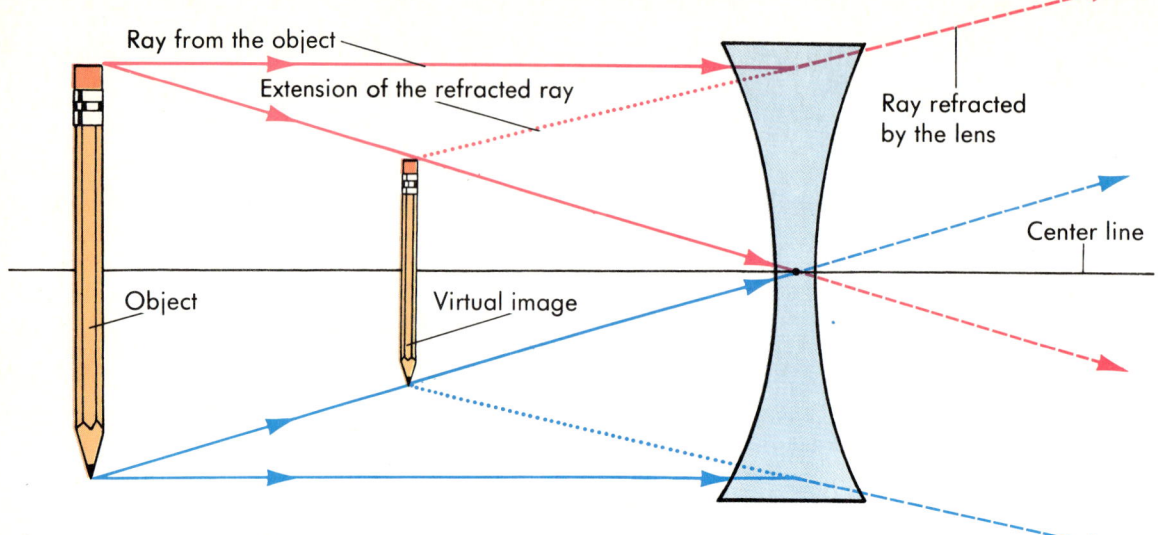

How a concave lens forms an image

Concave Lenses Spread Light Rays Apart

Another kind of lens, a **concave lens,** is thinner in its center than at its edges. A concave lens spreads light rays apart. Notice above how a ray parallel to the center line bends. It follows a line that passes through a focal point. But since the ray is bending outward, the focal point we use is on the near side of the lens. Each lens has a focal point on each side. Which way light passes through does not matter.

The other ray shown does not bend because it passes through the lens's center. You have to trace the two rays backward to find where they meet, as shown with dotted lines. The image appears right side up, smaller than the object, and on the same side of the lens. A paper would not pick up this virtual image because the light rays do not really meet at the image.

Challenge!

The large, flat lens often used in an overhead projector or as a window decoration is a Fresnel lens. Check an encyclopedia or a book on optical devices to find out how this lens works.

Review It

1. Explain how a convex lens forms an image.
2. What happens to light rays when they pass through a concave lens?

Activity

Lenses

Purpose
To investigate how lenses affect light rays.

Materials
- plastic wrap
- water
- piece of paper with words on it
- scissors
- paper towel
- color picture from a newspaper
- eye dropper (optional)
- ruler

Procedure
1. Cut out a 5 cm × 5 cm piece of plastic wrap.
2. Lay the plastic flat over the print on the paper. Note whether the plastic affects how the print appears.
3. Remove the plastic. Put a drop of water onto the plastic with your finger or an eye dropper, as in *a*. Make sure the drop is fairly large.
4. Blot any other water off the plastic with a paper towel.
5. Repeat step 2, but be sure that the drop is the only water on the plastic.
6. Move the plastic so the drop is over a small letter "e." Slowly move the plastic toward you, as in *b*. Observe and record what happens to the image of the "e."
7. Hold the plastic so the image of the "e" is inverted. Move the plastic, first to the left, then to the right, along a line of print. Record which way the image moves each time.

a

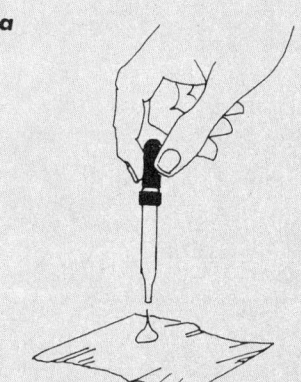

b

8. Repeat step 7, but move the plastic toward the top and bottom of the page.
9. Use your drop to examine the newspaper picture. Record your observations.

Analysis
1. Look at the pictures of lenses on the first page of section 19–3. Which lens is like your waterdrop lens?
2. What kind of image is formed when the water drop acts as a magnifying lens?
3. How are colors printed in the newspaper? What other objects do you know of that use this method to show color?

19-4
Eyes and Lenses

How do you see? Your eyes are amazing organs that adapt to many conditions. For example, you can see not only objects that are close by but also objects that are very distant. To understand how your eyes work, remember the following as you read:

a. How does the lens in your eye focus light?
b. How can vision be corrected?
c. How do other animals see?

The Lens in Your Eye

Your eye's transparent **cornea** (kor′nē ə) and lens focus light from an object and form an image. Notice in the picture below that they make a convex lens. They produce an upside down image on your retina, the part of your eye that is sensitive to light. The retina changes the image into electrical messages and sends them to your brain. Your brain interprets the image as right side up.

You see best when the image on your retina is in focus, regardless of the distance of an object from the lens in your eye. Your eye muscles change the shape of the lens to keep the image focused on your retina. Notice the lens's shape in the pictures. The muscles relax, allowing the lens to become fatter to see a close object clearly. To see a distant object clearly, the muscles tighten, making the lens thinner.

How an image forms in your eye

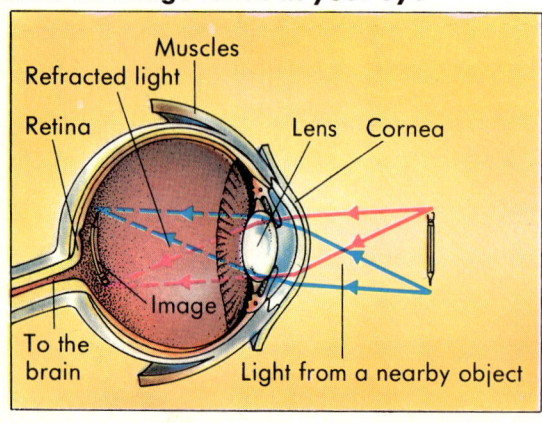
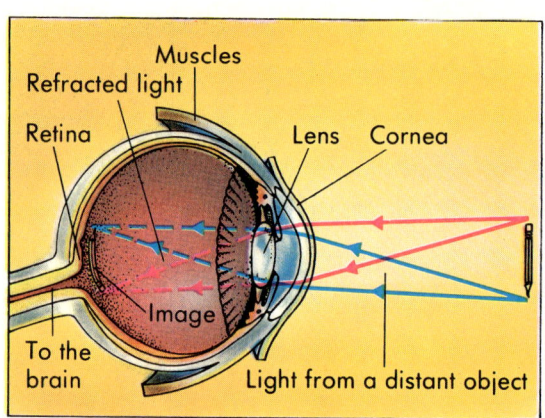

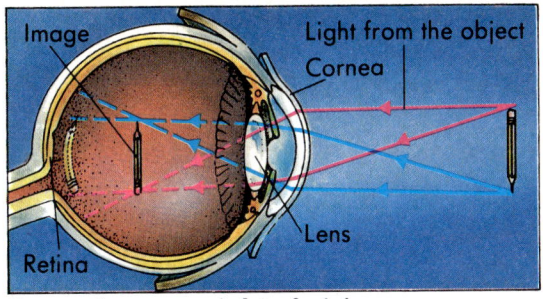
Correcting nearsighted vision

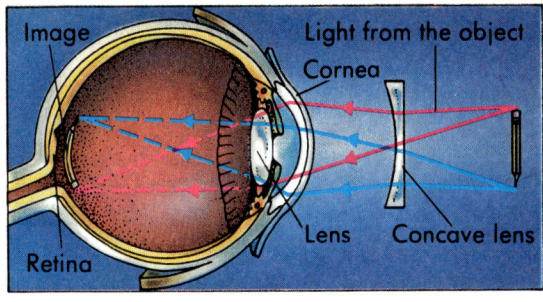

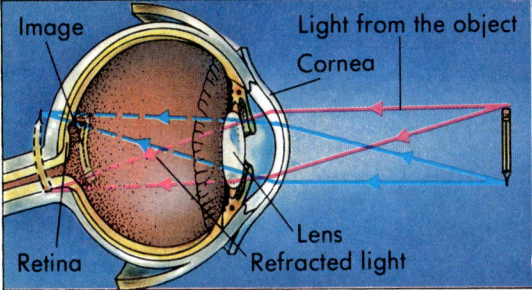
Correcting farsighted vision

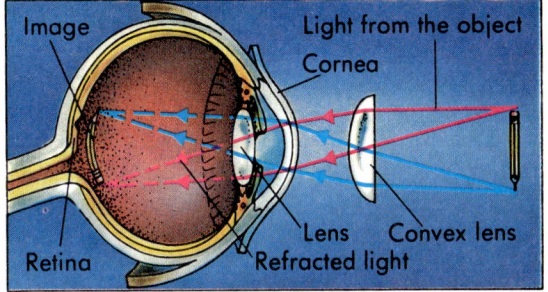

Correcting Vision

If the cornea and lens in your eye do not focus as they should, an extra lens can often solve the problem. People who cannot see distant objects clearly are **nearsighted.** The retina is too far from the eye's lens. In the first picture above, the lens focuses images of distant objects in front of the retina. The image appears fuzzy, because the light is spread over the retina. In the second picture, a concave lens corrects the focus.

People who cannot see nearby objects clearly are **farsighted.** The third picture above shows how a farsighted person sees. The retina is too near the lens. The images of nearby objects would appear behind the retina. The image appears fuzzy because the light hits the retina before it can form a good image. Notice in the fourth picture how a convex lens helps correct the focus.

People become more and more farsighted as they age. The muscles become weaker and so the lens does not thicken to see nearby objects well.

Challenge!

Glaucoma and cataracts are eye diseases. Find out how we can correct or cure them and how we can protect our eyes from disease.

The Eyes of Other Animals

Like the eyes of people, the eyes of many animals use lenses to form images. In many animals, however, the lenses are arranged quite differently.

Insects, lobsters, and crabs have many lenses in each eye. Each lens has six sides. The number of lenses in an insect's eye varies from 6 in some ants to 30,000 in some dragonflies. The photograph shows the eye of a kind of fly.

An insect's multiple lenses do not form very clear images. Since the lenses do not adjust to distance, objects more than one meter away from the insect appear fuzzy. Also, each lens refracts only a tiny part of the light from an object. So the insect does not see the whole object at one time.

The glowing eyes of the cat family have fascinated people for ages. In addition to a lens, cats have a mirrorlike surface inside their eyeballs. This surface reflects light onto the retina, which makes the cat's eyes appear to glow.

Review It

1. How does the human lens change shape to focus light?
2. Explain how lenses are used to correct poor vision.
3. Describe the eyes of insects.

Did You Know?

Mirages

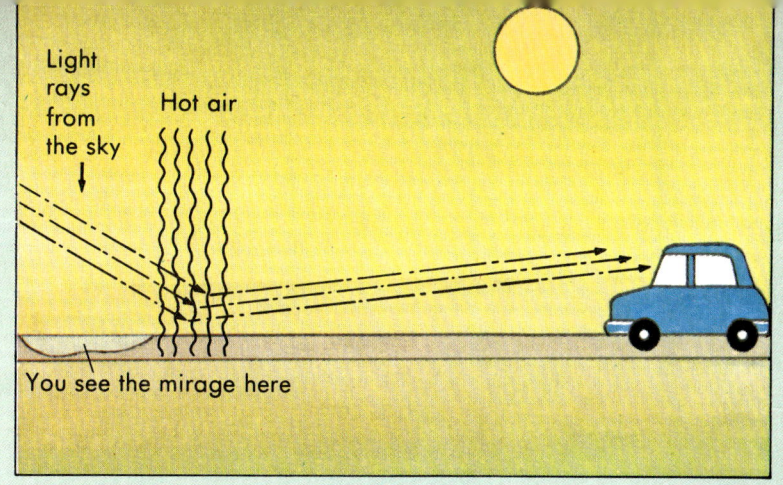

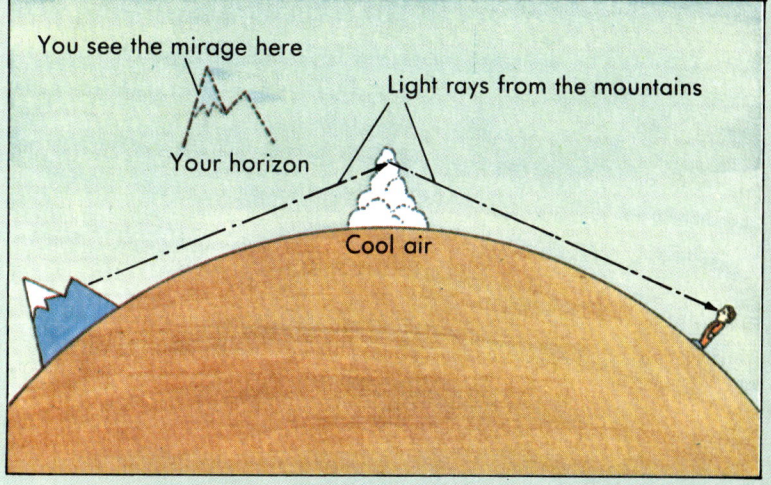

Sometimes what you see is not really there. It is an optical illusion—a trick played on your eyes. Since a camera works like your eyes, it can be tricked too. You can photograph opitcal illusions.

Refraction of light by air can produce a mirage (mə räzh′), a kind of optical illusion. A mirage is something you think you see that really is not at that place. Mirages happen because light refracts when it passes from air of one temperature and density to air of another temperature and density. The greater the change in temperature, the greater the amount of refraction of light.

Often, large differences in air temperature occur near the ground. So light passing near the ground can be refracted a great deal.

When you ride in a car on a hot day, you might see what seems to be a puddle of water ahead of you. However, the puddle disappears as you approach it. You are seeing light from the sky. The road heats up much more from sunlight hitting it than the air does. Heat from the road surface warms the air above it. This heated air refracts the light to cause the mirage. The first diagram shows how this mirage forms.

Mirages are sometimes caused by cold air instead of warm air near a surface. In this case, a distant object may appear very close and very large. On the way to the North Pole in 1906, the explorer Robert E. Peary thought he saw a beautiful mountain range. A few years later, other explorers thought they saw the same range. But when they drew closer, they realized they had seen a mirage. Light from the mountains had been refracted, as shown in the second diagram. The mountains were real. But they were actually far from the place where the explorers saw the mirage.

For Discussion
1. What is a mirage?
2. How does a difference in air temperature cause a mirage?

19-5
Using Lenses and Mirrors

Once the icy rings of Saturn, the red soil of Mars, and the tiny world of bacteria were sights people only imagined. Today we can see such objects with the help of instruments containing lenses and mirrors. Read on to find out how these devices work, and keep the following in mind:

a. How does a camera form an image?
b. How do telescopes use lenses and mirrors?
c. How do microscopes use lenses and mirrors?

How a Camera Works

A camera uses a lens to photograph objects. Light enters a convex lens in the front of the camera. We can control how large the opening is and how long it is open. The lens focuses the light on the film, which is coated with chemicals that are sensitive to light. As you see below, a real, upside-down image forms on the film. A viewfinder allows you to look through the camera's main lens or another lens to see what you want to photograph.

You can put different main lenses on some cameras to produce pictures magnified by different amounts. Each lens is really made of a series of several single lenses. A wide-angle lens is useful for photographing a large scene. A telephoto lens is useful for making objects, especially distant objects, appear larger and closer.

Inside a camera

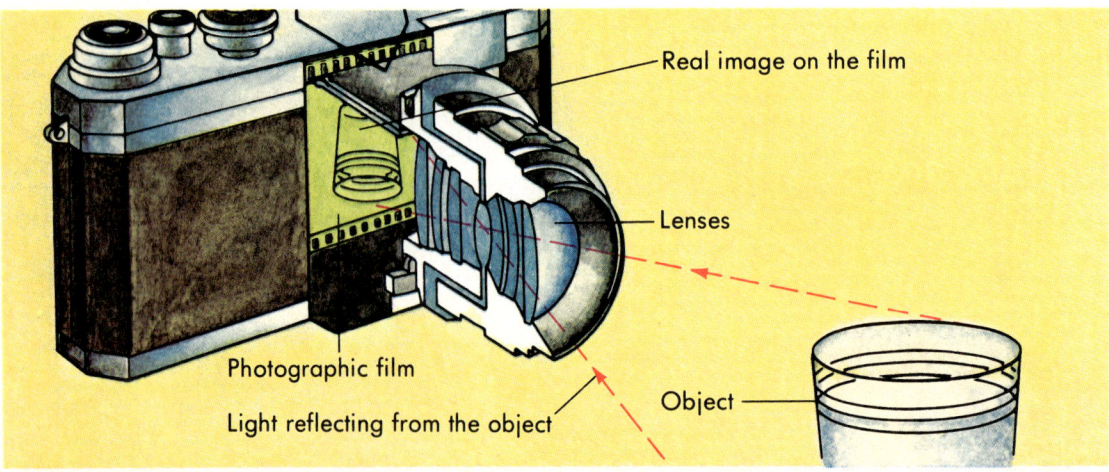

The Yerkes telescope of the University of Chicago

The Soviet telescope, largest in the world

Seeing Distant Objects

Around 1600, in the Netherlands, it was discovered that putting one lens in front of another made distant objects more visible. The scientist Galileo soon used this discovery to make his own telescope.

The first picture on the right shows a simple refracting telescope, which has convex lenses. Light enters a lens that bends light from a star to make an image. The other lens allows you to see this image. The picture at the left above shows the world's largest refracting telescope.

About seventy years later, Isaac Newton replaced the main lens in a refracting telescope with a concave mirror. He invented the reflecting telescope. The path of light in a reflecting telescope is different from that in a refracting telescope. The effect is similar, however. In the reflecting telescope shown on the right, a concave mirror reflects light to a place where you can see the image. Shown at the right above is the world's largest reflecting telescope. Its main mirror, which is 6 meters across, is at the bottom of the telescope.

The largest telescopes do not work by producing larger images. Instead, they make an object more visible because they collect more light. If the surface area of the main lens or mirror is greater, the image is brighter and clearer.

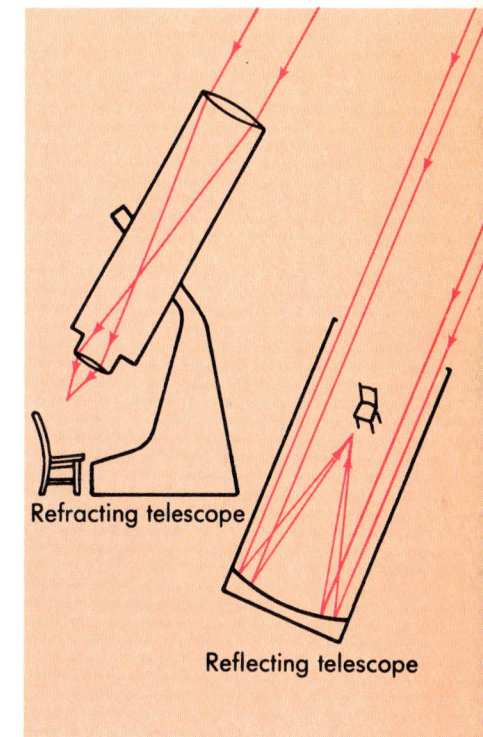

Refracting telescope

Reflecting telescope

Light rays in a microscope

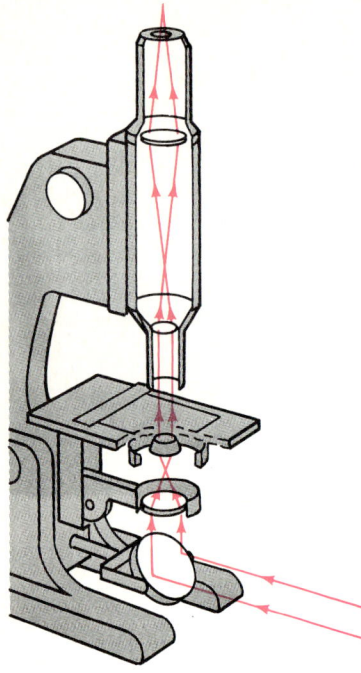

Seeing Tiny Objects

In the middle 1600s, again in the Netherlands, scientists put two lenses together to make nearby objects appear larger. They invented the microscope. The drawing at the left shows the path of light rays in a microscope.

Light hits the mirror at the bottom and is reflected up. It passes through a convex lens, which makes the rays parallel so they evenly light up the object you are viewing. After the light passes through the object, it passes through the microscope's main lens. This second convex lens makes a real image that is viewed with a third convex lens at the eyepiece. The total effect of the three lenses is to magnify the object's appearance.

At the lower left, pond water is being prepared for viewing in a microscope. The next picture shows how the tiny organisms in the water appear in the microscope.

Review It

1. How does a camera use a lens to produce a picture?
2. How do telescopes produce images?
3. How do microscopes produce images?

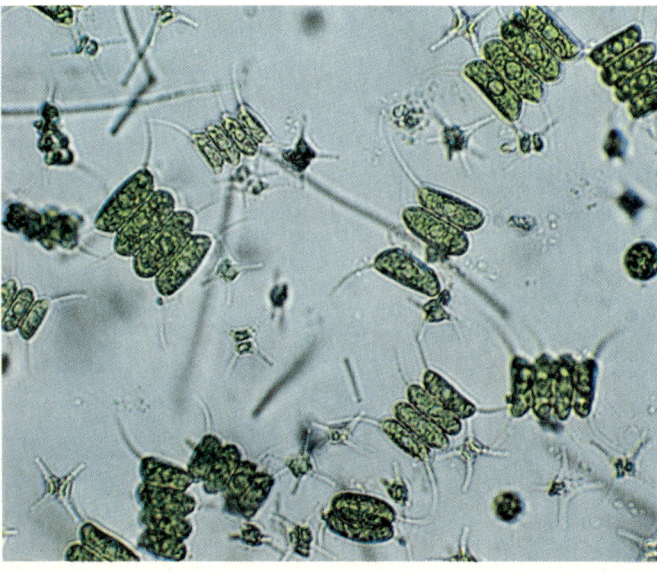

Activity

Making a Pinhole Camera

Purpose
To learn how a simple camera works.

Materials
- shoe box
- straight pin
- some heavy cardboard, somewhat wider, but not higher, than the end of the box
- adhesive or masking tape
- aluminum foil
- "slow" photographic film, cut in pieces and wrapped in aluminum foil

Procedure
1. Cut a 3 cm × 3 cm square in the center bottom of one end of the box.
2. Tape some foil over the square, as shown.
3. With a pin, punch a hole near the top of the foil over the square hole.
4. Bend the ends of the cardboard so that it stands in the box as shown. Cover the box tightly.
5. NOTE: Check the box to be sure that light enters only at the pinhole. Cover any cracks with tape.
6. Tape a cardboard flap over the foil to cover the pinhole, as shown.
7. Take the box, one piece of wrapped film, tape, and scissors into a pitch-dark room. *NOTE: Read step 8 before you enter the dark room.*
8. Unwrap the film, but save the foil. Tape the film to the heavy cardboard so it faces the pinhole. Leave the dark room with the foil and covered box.
9. Choose a dark object against a bright background to photograph. The distance between the film and the pinhole and the time the pinhole is open will affect your image.
10. To take a picture, uncover the pinhole for at least 10 minutes. Rest your camera on something steady while the pinhole is open.
11. Cover the hole and return to the dark room with your foil and box.
12. Open the box and remove the exposed film. Wrap it carefully in foil.
13. Repeat steps 8–11 using another piece of film. Choose a scene of a different brightness and distance from your camera.
14. When you are finished, give your exposed film to your teacher to be developed.

Analysis
1. What could happen to your image if light rays entered your camera from places other than the pinhole?
2. How does the pinhole affect light rays?

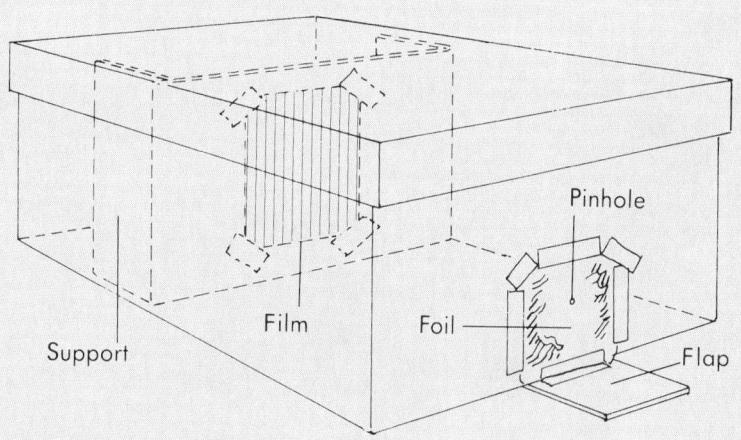

359

Chapter Summary

- Mirrors produce images by reflecting light. (19-1)
- Light rays only seem to meet at a virtual image. The image will not appear on a screen. (19-1)
- A plane mirror forms a virtual image that is as far in back of the mirror as the object is in front of it, right side up, and flipped left to right. (19-1)
- Light rays meet at a real image. The image will appear on a screen. (19-2)
- A concave mirror focuses light rays to produce a real or a virtual image. (19-2)
- A convex mirror spreads light rays to produce a virtual image. (19-2)
- Lenses produce images by refracting light. (19-3)
- A convex lens focuses light rays to produce a real or a virtual image. (19-3)
- A concave lens spreads light rays to produce a virtual image.
- Many animals, including people, have lenses in their eyes to form images. (19-4)
- Many instruments, including cameras, telescopes, and microscopes, make use of mirrors, lenses, or both. (19-5)

Interesting Reading

Laycock, George. *The Complete Beginner's Guide to Photography*. Doubleday, 1979. A discussion of everything involved in taking pictures, from choosing a camera to making a career out of photography.

Rahn, Joan. *Eyes and Seeing*. Atheneum, 1981. A description of the eyes of animals, from flatworms to humans. Simple experiments are included.

Simon, Seymour. *Look to the Night Sky*. Viking, 1977. A good explanation of how to use telescopes to explore the night sky.

Questions/Problems

1. How can you determine whether an image is a real image or a virtual image?
2. Compare the images a plane mirror produces with the images a curved mirror and a lens produce.
3. Compare the ways a concave mirror and a convex lens affect light.
4. Why does a concave lens not form a real image?
5. Explain how the human eye is like a camera by comparing the parts of each.
6. Compare the human eye with the eyes of other animals.
7. Contrast the ways a reflecting and a refracting telescope produce images.

Extra Research

1. Hold this book up to a mirror and read the words, or write something on paper as you watch the mirror instead of the paper. Explain your observations.
2. Examine the print in this book through a magnifying glass. Hold the magnifying glass very close to and very far from the page. Notice when the image is upside down and right side up. Compare this lens to your water drop lens. When is the image real? virtual?
3. If the front door of your house has a peephole, find out which kind of lens it has. Shine a flashlight through the hole to see whether the light spreads apart or comes together at the other side.
4. Make a periscope by taping a mirror diagonally across each of two opposite corners inside a shoe box. Cut holes in the sides of the box in front of each mirror and tape the cover to the box. Use your periscope to see around corners. Then diagram the path of light rays in your periscope.

Chapter Test

A. Vocabulary   Write the numbers 1–10 on a piece of paper.
Match the definition in Column I with the term it defines in Column II.

Column I

1. flat mirror
2. mirror that curves away from the light source
3. mirror that curves toward the light source
4. image formed by light rays coming together
5. image that does not appear on a screen
6. point to which a mirror or lens focuses parallel light rays
7. lens that focuses light
8. lens that spreads light apart
9. the eye's lens cannot form a sharp image of a nearby object
10. the eye's lens cannot form a sharp image of a distant object

Column II

a. concave lens
b. concave mirror
c. convex lens
d. convex mirror
e. farsighted
f. focal point
g. nearsighted
h. plane mirror
i. real image
j. virtual image

B. Multiple Choice   Write the numbers 1–10 on your paper.
Choose the letter that best completes the statement or answers the question.

1. Mirrors affect light by   a) changing its speed.   b) refracting it.   c) reflecting it.   d) changing its color.

2. Your image in a plane mirror is
a) magnified.   b) flipped from left to right.
c) exactly like you.   d) refracted.

3. Nearsighted vision can be corrected with a
a) concave lens.   b) convex lens.
c) combination of both lenses.
d) combination of a lens and a mirror.

4. A light source stands at a concave mirror's focal point. The mirror   a) reflects the light to a point.   b) reflects the light in parallel rays.
c) refracts the light to a point.   d) refracts the light in parallel rays.

5. Both cameras and magnifying glasses use
a) concave lenses.   b) convex lenses.
c) concave mirrors.   d) convex mirrors.

6. Convex mirrors are useful because they
a) give a wide field of view.   b) flip images from left to right.   c) form real images.
d) focus light to a point.

7. The microscope   a) uses mirrors to produce an image.   b) is the opposite of a telescope.
c) makes large objects appear smaller.
d) makes small objects appear larger.

8. All lenses   a) spread light apart.   b) focus light.   c) are usually curved and transparent.
d) magnify objects.

9. All telescopes   a) make distant objects more easily visible.   b) make nearby objects seem polarized.   c) are reflecting telescopes.
d) are refracting telescopes.

10. In the eye, images are usually formed on the   a) muscle.   b) lens.   c) cornea.
d) retina.

Chapter 20
Sound

We are surrounded by sounds all the time. Sounds can be very different from each other. Different instruments in the marching band in the picture, for example, produce different sounds. A flute makes high sounds. A tuba makes low sounds. Each sound is a kind of music.

Sound is a wave, so you already know something about it. The first section of this chapter will demonstrate the wave properties of sound. The next section will discuss how fast sound travels and how noise affects us. The last section will use what you know about sound to describe its application in music.

Chapter Objectives

1. Describe sound using wave properties.
2. Describe the effects of sound waves and materials on each other.
3. Explain how instruments make music.

20–1
Sound as a Wave

A ringing bell, a thunderclap, laughter, and rock music are sounds that seem very different to us. However, all sounds are alike because they are waves. To learn how wave properties apply to sound, explain the following:

a. What causes sound?
b. What is a reflected sound wave?
c. How do sound intensity and loudness differ?
d. What is the pitch of a sound?
e. What is the Doppler effect?

Making Sound

If you hum holding the front of your neck, you feel your vocal cords moving. The strings on a guitar also move back and forth to make music. **Vibrations** (vī brā′shəns) are quick, back-and-forth motions that cause sound. A vibrating object makes the matter around it vibrate. The tuning fork shown is an example. When it is struck, its vibrations move through both the water and the air as waves.

Sound waves travel through matter as compressional waves. They travel through matter just as other compressional waves move through a spring. Because sound is a vibration of molecules, matter must be present to provide the molecules. So sound waves can only move through matter. The astronauts walking on the moon needed radios to talk to each other. Outer space has no air to carry the sound waves of their voices.

You can hear because, when sound waves reach your ears, the waves make your eardrums vibrate. Nerves then send your brain messages about the vibrations. The brain interprets the messages as sound.

Sound Waves Reflect

Like all waves, sound waves can be reflected. An echo is a reflected sound wave. Animals and people can use echoes to find distances to objects. For example, bats use echoes to fly in the dark and to find food.

People use the echoes of sonar (*so*und *n*avigation *a*nd *r*anging) to find schools of fish and to chart the oceans. In the drawing, a sound wave sent from a ship hits the ocean floor and reflects back to the ship. The distance that the sound travels is calculated using the speed of the sound wave and the time the wave needs to travel to and from the ocean floor. This calculation tells the depth of the ocean at that place.

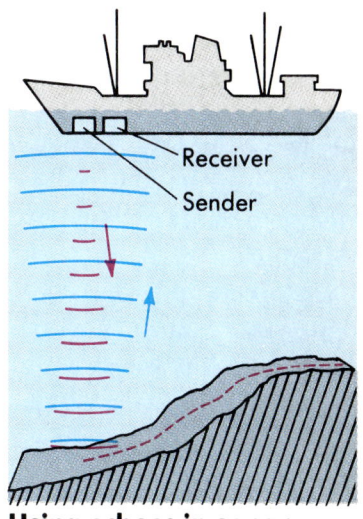

Using echoes in sonar

Sound Intensity and Loudness

Sound waves spread out much as light from a lamp spreads. Sound waves move away from their source in all directions. Therefore, you can hear sounds in any direction from their source. As the waves move farther out, they spread over a greater region. As with light, sound intensity decreases if you move farther away from the source.

Intensity and loudness are not the same. We can measure intensity in units called **decibels** (des′ə belz). Every increase of ten decibels means that the intensity of a sound is ten times as great. Loudness is more difficult to measure, because what seems loud to one person may not seem loud to another. However, most people agree that sounds of more than 120 decibels are painful.

A very intense sound has a lot of energy and a large amplitude. It seems loud. A less intense sound has less energy and a smaller amplitude. It seems soft. The chart lists some common sounds and their intensities. Listening too often to sounds above 85 decibels in intensity may cause a hearing loss.

Some Common Sounds

Sound	Decibels
Threshold of hearing	0
Rustle of leaves	15
Whisper	20
Light traffic	50
Normal conversation	55
Busy street traffic	65
Typewriter	75
Alarm	80
Vacuum cleaner	85
Motorcycle	90
Subway train	95
Jackhammer	110
Rock band	115
Thunder overhead	120

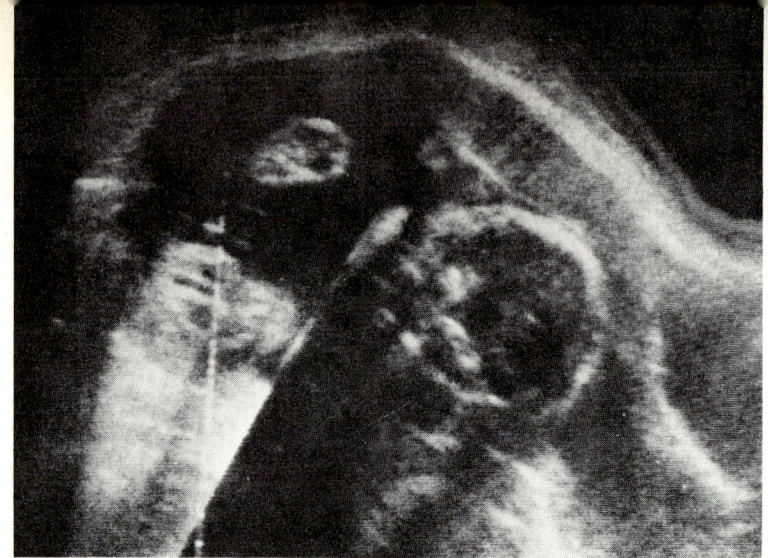

We Can Hear Some Sounds

Pitch is how high or low a sound seems. Pitch depends on how quickly the sound source vibrates, which is its frequency. A high frequency has a high pitch. A low frequency has a low pitch.

Our ears can detect sounds between frequencies of 20 and 20,000 hertz. Sounds above this range are **ultrasonic** (ul′trə son′ik). Some animals hear such sounds. Dogs hear frequencies up to 25,000 hertz. The echoes bats use are about 80,000 hertz.

Although we cannot hear ultrasonic waves, we can use them. One kind of camera sends out waves of about 50,000 hertz. The object to be photographed reflects the waves to the camera. A device in the camera times the trip back and forth and figures out the object's distance. The camera focuses for that distance and takes a picture.

Industrial testing companies use ultrasonic waves to find fine cracks and breaks in parts. Doctors use ultrasonic waves to "photograph" a fetus, shown above, and learn more about the fetus's condition.

Have You Heard?
Hummingbirds are named for the sound of their beating wings. This vibration has a frequency of about 50 hertz.

The Doppler Effect Changes Pitch

You may have noticed how the pitch of a car's engine drops as it passes you. Since pitch depends on frequency, something must happen to the frequency to cause this effect.

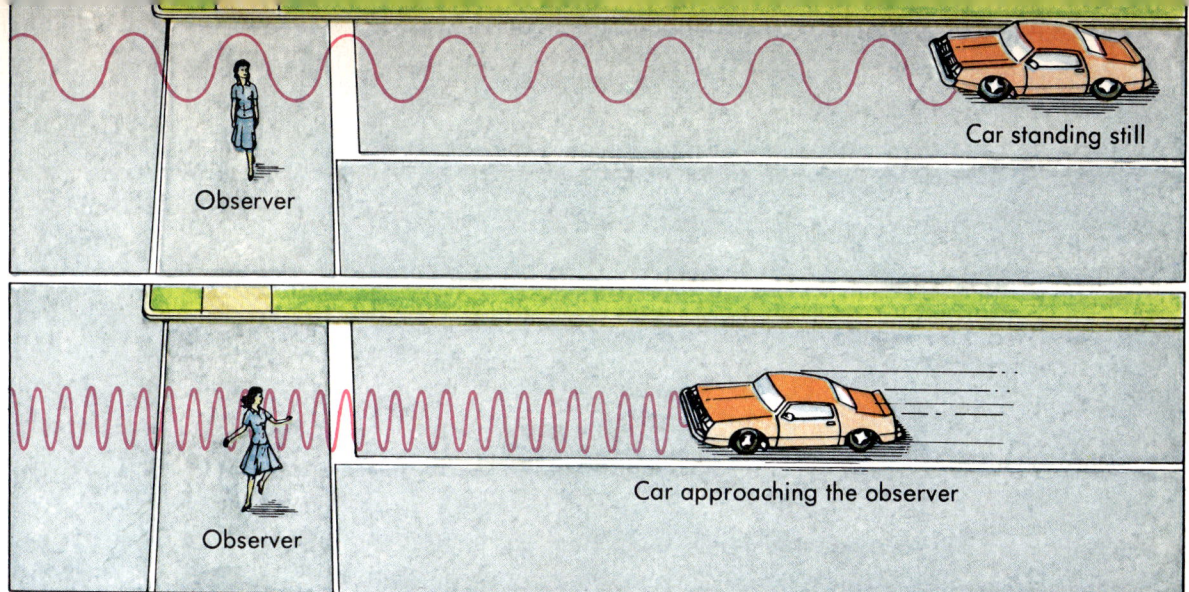

The situation pictured above shows what happens. As the car approaches, each sound wave travels less distance to reach you than earlier waves did. Therefore, the wavelength is shortened. The car seems to chase its own waves. However, the car never catches up, because sound moves much faster than the car. Since more waves reach you every second that the car nears you, the frequency and pitch are higher than normal. As the car passes, the pitch drops as the wavelength returns to its true length. When the car moves away, fewer waves reach you each second, so the pitch is lower.

A similar change happens when you move instead of the sound source. Any change in frequency when you or a wave source moves is named the **Doppler** (dop′lər) **effect** for its discoverer, Christian Doppler.

All waves show the Doppler effect. In the drawing, police radar uses the Doppler effect to measure changes in the reflected radio waves and determine the car's speed.

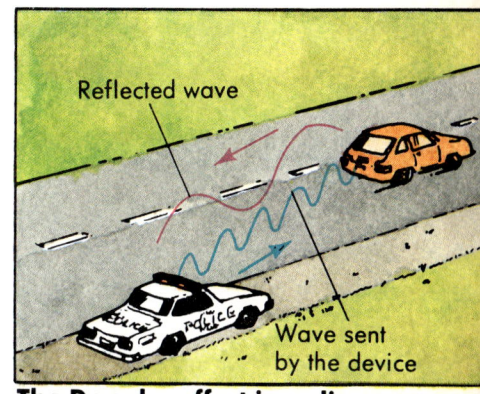

The Doppler effect in police radar

Review It

1. How are all sounds produced?
2. Explain one way echoes are used.
3. How do we measure the energy of sound waves?
4. Which frequencies can people hear?
5. Name two waves that show the Doppler effect.

Activity

Properties of Sound

Purpose
To show how matter affects sound.

Materials
- Erlenmeyer or Florence flask
- stopper to fit flask
- wire
- small bell
- Bunsen burner or hot plate
- matches
- ring stand and ring
- wire gauze
- tongs
- safety goggles

Procedure
1. Attach the bell to the wire and insert the wire into the stopper, as in *a*.
2. Stopper the flask so that the bell hangs in it, as in *b*.
3. Shake the flask and record whether you can hear the bell.
4. Unstopper the flask and put a small amount of water in it.
5. Arrange the flask on the hot plate or on the wire gauze on the ring and ring stand. CAUTION: Do not touch any hot surfaces. Wear your safety goggles.
6. Boil the water in the flask until steam comes out of it.
7. Remove the flask from the heat. CAUTION: Use the tongs and do not touch the hot glass or water.
8. Stopper the flask and let it cool for a few minutes.
9. Repeat step 3 and record whether the sound is louder or softer than it was before.

a

b
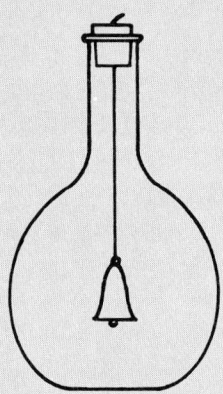

Analysis
1. After you stoppered the heated flask, what happened to the gases in it?
2. What property of sound caused the result you observed in step 9?

Did You Know?

Designs for Better Listening

"You could have heard a pin drop!" This familiar phrase describes excellent conditions for listening. Exactly what factors make good listening conditions? The science of acoustics attempts to answer this question.

For centuries, a good concert hall or auditorium resulted mostly from luck, not science. Builders commonly used superstitions and folk wisdom. Some builders stretched wires across cathedrals to improve the sound. Others thought the walls of a concert hall had to "age" like the wood in a fine violin.

The first scientific studies in acoustics were made in the early 1900s by the American physicist Wallace Clement Sabine. Sabine found that the size, shape, and construction of a hall determined the quality of listening conditions. He developed the idea of reverberation time, which is the time a sound takes to die out in a room. Reverberation time is determined by the size of a room and the materials in it.

Sabine learned that long reverberation times give musical sounds a chance to blend properly. However, spoken words need only a short reverberation time. These discoveries tell us that rooms good for speeches might be terrible for music.

Designing an auditorium today requires many calculations. Reverberation time must be considered. The correct reflection of sound waves from the ceiling, walls, and floor is also a factor. The size and location of amplifiers is important. Finally, the proper balance between a sound and its echo must be calculated. Designs are often tested in echo-free rooms, called anechoic (an/ē kō/ik) rooms, such as the one shown.

Acoustical engineers are just learning that personal preferences are important too. No mathematical equations show which music halls people prefer. We like certain sounds more than others. People prefer to be surrounded with sound. Narrow halls seem to produce better sounds than wide halls. The proper blend of all factors, both scientific and personal, is needed to produce just the right auditorium.

For Discussion
1. What is reverberation time?
2. Designing an auditorium today presents problems not known 100 years ago. What are some of these problems?

20-2 Characteristics of Sound

Fifty years ago, pilots died trying to fly their planes faster than the speed of sound. Their planes nosed over or fell apart from the powerful waves they caused. Then engineers redesigned the planes' wings to sweep back. This new design smoothed the flow of air over the wings. In 1947, the first faster-than-sound flight took place. Now even commercial airplanes fly this fast. Keep reading to learn about the speed of sound, and ask yourself:

a. What determines the speed of sound?
b. How can noise be a kind of pollution?

The Speed of Sound

Sound moves at different speeds through different materials. As the table at the left shows, sound travels fastest in solids and slowest in gases.

The speed of sound depends on how easy it is to squeeze the material's molecules closer together. Sound travels faster through materials that are harder to compress and slower through those that are easier to compress. To understand why, look at the drawings below.

In a solid, molecules are held in place close to each other. When a wave disturbs some molecules, it disturbs all of them. The wave can pass quickly from one molecule to another. In a gas, molecules are not held in one place. They can move in any direction. When a wave disturbs some gas molecules, they can move without affecting other molecules as directly. Therefore, the wave takes more time to pass through the gas.

Speed of Sound in Various Materials

Material	Speed (m/s)
Air (20°C)	344
Water (15°C)	1,450
Brick	3,650
Oak	3,850
Aluminum	5,100
Steel	5,200
Granite	6,000

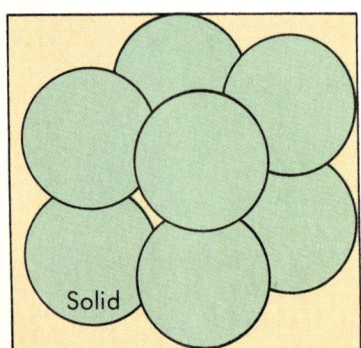
Solid

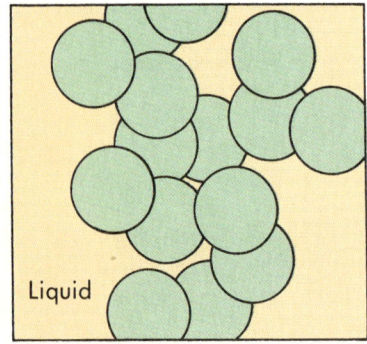
Liquid

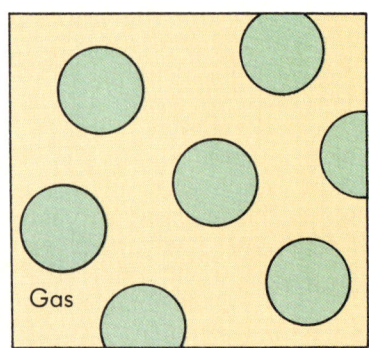
Gas

Moving slower than the speed of sound

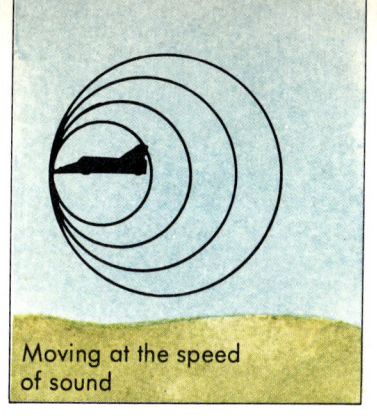
Moving at the speed of sound

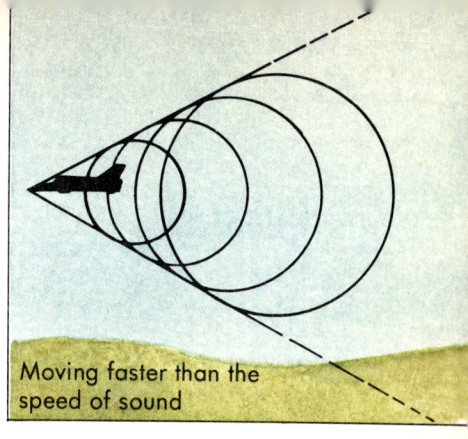
Moving faster than the speed of sound

People once thought an airplane could not fly faster than the speed of sound in air without destroying itself. As a plane flies, it compresses air in front of it, as shown. This wave travels in front of the plane at the speed of sound. If the plane flies more slowly than the speed of sound, the plane never catches up with the wave. To fly faster than the speed of sound, the plane must pass through a "wall" of compressed air—the sound barrier.

While the plane flies faster than sound, the air is compressed along a cone, as shown. This air hitting the ground causes a sonic boom, which is as loud as thunder. The boom trails behind the plane as long as it flies faster than sound.

Noise Pollution

Sonic booms are just one form of **noise,** which is unwanted or unpleasant sound. Noise is pollution because it harms people and property. Noise can break plaster and glass and harm your hearing. Such noises are harmful because of their high intensities. The pressure they cause on your eardrum or on an object is high enough to break it. All noise raises your blood pressure and makes you irritable.

We can lessen the effects of noise by muffling noisy devices, such as cars and planes. Special ceiling materials for rooms also absorb sound and lessen noise.

Challenge!
Use an encyclopedia to find out how the human ear works.

Review It

1. How does a material affect the speed of sound in it?
2. State two examples of noise pollution.

20–3 The Sound of Music

A piano looks and sounds very different from a drum or a violin. Yet all three instruments are similar in the way they make sounds. Apply what you know about sound to instruments, and ask yourself:

a. How do instruments produce music?
b. How do instruments increase vibrations?

Music from Instruments

Music is pleasant sound with regular wave patterns. Musical instruments usually make sounds that go well together. Three basic kinds of musical instruments are the string, percussion (pėr kush′ən), and wind instruments.

String instruments, such as the guitar below, have strings stretched over a box. A player vibrates the strings with a finger, a bow, or a pick.

To play the correct notes, the player must adjust the pitch of the strings by tightening them. Tighter strings make higher sounds. A player can then shorten the vibrating section of a string by pressing down on it. Shorter strings also make higher sounds.

Trombone

A string's thickness also affects its pitch. A thick string sounds lower than a thin one of the same length and tightness. Each instrument has strings of different thicknesses.

Percussion instruments produce sounds when a hand or stick strikes them. For example, the flexible, stretched material on a drum vibrates when struck. The piano is both a percussion and a string instrument. Striking a piano key makes a wood hammer hit a wire inside the piano. The vibrating wires make the sounds.

A wind instrument produces sound when the air column inside it vibrates. The clarinet at the right is a wind instrument. The player blows air onto a reed in the clarinet. The reed vibrates and makes the air inside the instrument vibrate. To change the pitch, the player presses the keys or covers the holes. This action changes the length of the air column, which changes the pitch. As with the strings, short air columns sound high, and long ones sound low.

The trumpet at the lower right and the trombone above work in a similar way. However, instead of a reed, the player's lips must vibrate to vibrate the air column. Pushing down on the buttons, called valves, lengthens the trumpet's air column. Pulling the trombone's slide shortens its air column.

No matter how hard you strike a piano key, it will always play that same frequency. Only the loudness changes. Every object has its own frequency of vibration. This **natural frequency** depends only on the object's size, shape, and material.

Clarinet

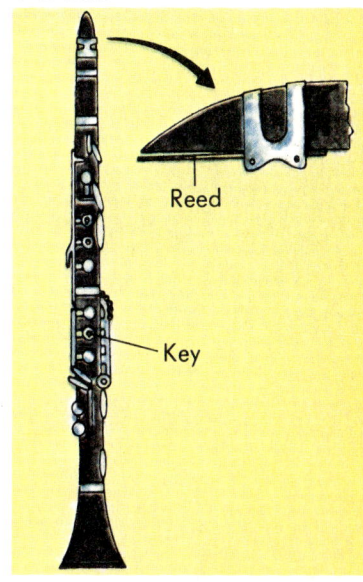

Trumpet

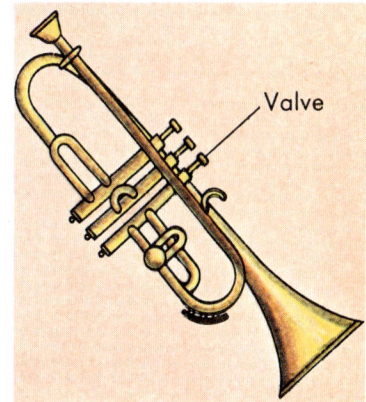

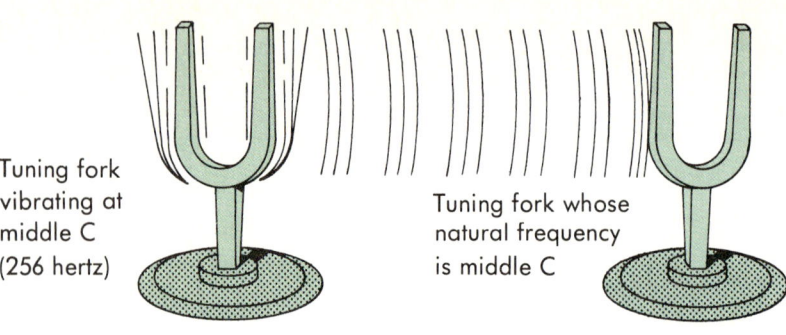

Tuning fork vibrating at middle C (256 hertz)

Tuning fork whose natural frequency is middle C

Increasing Vibrations

The vibrations of a reed, a string, or a player's lips are actually very faint. Therefore, musical instruments are built to increase the original vibrations. **Forced vibrations** happen when one object causes another to vibrate at the original frequency, which is not necessarily the natural frequency of the second object. For example, a piano's wires are attached to a sounding board, as shown. The vibrating wires make the board vibrate, which makes the sounds louder. A similar effect happens in a string instrument. The vibrating string causes the instrument and the air inside it to vibrate.

A different kind of vibration happens in wind instruments, which are built so that each length of the air column produces only its natural frequency. When the player blows into the instrument, the air column reinforces the original vibration. **Resonance** (rez′ə nəns) occurs when one vibration reinforces another because they have the same natural frequency.

Still another effect is shown above. Both tuning forks have the same natural frequency. When one fork vibrates, the second fork starts vibrating without being struck. These **sympathetic vibrations** happen when waves from one vibrating object cause another to vibrate because they both have the same natural frequency.

Have You Heard?

Sympathetic vibrations can cause an avalanche. A loud sound, such as a shout, can cause loosely packed snow to vibrate sympathetically. When it vibrates, it moves and may tumble down a mountain.

Review It

1. Explain how each kind of instrument makes music.
2. What are three ways to strengthen a vibration?

Activity

Making Music

Purpose
To observe the factors affecting both the vibrations and pitch of strings and air columns.

Materials
- 4 rubber bands of different thicknesses
- ruler
- 2 pencils
- empty soda pop bottle
- tuning fork
- water

Procedure

Part A
1. Stretch a rubber band around the length of the ruler.
2. Note the rubber band's tightness by pulling on it.
3. Insert the 2 pencils, as in *a*.
4. Pluck the rubber band. Note its pitch.
5. Repeat steps 1–4 for the other rubber bands.
6. Record how the band's thickness and tightness affect its pitch.

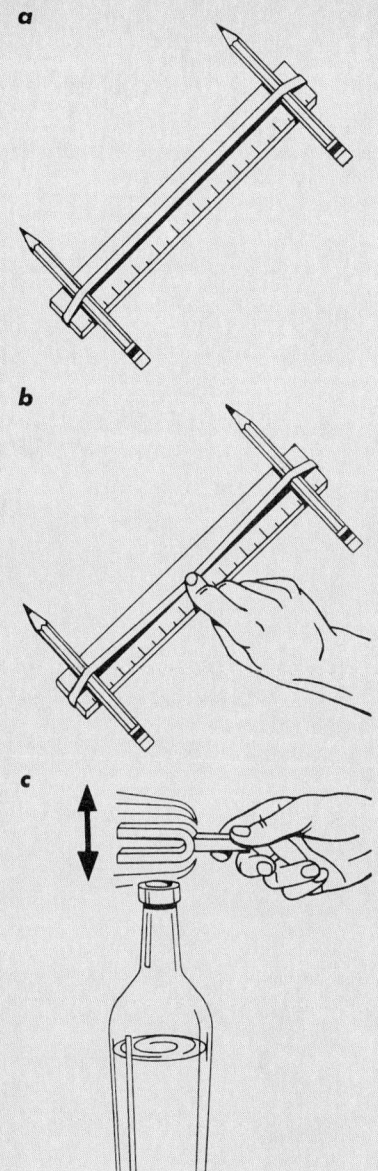

Part B
1. Pluck one band.
2. Hold down the middle of the band.
3. Pluck each half of the band, as in *b*.
4. Record how the pitch of the shortened rubber band compares with the pitch of the whole rubber band.

Part C
1. Put some water in a bottle.
2. Blow across the top and listen.
3. Hit a tuning fork on something soft, such as the rubber heel of your shoe.
4. Move the vibrating tuning fork up and down over the bottle's mouth, as in *c*.
5. Record what happens to the loudness as you move the tuning fork.
6. Repeat steps 1–5 with different amounts of water in the bottle.

Analysis
1. What factors affect the pitch of the rubber bands?
2. Why does the tuning fork sound louder at certain positions than at others?

Chapter Summary

- Vibrations cause all sound waves. (20–1)
- Sound must move through matter. (20–1)
- Echoes can be used to determine the distance to an object. (20–1)
- A sound wave's energy determines its loudness and intensity. (20–1)
- Pitch depends on the frequency of a sound wave. (20–1)
- The Doppler effect occurs either when a source of waves or when the observer moves toward or away from each other. (20–1)
- The speed of sound in a material is determined by how easily the material's molecules are compressed. (20–2)
- A sonic boom occurs whenever an object moves faster than the speed of sound. (20–2)
- Too much noise can harm people and property. (20–2)
- Musical instruments use vibrating air columns, strings, or other objects to produce different sounds. (20–3)
- Every object has its own natural frequency of vibration. (20–3)
- Forced vibrations, sympathetic vibrations, and resonance transfer vibrations from one object to another. (20–3)

Interesting Reading

Grey, Jerry. *Noise, Noise, Noise!* Westminster, 1975. Describes the effects of noise and how to reduce them.

Knight, David C. *Silent Sound.* Morrow, 1980. Discusses ultrasonics and its uses.

Tannenbaum, Beulah, and Stillman, Myra. *Understanding Sound.* McGraw-Hill, 1973. Describes sound, sound-producing devices, music, and hearing.

Questions/Problems

1. Suppose a music box were placed in an airtight room with no air. Could its song be heard inside the room? outside the room? Explain your answers.
2. You hear the pitch of a car horn become lower. Explain how you know whether the car has started moving toward or away from you.
3. The sound of a train reaches you more quickly through the rails than through the air. Why?
4. How would you adjust a single wire to produce notes of different frequencies?
5. If you hold a glass over your ear and play a few notes on a piano, some notes sound louder than others. Explain why.

Extra Research

1. Ask a friend to stand about 25 m away from you down the length of a chain link fence. Tap a post with a hammer. Explain what you and your friend observe.
2. Find out if your town has laws about how loud sounds can be. How do these rules fit the decibel chart in this chapter?
3. Make a musical scale by filling at least eight empty soda pop bottles with different amounts of water. Use a tuning fork or pitch pipe to get started. Play a song on your "instrument."
4. Open the top of a piano to expose the wires. Listen during and after you hum or sing a tone into the piano. Explain what you observe. Repeat with other notes.

Chapter Test

A. Vocabulary Write the numbers 1–10 on a piece of paper.
Match the definition in Column I with the term it defines in Column II.

Column I

1. unit for comparing intensity of sounds
2. back-and-forth motion
3. highness or lowness of a sound
4. too high for people to hear
5. one source makes another vibrate at the original frequency
6. unwanted or unpleasant sound
7. a vibration reinforces another of the same natural frequency
8. apparent change in pitch
9. depends on an object's size, shape, and material
10. a vibration causes another of the same natural frequency

Column II

a. decibel
b. Doppler effect
c. forced vibrations
d. natural frequency
e. noise
f. pitch
g. resonance
h. sympathetic vibrations
i. ultrasonic
j. vibration

B. Multiple Choice Write the numbers 1–10 on your paper.
Choose the letter that best completes the statement or answers the question.

1. A sound's frequency determines its
a) amplitude. b) energy. c) intensity.
d) pitch.

2. Sound travels fastest through a) solids.
b) liquids. c) gases. d) a vacuum.

3. Vibrations causing sound a) move molecules back and forth. b) spread molecules apart. c) pack molecules together. d) a, b, and c.

4. A very intense sound a) has little energy.
b) has a small amplitude. c) seems loud.
d) seems soft.

5. The Doppler effect can happen to
a) compressional waves only. b) sound waves only. c) water waves only. d) all waves.

6. As a sound source moves past a stationary observer, the observer may hear (a) a) rise in pitch. b) drop in pitch. c) rise in pitch, then a drop. d) no change in pitch.

7. A sonic boom occurs when a(n) a) sound becomes ultrasonic. b) echo happens.
c) object moves faster than the speed of sound. d) noise is made.

8. Shortening a vibrating string causes the pitch of the sound produced to go a) up then down. b) down. c) up. d) remain the same.

9. Wind instruments increase vibrations with
a) forced vibrations. b) resonance.
c) sympathetic vibrations. d) none of the above.

10. Sympathetic vibrations can occur when the natural frequency(-ies) a) of both objects are the same. b) of both objects are any number.
c) of one object is one-half that of the other.
d) of one object is one-fourth that of the other.

377

Careers

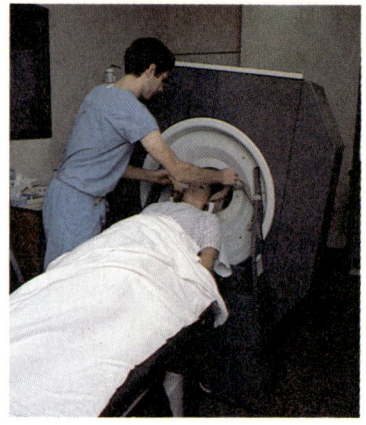

Studio engineer
When you watch a movie with an old sound track, you might see the actors move their mouths before or after you hear their voices. Today, such delays would be fixed by a studio engineer.

Studio engineers work wherever sound is reproduced for movies, television, radio, and records. They study the acoustics of recording studios and decide where to place microphones or speakers. Studio engineers work to cut down echoes, reverberations, and delays. They control sound volume and quality to improve recorded sound tracks.

Studio engineers learn acoustics in college or technical school. They also study the electronics of complex audio equipment.
Career Information:
National Association of Broadcasters, 1771 N St., NW, Washington, DC 20036

X-ray technician
Without an X ray, injuries inside your body might not be found. Technicians take the pictures that uncover things such as tiny, hair-thin cracks in bones.

Hospitals and clinics hire special technicians to take X-ray pictures for both routine checkups and emergency cases. Technicians help patients pose for their X rays. They shield areas that are not being examined, and then take the picture. Finally, technicians develop X rays so that doctors can use them to diagnose broken bones, tumors, and lung diseases.

X-ray technicians train for 1-2 years in hospitals or at technical schools. There, they learn about anatomy and how to take and develop X rays.
Career Information:
The American Society of Radiologic Technologists, 55 E. Jackson Blvd., Chicago, IL 60604

Photographer
Light, lenses, chemically treated paper, and a camera can produce a picture. Photographers use special skills to create these pictures.

Photographers use their knowledge of light, lenses, and film to produce the effect they want. Then, by doing some special chemistry, photographers develop and print the pictures they have taken.

Photographers work in many places. Taking pictures for newspapers, magazines, science and medical journals, and portraits of people are just a few examples of the work photographers do.

Beginning photographers need to master many kinds of cameras, lenses, and developing techniques. They learn and practice these skills as they train with experienced photographers.
Career Information:
Professional Photographers of America Association, 1090 Executive Way, Des Plaines, IL 60018

  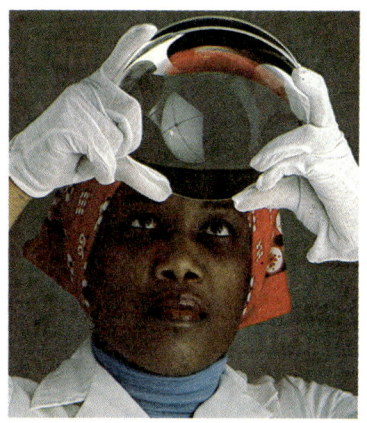

Laser technician

Laser beams play an important role in medicine and industry. Their use is becoming more widespread every day. A laser technician oversees the production of laser equipment and makes sure it is built exactly as planned.

Laser technicians study for two years to get a degree, and with on-the-job training learn about both the design and the mechanics of a laser. This preparation helps the technician when he has to explain design ideas to shop workers and mechanical problems to people who design lasers.

Career Information:
Laser Institute of America, 5151 Monroe St., Suite 103-West, Toledo, OH 43623

Research tester

If an airplane crashes, examiners inspect the wreck to learn why it went down. Testing planes and other products before they are used can prevent accidents. Research testers perform such tests.

Almost every industry hires research testers who use X-ray equipment, microscopes, lasers, photographs, and chemicals to check products. These testing methods let the researchers get a close look without taking the machine apart. If they see something wrong with a part, they stop production and have the part replaced.

Research testers often learn one specific skill by training on the job. More complex testing methods are taught in technical schools.

Career Information:
American Society for Nondestructive Testing, 4153 Arlingate Plaza, Caller no. 28518, Columbus, OH 43228

Lens maker

Do you have perfect eyesight? If not, you probably wear eyeglasses. The lenses in glasses correct problems with the lenses in your eye. A lens maker makes corrective lenses. Some lens makers make lenses for telescopes, cameras, and binoculars.

If you need glasses, your eye doctor writes you a prescription. Then you choose the eyeglass frames you want. Both the prescription and the frames go to the lens maker, who grinds a lens to fit your frames and your vision. Lens makers use careful measurements to make sure you get a perfect fit.

Lens makers train as apprentices for three years after high school. They study optical mathematics and physics. They also learn how to grind lenses.

Career Information:
Optical Laboratories Association, 6935 Wisconsin Ave., Suite 200, Chevy Chase, MD 20815

UNIT SIX
ELECTRICITY AND MAGNETISM

This photograph might remind you of a street map. Or maybe you think it is a new kind of machine part. What else could it be?

The object is smaller than your fingertip. But do not let its size fool you. Actually, the object is a tiny part used in certain devices that use electricity. Such parts are very important to the operation of calculators and televisions. As you study Unit Six, you will learn how electricity and magnetism are related and how they affect computer technology.

Chapter 21 Electricity
The electric force is a basic force in nature. Electricity is affected differently by different materials.

Chapter 22 Magnetism
A material's magnetic nature depends on its electron structure. Electric and magnetic forces are parts of the same force.

Chapter 23 The Electronic Revolution
Our need to process information quickly has resulted in a new technology based on the flow of electrons.

Chapter 21
Electricity

Electricity can harm us as well as help us. Lightning is a dangerous form of electricity. Yet electricity can be tamed and carried over wires to our homes. Imagine how difficult our lives would be without electricity.

This chapter discusses how objects can have electric charge, how those charges move, and how we use electricity every day.

Chapter Objectives
1. Explain how electric charge builds up.
2. Explain how voltage, current, and resistance are related to each other.
3. Describe the two kinds of circuits.
4. Explain how we use electricity in our homes.

21–1
Electric Charge

What do lightning, a shock from a metal doorknob, and fly-away hair have in common? As you read, you will understand how they happen. Keep the following questions in mind:

a. In what way is all matter electrical?
b. How does charge build up?
c. What is a spark?

Matter Is Electrical

The atoms in all matter contain small particles. Two of these particles, the electron and proton, have an electric charge. A third particle, the neutron, has no electric charge. It is neutral. Each proton has a positive charge. Each electron has a negative charge. The movement of electrons makes the electricity we use daily.

An atom has the same number of electrons and protons. The electrons' negative charge balances the protons' positive charge. The sum of the negative and positive charges is zero, so the atom is neutral.

Protons and neutrons are not completely free to move, but sometimes the electrons are. Often a neutral atom loses some of its electrons. It then has more protons than electrons. The charged atom now has a positive charge. Sometimes a neutral atom gains one or more electrons. Then the charged atom has a negative charge. The charged atoms shown below have become charged by losing or gaining electrons.

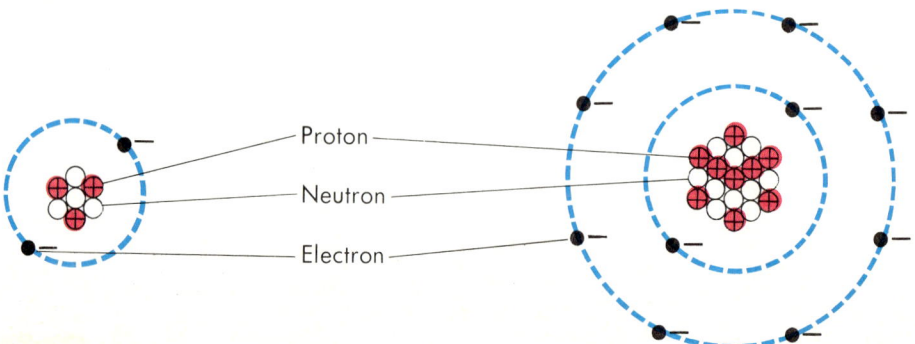

Positively charged lithium

Negatively charged fluorine

Atoms with unlike charges attract (pull on) each other. Atoms with like charges repel (push away from) each other.

Scientists know that any object, including an atom, can change its motion only if a force is applied. Therefore, an **electric force** must exist between all charged atoms. This electric force holds the atoms of matter together.

How Charge Builds Up

If a neutral object is rubbed against another neutral object, some of the electrons may move from object to object. An electric charge is produced on both objects.

You may feel the effects of electric charge when you comb your hair. Like the girl's hair in the first picture, your hair is usually neutral. In the second picture, the girl pulls the comb through her hair. The comb rubs against the hair and removes electrons. Now her hair has more protons than electrons, so its surface has a positive charge. The extra electrons give the comb a negative charge. Because unlike charges attract, the positively charged hair sticks to the negatively charged comb in the third picture. The positively charged hair moves toward the negatively charged comb even before they touch.

Building up charge

  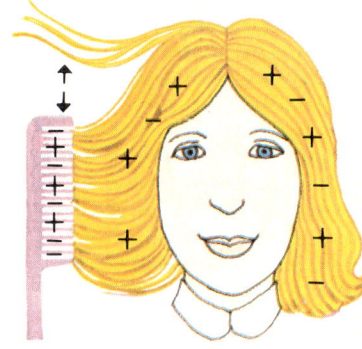

Because like charges repel, the hairs move away from each other. Charged objects do not have to touch for the electric force to exist between them.

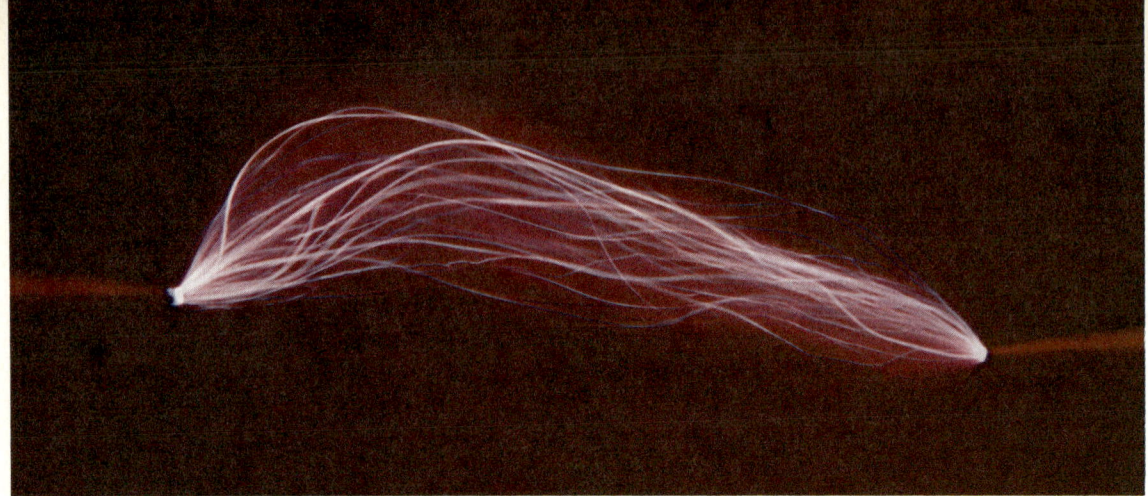

Have You Heard?

The saying "Lightning never strikes twice in the same place" is not true. In just one storm, the Empire State Building in New York City was struck by eight bolts of lightning in twenty-four minutes.

Challenge!

Find out how lightning rods protect your home by reading about them in reference books.

Sparks Are Moving Charges

Sometimes a lot of charge builds up on an object. If that charge jumps from one place to another, a spark is produced. The picture shows a spark made by charges jumping from one object to another.

Lightning is a big spark. Before and during a thunderstorm, some parts of clouds become positively charged. Other parts become negatively charged. Scientists do not know how this process happens. They do know, however, that the negative electrons move away from each other. When the electrons become too crowded on a cloud, they move suddenly to another part of the same cloud, to another cloud, or to the ground. When the electrons move, lightning results.

You may also notice sparks when you walk across a rug and touch a metal doorknob. Your shoes rub against the rug and pick up additional electrons. As the electrons repel each other, they spread out over your body. When you touch the metal doorknob, you feel a shock and may see a spark as the electrons jump from you to the doorknob.

Review It

1. What makes an atom neutral?
2. How does an object become charged?
3. Explain what can happen when you rub your feet on a carpet and then touch a doorknob.

Activity

Electric Charge

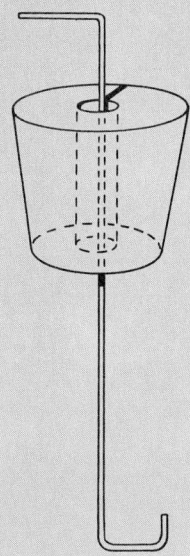

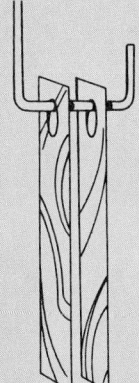

Purpose
To produce and observe the effects of electric charge.

Materials
- Erlenmeyer flask
- 1-hole cork or rubber stopper to fit flask
- 20 cm of copper wire
- two 3 cm × 1 cm aluminum foil strips
- plastic or rubber comb
- piece of wool

Procedure
1. Put the wire through the hole in the stopper.
2. Bend the wire's top end so it does not slide through the hole.
3. Bend the wire's bottom end into a hook. The wire should look like *a*.
4. Hang the foil strips loosely on the hook, as in *b*.
5. Insert the stopper into the flask. You have made an electroscope, which you can use to detect charge.
6. Rub the comb with the wool cloth about 25 times, or comb your hair about 25 times.
7. Bring the comb near the wire sticking out of the stopper.
8. Observe and record what happens to the foil strips.
9. Move the comb away from your electroscope.
10. Repeat step 8.
11. Repeat steps 6–7.
12. Touch the wire.
13. Repeat step 8.

Analysis
1. Use the idea of building up charge to explain your observations.
2. What caused the foil strips to move? (What causes a change in any object's motion?)

21–2
Electric Current—Charges on the Move

How is the electricity in your home similar to a spark or to the electricity in a flashlight battery? Sparks jump and home appliances work because electricity is moving charges. As you read about how charges move, think about the following:

a. Why does electric current flow more easily through some substances than through others?
b. How does a dry cell produce current?
c. How is current related to voltage and resistance?
d. How must current move?

Charges Move Through Conductors

Charges build up on some materials. But charges can also move through matter. The flow of electric charges is electric **current** (kėr′ənt). A material that allows electric current to pass through it easily is a good electrical **conductor.** Current moves through conductors because electrons easily escape from one atom of a conductor to another. Many metals, such as gold, silver, copper, and aluminum, are good conductors of electricity.

An **insulator** (in′sə lā′tər) is a material that does not allow electricity to flow through it easily. Insulators do not conduct well because their atoms hold electrons tightly. Rubber, plastic, and glass are good insulators.

The electric cords shown below carry electricity in copper wires. The insulator stops charges from flowing from one wire to another or to your hand.

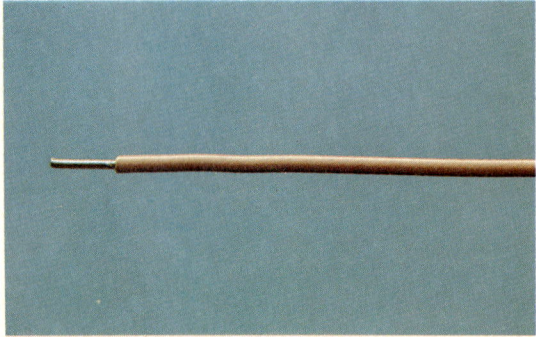

Even the best conductor slows the flow of electrons a little. Atoms in a conductor scatter electrons, which slows the electrons down slightly. Atoms in an insulator hold electrons so strongly that little or no current flows. Anything that slows or stops the flow of electrons is **resistance** (ri zis′təns). Insulators have high resistance. Conductors have low resistance. The **ohm** (ōm) is the unit for measuring resistance.

Have You Heard?

Resistance converts electricity into light and heat. Many everyday appliances apply this fact and use poor conductors. A thin wire in the light bulb glows because it resists the current passing through it. The wires in a toaster are also poor conductors. These wires get red-hot as they resist the passage of current. This heat toasts the bread.

How a Dry Cell Produces Electric Current

In the pictures below, the flow of water through the pipe is like the flow of electrons through wire. To keep the water flowing, something must push the water. A pump does work on the water by pushing it through the pipes. In the same way, the battery provides an electric "push." The battery does work on the charge by pushing it through the wire. The push the battery gives to each electron is the **voltage** (vōl′tij). The **volt** is the unit for measuring voltage.

Comparing a water system and an electrical system

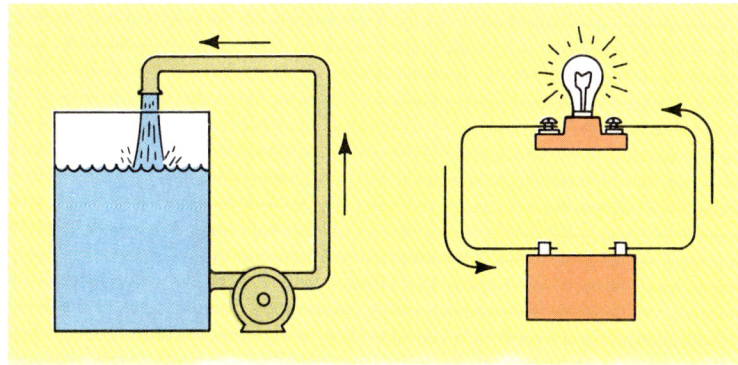

One source of voltage is a **dry cell,** such as the flashlight battery at the right. This dry cell is made of a carbon rod surrounded by an acid paste. A zinc case encloses the rod and the paste. A chemical reaction between the acid paste and the zinc releases electrons. The flow of electrons to the carbon rod makes an electric current. A typical flashlight battery operates at only 1.5 volts.

A dry cell

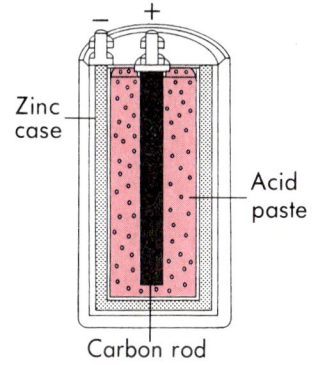

How Current Is Related to Resistance and Voltage

Resistance and voltage determine how much current moves through a conductor. To understand this idea, compare the pictures of the pipes. The only difference in the pipes is their diameters. The pump pushes water equally in both cases. But more water passes through the pipe with the large opening. The larger pipe offers less resistance to the flow of water than the smaller pipe. In a similar way, more current flows through a conductor with less resistance than through a conductor with more resistance.

The pictures below show how the amount of resistance and voltage affect the current. In the first picture, copper wires connect a bulb to a dry cell. The bulb glows brightly. If you replace one or both wires with a conductor of higher resistance, as in the second picture, the total resistance is raised. As a result, the amount of current going through the bulb decreases. The bulb dims. If a second dry cell is added, as in the third picture, the voltage increases. Now the current increases, and the bulb glows brightly again.

In the 1800s, Georg Ohm defined the relationship between current, voltage, and resistance. According to **Ohm's law,**

$$\text{current} = \frac{\text{voltage}}{\text{resistance}}.$$

Remember that voltage is measured in volts, and resistance is measured in ohms. Current is measured in **amperes** (am′pirz), the number of electrons moving past a point in a certain time. Ohm's law is important for designing any electric device, such as a toaster.

You can use Ohm's law to figure out how much current runs through a toaster. A wall outlet's voltage is about 110 volts, and the toaster's resistance is about 10 ohms. Using Ohm's law,

$$\text{current} = \frac{110 \text{ volts}}{10 \text{ ohms}}$$
$$= 11 \text{ amperes.}$$

This equation means that, to toast your bread, 110 volts "push" 11 amperes of current through a 10 ohm wire.

Current Moves in a Circuit

A current flows only when it can follow a complete path back to its starting point. Such a path is called a **circuit** (sėr′kit). If the path has a gap, the circuit is open and the current cannot flow. When the path has no gap, the circuit is closed and the current can flow.

A circuit has three basic parts, as the drawing shows. First, a source of voltage, such as a battery, pushes the electric charges. Second, some object serves a purpose when the current passes through it. This object may be a lamp or motor. Third, wires provide a complete, closed path from the battery to the object and back to the starting point.

Review It

1. Name three conductors and three insulators.
2. How does voltage affect current in a dry cell?
3. What is Ohm's law?
4. Describe the path of current through a circuit.

For Practice

Use Ohm's law to solve these problems.
• A toaster with a resistance of 10 ohms is plugged into a 220-volt electric line. How much current flows through the toaster?
• The two dry cells in a flashlight put out a total of 3 volts. What is the bulb's resistance if 0.5 ampere flows through the circuit?

A simple circuit

21-3
Two Kinds of Circuits

Some strings of holiday lights go out when one bulb burns out. Other strings of lights remain lit, even though a bulb burns out. The difference between the strings of lights is the kind of circuit connecting them. Read about the two kinds of circuits, and ask yourself:

a. What is a series circuit?
b. What is a parallel circuit?

Series Circuits Have One Path

A simple circuit is shown below. It uses a dry cell, a flashlight bulb, a socket, a switch, and a wire to form a complete path. The switch makes opening and closing the circuit easy.

You can add lights or appliances to a circuit in one of two ways. In one kind of circuit—a **series circuit**—all the parts are connected one after the other. Notice the connection between the two bulbs in the second picture. Both are in the same path with the other parts of the circuit.

Removing any part of a series circuit leaves a gap in the current's path. For instance, removing a bulb in the third picture opens the circuit. Because current cannot flow to reach the second bulb, it goes out too.

Current path through a series circuit

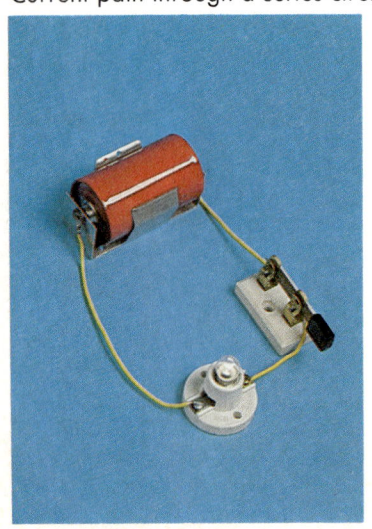
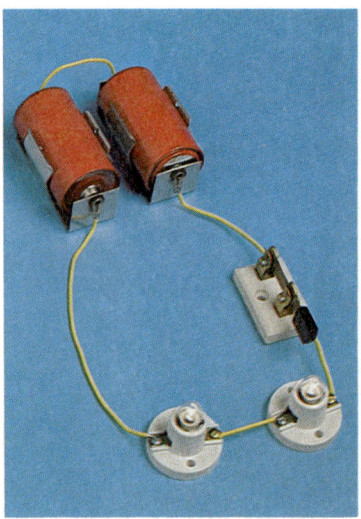
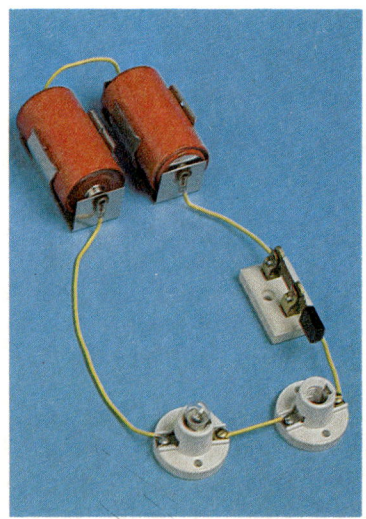

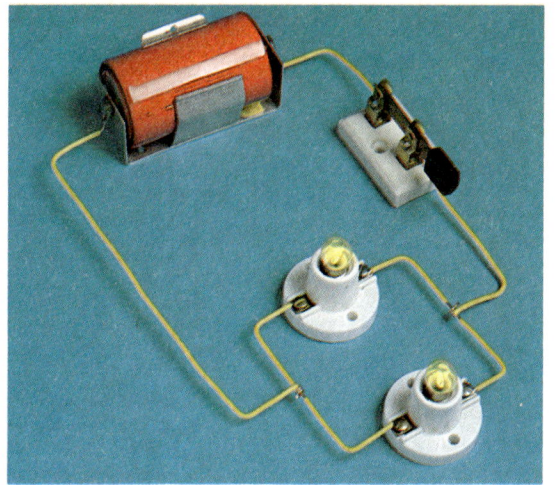
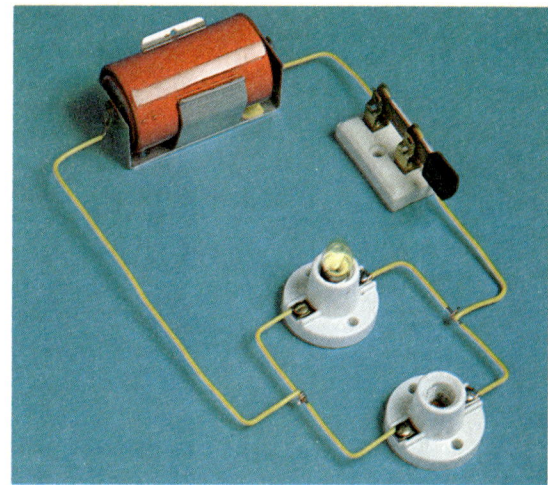

Current path through a parallel circuit

Parallel Circuits Are Practical

The circuit pictured above also contains two bulbs. This kind of circuit is a **parallel circuit** because a different path connects each appliance to and from the source of voltage.

One of the bulbs is unscrewed in the second picture. However, the current path has no gap. The current can take the path to the second bulb. The second bulb remains lit.

If homes were wired in series, all the lights and appliances would have to be on at the same time for any of them to work. Parallel circuits provide more than one path for the current. Your home is wired in parallel to allow you to use only the lights and appliances you need.

Review It

1. How are the parts of a series circuit connected?
2. Why are our homes wired in parallel?

21–4
Using Electricity

You flip a switch, and a lamp lights. You push a button, and a stereo plays. You turn a knob, and the television comes on. Consider these questions as you learn about the uses of electricity:

a. What are the two kinds of current we use?
b. How do we control and use electricity in our homes?

Two Kinds of Current

Current in a battery differs from current in your home. Batteries produce **direct current** (d.c.), which always flows in one direction. Current reaching your home—**alternating current** (a.c.)—changes direction many times each second.

Electric companies use a.c. instead of d.c. because a.c. is cheaper to produce and to send over long distances. A device that changes the voltage—a **transformer** (tran′sfôr′mər)—raises the voltage. This voltage travels farther before dying down.

The voltage of electric lines is too high for use in homes. Another transformer, like those shown, lowers the voltage. Large appliances, such as clothes dryers, use 220 volts. Small appliances, such as lamps, use 110 volts.

The electric company measures the power—the energy per time—you use in your home. The unit for power is the kilowatt (1,000 watts). For example, 100 kilowatts used for one hour equals 100 kilowatt-hours of electricity. Your electric bill would tell you the cost of this energy.

Electricity at Home

We must control electricity so that we use it only when we need it. A switch controls electricity by opening and closing the circuit. Other devices also open and close circuits to control electricity.

Fuses or **circuit breakers** open a circuit when too much current flows through it. If too much current flows through a wire, the wire may overheat and cause a fire. The pictures show fuses and circuit breakers. Too much current heats and melts a piece of metal in the fuse. The circuit opens, and the flow of current stops. A circuit breaker switches off automatically if too much current flows. In either case, you must find and correct the cause of the extra current. The cause may be that too many appliances are on one circuit. When the problem is corrected, you close the circuit by replacing the fuse with a new one or by flipping the switch to reset the circuit breaker. The circuit is closed, and current flows again.

Fuses

Making sure wires are properly insulated can protect our homes from fire. For example, the insulation on a wire can wear away. Then current can pass out of the original circuit and into something else. Instead of going through the circuit, the current takes a shorter path—a **short circuit.** The device cannot work because not enough current reaches it, and the current flowing through the short circuit might spark a fire.

Circuit breakers

Review It

1. Explain why a.c. is sometimes more useful than d.c.
2. Describe ways to prevent fires by controlling current.

395

Activity

Switches and Fuses

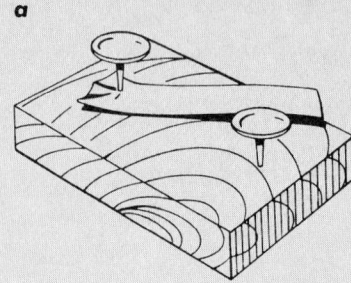

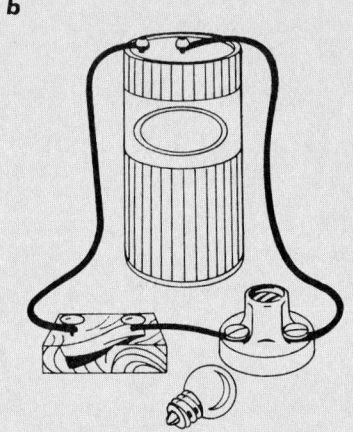

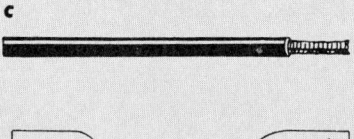

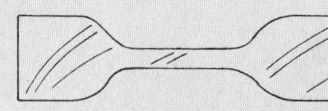

Purpose
To observe the effects of a simple switch and fuse.

Materials
- 1.5-volt dry cell
- flashlight bulb
- miniature socket
- block of wood
- 2 thumbtacks
- piece of aluminum foil, 10 cm × 2 cm
- three 15-cm pieces of insulated wire, with ends stripped
- scissors

Procedure

Part A
1. To make a switch, place the tacks and foil on the wood, as in *a*. Do not press the tacks all the way in.
2. Current leaves one dry cell terminal and returns to the other. Either both terminals are on the cell's top or one is on top and one is on the bottom. Connect a wire from a terminal to a tack.
3. Connect another wire to a socket terminal and the other dry cell terminal.
4. Connect the third wire to the other socket terminal and the other tack, as in *b*.
5. Screw the bulb into the socket.
6. *CAUTION: Do not touch any metal parts.* Place the loose foil end on the tack.
7. Record what happens.

Part B
1. Remove the bulb.
2. Remove the foil.
3. Cut the foil, as in *c*.
4. To make a fuse, tack the foil to the wood at both ends. Do not push the tacks all the way in. Make sure the wires still touch the tacks.
5. *CAUTION: Do not touch any metal parts.* Replace the bulb.
6. Record what happens.

Part C
1. Unscrew the bulb.
2. *CAUTION: Do not touch any metal parts.* Remove both wires from the socket.
3. Touch the stripped ends to each other.
4. Record what happens.

Analysis
1. How did the switch make the light go on?
2. How did the foil fuse protect the circuit?

Breakthrough

Superconductivity

Electricity travels more easily in cold metal than in the same metal when warm. The Dutch scientist Heike Kamerlingh Onnes studied this property seventy years ago. He cooled mercury to a very low temperature. Suddeny, the mercury's resistance to the flow of electricity disappeared entirely. The mercury could conduct an electric current without any loss of energy to resistance. The metal had become a superconductor of electricity.

The possible uses of this discovery were breathtaking. Only a very small amount of energy would be needed to supercool the metal. The wires would offer no resistance to the flow of current. So additional energy would not be needed to boost the voltage over long distances.

Electric power companies around the world are using superconducting materials in different ways. They are constructing supercooled machines and power cables that will provide cheaper electric power in the future.

Computer manufacturers have devised yet another way of using superconductors. They are studying supercooled circuits called Josephson junctions, shown in the photograph. Due to their lack of electrical resistance, these devices conduct electrical messages very quickly. They are reliable and require very little energy. These characteristics are used to design high-speed computers with the ability to store huge amounts of information.

Each year brings more and more advances in the uses of superconductivity. These advances should improve living conditions in a world looking to conserve energy.

For Discussion
1. How does a lack of resistance make superconductors more efficient than other conductors?
2. What are some possible uses of superconductors?

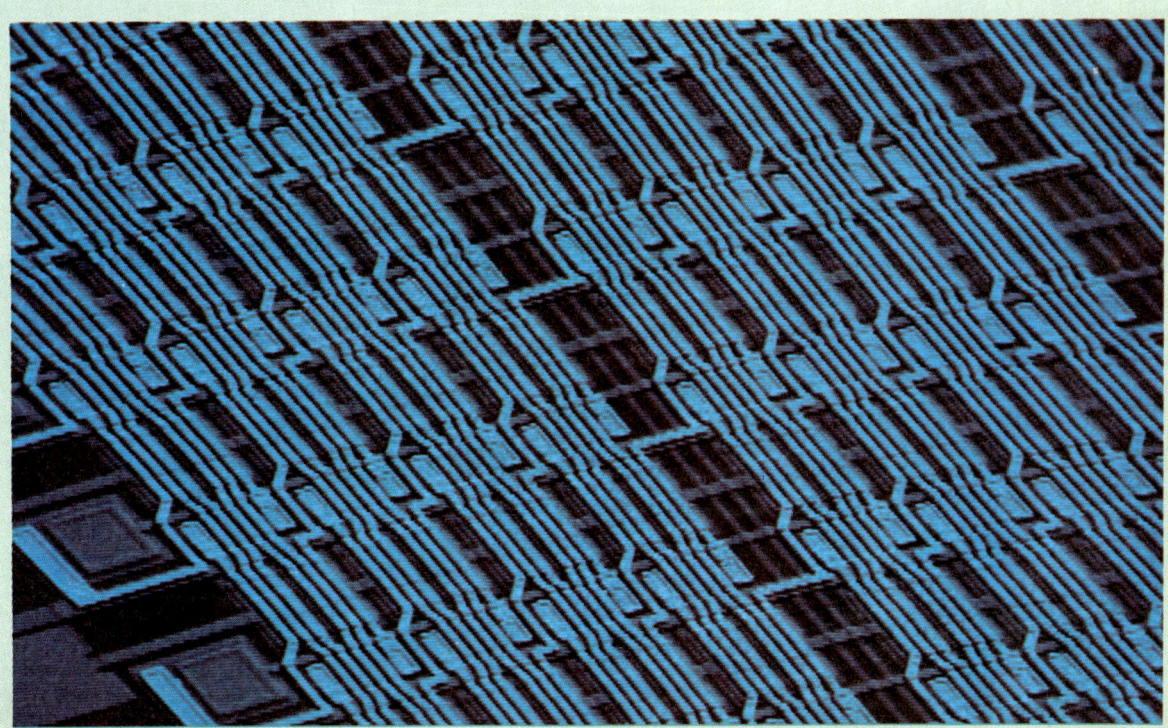

Chapter Summary

- Atoms become electrically charged when they gain or lose electrons. (21–1)
- Like charges repel. Unlike charges attract. (21–1)
- A spark occurs when electric charge jumps from one place to another. (21–1)
- Moving charges are electric current. (21–2)
- Some materials resist current flow more than other materials. (21–2)
- A source of voltage pushes charges. (21–2)
- Ohm's law relates current to resistance and voltage: current = voltage ÷ resistance. (21–2)
- Current must move through a circuit. (21–2)
- The two kinds of circuits are parallel and series circuits. (21–3)
- Alternating current is used more often than direct current. (21–4)
- Switches, fuses, and circuit breakers control current in circuits. (21–4)

Interesting Reading

Asimov, Isaac. *How Did We Find Out About Electricity?* Walker, 1973. Traces the discovery and uses of electricity from ancient Greece to the 19th century.

Epstein, Sam, and Epstein, Beryl. *The First Book of Electricity*. Watts, 1977. Describes the everyday uses of electricity, from TV to toasters.

Liebers, Arthur. *You Can Be an Electrician*. Lothrop, 1974. Tells what electricians do and describes the training needed to become one.

Questions/Problems

1. If you rub a balloon on your hair or sweater and touch the balloon to a wall, it will stay there. Explain how.
2. Explain how doubling the number of batteries in a circuit increases the current. The resistance remains the same.
3. If 10 amperes of current flow through your home wiring, use Ohm's law to determine the overall resistance of all the appliances and wires in your home's 110-volt circuit.
4. Explain why the flow of current needs a complete circuit.
5. Discuss why series circuits would be unacceptable for household wiring.
6. How can you prove that your home is wired in parallel?
7. Explain why it is dangerous to use damaged wiring in your home.

Extra Research

1. Look on the bottoms or backs of appliances in your home. List those that use the most current.
2. Ask an adult to help you find the circuit breaker cabinet or the fuse box in your home. Note how many different circuits your home has.
3. Charge a plastic comb by rubbing it with wool cloth. Then slowly move the comb near a small, steady stream of water from a faucet. Explain what happens.
4. Take apart a flashlight. Draw the parts of its circuit. Mark the path of the current flowing through this circuit. (The circuit may be built into the case.)

Chapter Test

A. Vocabulary Write the numbers 1–10 on a piece of paper.
Match the definition in Column I with the term it defines in Column II.

Column I

1. current supplied to homes
2. allows the use of more than one voltage in a house
3. more useful kind of wiring
4. kind of circuit in which current stops flowing when any part of the circuit is turned off or removed
5. lessens the flow of electric current in a conductor
6. kind of current that a dry cell produces
7. unit used to measure resistance
8. substance through which electrons easily pass
9. substance with high resistance
10. unit used to measure amount of current

Column II

a. alternating current
b. ampere
c. conductor
d. direct current
e. insulator
f. ohm
g. parallel circuit
h. resistance
i. series circuit
j. transformer

B. Multiple Choice Write the numbers 1–10 on your paper.
Choose the letter that best completes the statement or answers the question.

1. An object with the same amount of negative charge as positive charge is a) neutral. b) opposite. c) unlike charged. d) like charged.

2. A charged object will cause another charged object to a) be attracted. b) be repelled. c) be attracted or repelled. d) remain unchanged.

3. The push given to a charge by a battery is measured in a) amperes. b) fuses. c) volts. d) ohms.

4. A pump that pushes water through a pipe is similar in function to a a) dry cell. b) light bulb. c) toaster. d) wire.

5. Current follows a complete, closed path in a(n) a) wire. b) insulator. c) fuse. d) circuit.

6. If a gap develops in the path of an electric charge, the circuit is a) open. b) fused. c) closed. d) short.

7. A device that opens a circuit in case of overheating is a(n) a) ampere. b) ohm. c) dry cell. d) fuse.

8. The kilowatt measures a) energy. b) power. c) current. d) resistance.

9. When four light bulbs are wired in series, if one goes out a) the second bulb goes out, but the other two bulbs continue to shine. b) all the remaining bulbs go out. c) the three remaining bulbs continue to shine. d) the remaining bulbs flicker on and off.

10. When four light bulbs are wired in parallel, if one goes out a) the second bulb goes out, but the other two bulbs continue to shine. b) all the remaining bulbs go out. c) the three remaining bulbs continue to shine. d) the remaining bulbs flicker on and off.

Chapter 22
Magnetism

The beautiful scene in the picture is not a fireworks display. It is a natural light show that occurs over the very northern and southern parts of the earth. These lights brighten the night sky with their dazzling colors and patterns. Both the earth and the sun contribute to this effect.

The northern or southern lights are only one result of the force which is the subject of this chapter. Your study will begin with properties of magnets. It will continue with how objects become magnets and will end with how electricity and the magnetic force affect each other.

Chapter Objectives

1. List the properties of magnetic fields and poles.
2. Explain how the atoms of magnetic substances behave.
3. Describe the properties of permanent magnets.
4. Explain how electromagnets work and list some devices that use them.
5. Explain how magnetism and electricity are related.

22–1
Magnetic Properties

The ancient Greeks were familiar with a hard black stone that attracted iron. Much later, the English named this stone "lodestone," because it "led" travelers in the direction of the North Star. Today we know that this stone is magnetite, the magnetized ore of iron. Keep reading to learn more about magnets, and ask yourself:

a. How do magnetic poles behave?
b. What is a magnetic field?
c. What does earth's magnetic field resemble?

A Magnet Has Two Poles

Magnets are objects that attract objects made of iron and a few other elements by a force called **magnetism.** Only a few elements can take on permanent magnetism. Iron does not always attract other iron, but when it does, we say it is magnetic. Nickel, cobalt, and gadolinium are other elements that can be made into permanent magnets. The bar magnet shown at the left picks up steel paper clips because steel contains iron.

Notice that more paper clips cling to the magnet's ends than to its middle. Magnetism is strongest at a magnet's **poles.** The ends of a magnet are usually its poles.

Even though magnets come in many shapes and sizes, every magnet has at least two poles. If you cut a magnet into pieces, each piece will still have at least two poles.

A magnet will turn so that the same pole always points north. You can observe this fact by letting a bar magnet turn freely. If you hang the magnet from a string tied around its middle, the magnet will turn until one end points north. This end is the magnet's **north pole.** The end that points south is its **south pole.** If you move the magnet, its north pole will move until it again points north.

When people discovered that a magnet always points north, they used magnets to find directions. Over 900 years ago, sailors used a small, free-swinging magnet—or **compass**—to navigate. Since the magnet pointed north, sailors knew in which direction they were sailing. Navigators on planes and ships still use compasses.

A north pole of one magnet and a south pole of another magnet attract each other. Two north poles push each other away. Also, two south poles push each other away. As with electric charges, like poles repel and unlike poles attract. Because they attract and repel, we know that a force acts between the poles.

A Magnetic Field Surrounds Every Magnet

Iron filings arrange themselves around a magnet as shown at the right. The filings form a pattern around the magnet. Yet many filings do not touch the magnet. The region around a magnet where its magnetism acts is the **magnetic field.** Notice that the filings are densest at the magnet's poles, where the magnetic field is strongest.

When you place like poles near each other, the filings line up differently. The right picture shows how the magnetic field acts. When you place unlike poles near each other, the filings line up in still another way. The picture below shows how the field acts in this situation.

The magnetic field between unlike poles

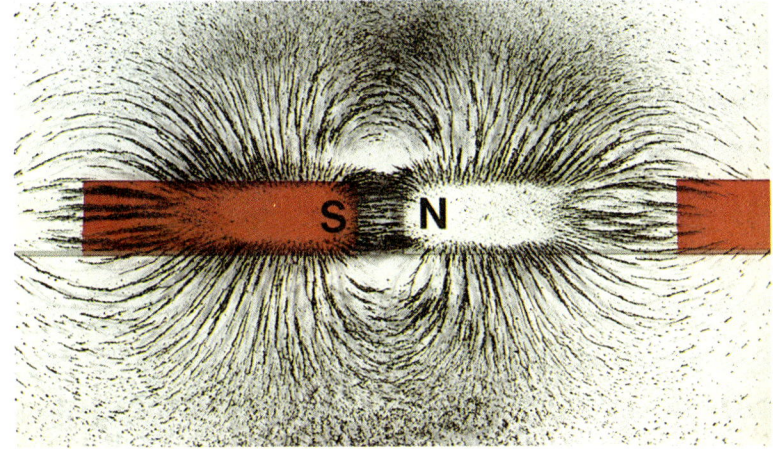

Challenge!

A navigator must be able to tell the difference between geographic north and magnetic north to plot a ship's course. Look in an encyclopedia or a book about navigation to find out how.

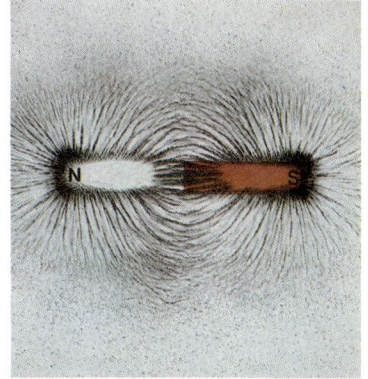

The magnetic field around a magnet

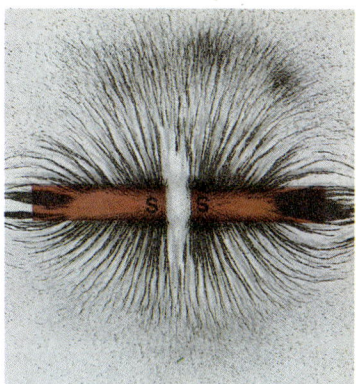

The magnetic field between like poles

403

Have You Heard?

Spacecraft have detected magnetic fields around Mercury, Jupiter, and Saturn. Studying these magnetic fields, and why Mars and Venus have such weak fields, helps scientists learn more about Earth's magnetic field.

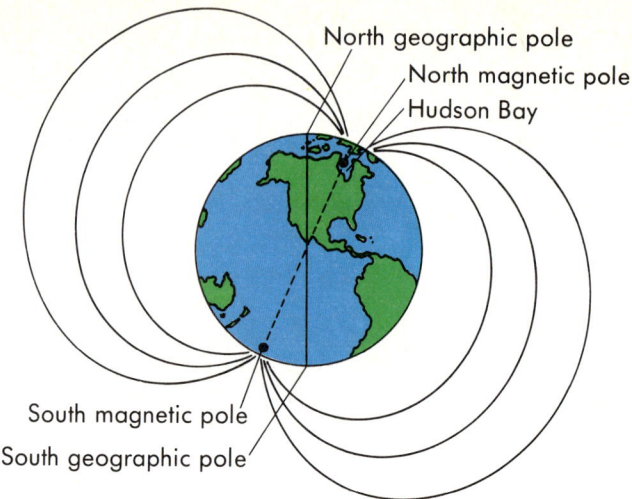

Earth's magnetic field

Earth's Magnetism

A magnetic field surrounds the earth. This field acts much like the field of a bar magnet. One proof that the earth has a magnetic field is the beautiful display of lights at night, called the aurora. Charged particles from the sun enter earth's magnetic field and hit molecules in the atmosphere. The molecules then radiate light. Auroras such as the one shown are most visible near the poles because the magnetic field guides the particles toward the poles.

Scientists think that the earth's spinning and the movement of molten iron inside the earth cause the earth's magnetic field. The "bar magnet" inside the earth lines up approximately with earth's axis of rotation. The north geographic pole does not coincide with the north magnetic pole. The same is true for the south geographic and magnetic poles. The drawing at the top shows the difference between the positions of the poles. A compass points to magnetic north, not geographic north.

Review It

1. Where is a magnet's force strongest?
2. Where does a magnetic field exist?
3. How do we know that the earth has a magnetic field?

Activity

Magnetic Poles

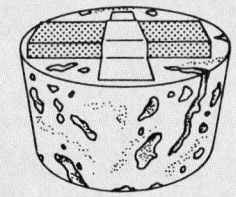
a

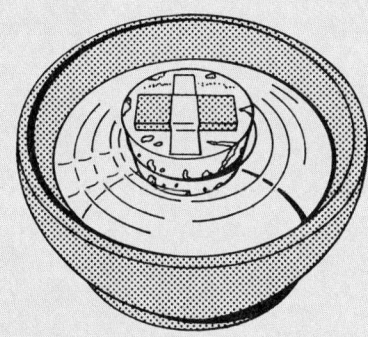
b

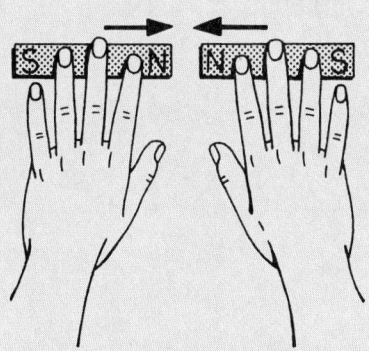
c

Purpose
To demonstrate some properties of magnets.

Materials
- 2 small bar magnets
- 2 wide corks
- bowl of water
- masking tape

Procedure

Part A
1. Mount one bar magnet on top of each cork with masking tape, as in a.
2. Float one of the corks with the magnet in the bowl of water. Keep the magnet out of the water, as in b.
3. Note which end of the magnet points north. If you do not know which way is north, ask your teacher.
4. Take the magnet and cork out of the water. Stick a piece of masking tape on this end and mark it with an "N" for "north pole."
5. Mark a piece of masking tape on the other end with an "S" for "south pole."
6. Put the magnet and cork back in the water pointing in a different direction.
7. Allow them to come to rest and note which way they point.
8. Write down whether the north-south directions in steps 4 and 8 are the same or different.
9. Repeat steps 2–5 for the second magnet and cork.

Part B
1. Remove the bar magnets from the corks, and place them flat on a desk with the marked sides up.
2. Move the north pole of one magnet next to the north pole of the other, as in c.
3. Record what happens.
4. Move the south pole of one magnet next to the south pole of the other.
5. Record what happens.
6. Move the north pole of one magnet next to the south pole of the other.
7. Record what happens.

Analysis
1. How do the results of Part A make a compass useful?
2. How do like poles act? unlike poles?

22–2 Magnets from Magnetic Substances

Why can some substances be magnetized while others cannot? Perhaps something inside a magnet causes its behavior. To learn why some substances can be magnetized, think about:

a. What are magnetic domains?
b. What kinds of substances become permanent magnets?

Magnetic Domains Line Up

Every atom has a bit of magnetism. Usually, though, their magnetism points in many different directions when many atoms are present. Their magnetism balances out. In magnets, however, the magnetism of all the atoms lines up in a region called a magnetic **domain.** Each domain contains billions of atoms. Because atoms are tiny, the size of a domain is small. A domain is only about a millionth of a meter across. Each domain acts like a little magnet inside the substance.

When you place certain substances in a strong magnetic field, the domains line up and the substance becomes a magnet. We know that a sewing needle can be made into a magnet, because a magnet attracts it. However, the needle does not start out as a magnet because it does not attract other needles. Its domains line up in many different directions, as shown. Placing the needle in a magnetic field makes most of the domains line up in the field's direction. The photograph shows the domains in a magnet.

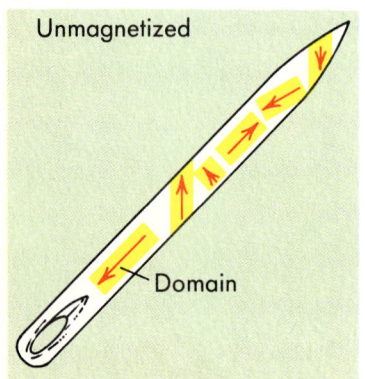
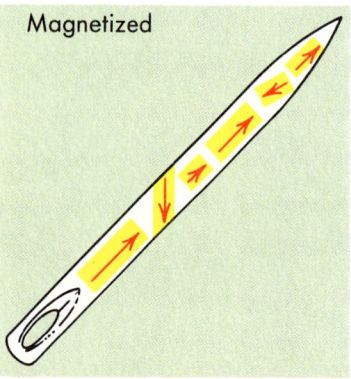
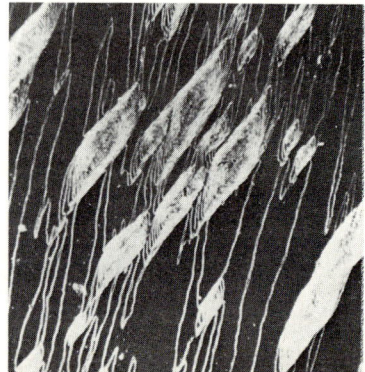

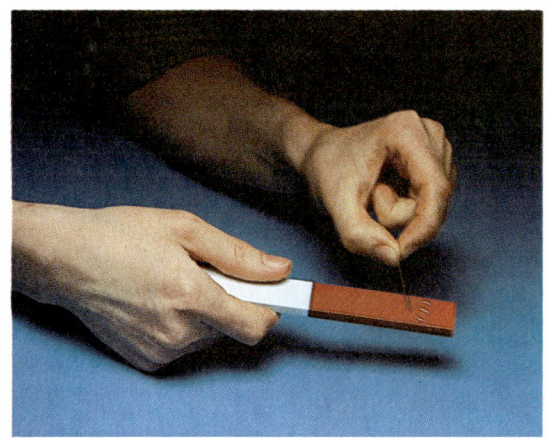 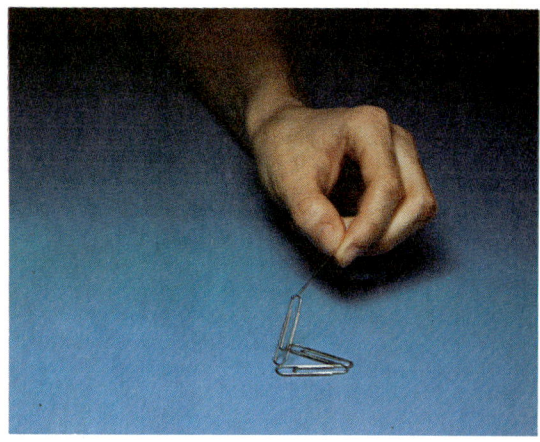

Permanent Magnets

Even when you remove the outside magnetic field from certain substances, some of the domains remain lined up. The object is a permanent magnet. These magnets are made of substances that keep their magnetism for a long time. Most bar magnets are permanent magnets made of steel.

The pictures show how to make a steel sewing needle into a permanent magnet. Rub a bar magnet many times along the needle in one direction. The magnet's field makes the steel's domains line up. When the field is removed, the needle is still a permanent magnet. The second picture shows that the needle now picks up paper clips.

A strong magnetic field is needed to force domains to line up. Once these domains line up, they stay in place under normal circumstances. You can force the domains out of line by heating or striking the magnet. Then its atoms move enough to move the domains out of line. The substance is no longer magnetized. Even a permanent magnet does not stay magnetized forever. After many years, it loses its magnetism as the domains slip out of line.

Review It

1. How do magnetic domains act in a magnet?
2. How are the domains of permanent magnets special?

22–3
Changing Electricity into Magnetism

People knew about magnets long before they understood them. How magnets work is based on discoveries made less than two hundred years ago. As you read about these discoveries and how they help you, consider:

a. How does electricity affect magnetism?
b. How does an electromagnet work?
c. How does an electromagnet help a motor work?

Electric Currents Cause Magnetism

In the 1820s, scientists showed that magnetism is closely related to electricity. The picture demonstrates that electric current produces a magnetic field. Current starts through a wire coil. Nearby, a compass needle suddenly turns away from earth's magnetic pole and points to the coil. The coil is now a magnet with a stronger field than earth's. The poles reverse if the current flows in the opposite direction. When the current stops, the coil loses its magnetism.

We now know that electricity causes all magnetism. Inside an atom, each electron moves and so gives the atom a magnetic field. These tiny magnetic fields are the ones that line up to make domains.

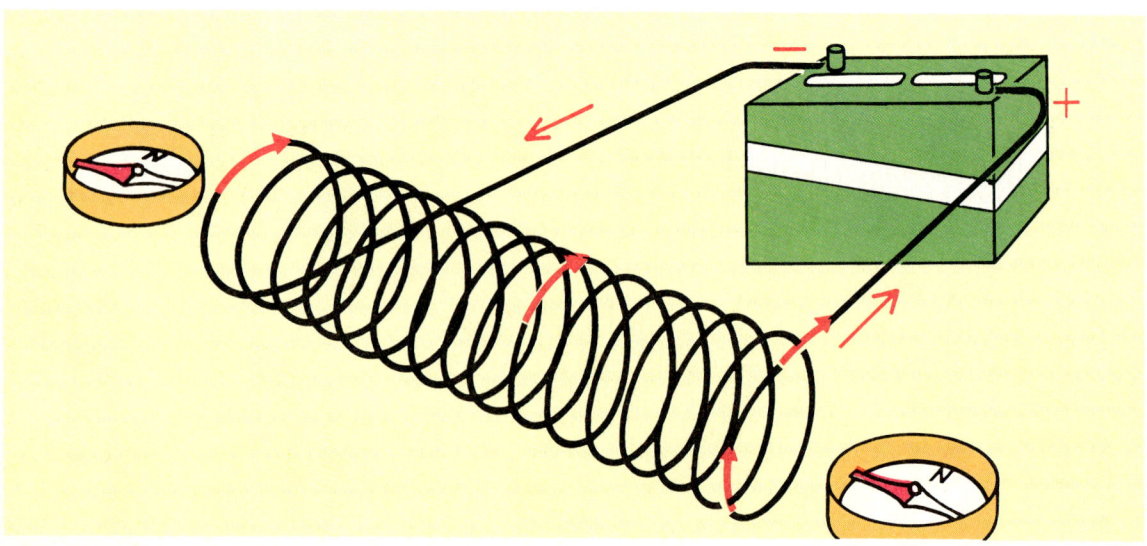

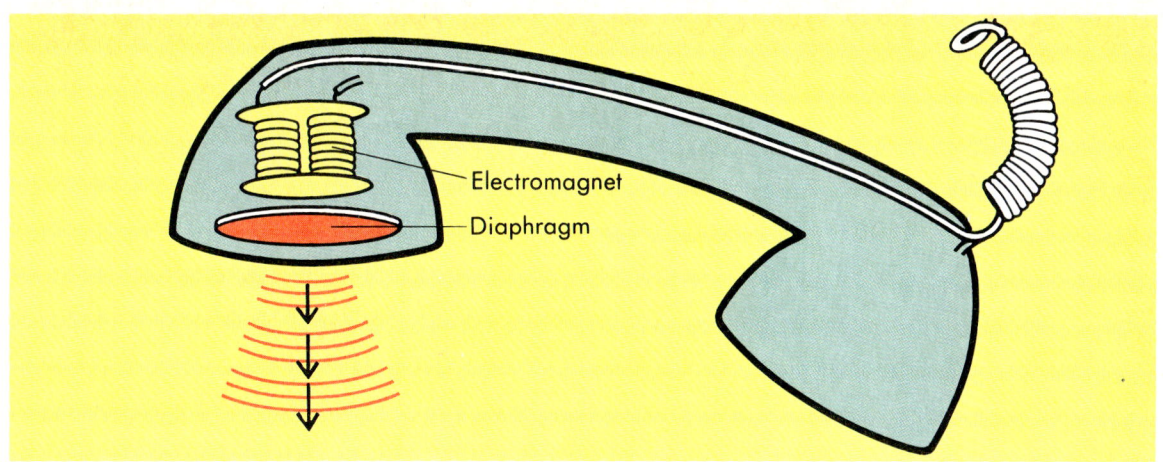

Using Electricity to Cause Magnetism

The discovery that electric current causes magnetism helps us every day. Every time you listen to a record or tape, dry your hair with an electric dryer, ring a doorbell, or talk on a telephone, you use electricity and magnetism. These devices contain an **electromagnet,** which is a coil magnetized by an electric current passing through it. Electromagnets are very useful because their magnetism can be switched on and off.

All you need to make an electromagnet is a wire coil and an electric current. The coil does not have to be magnetic when the current is switched off. The picture above shows such an electromagnet in your telephone. The crane at the right uses a strong electromagnet to move metal scrap. When the current is on, the electromagnet attracts the metal, and the crane lifts it. Turning the current off makes the electromagnet lose its magnetism. The crane drops the scrap.

Strong electromagnets are made by putting iron or another substance that can be magnetized in the center of the coil. The coil magnetizes the iron, and the fields from the coil and the iron are combined. You can also strengthen an electromagnet by increasing the number of turns of wire in the coil or the current in the coil. All these methods increase the coil's magnetic field. Therefore, the coil attracts more or larger objects.

Hair dryer motor

Toy car motor

Electric Motors Use Electromagnets

An electric **motor** uses electromagnetism to change electric energy into mechanical energy. Then the mechanical energy can be used to perform work. Electric motors power many kinds of machinery, as shown at the left.

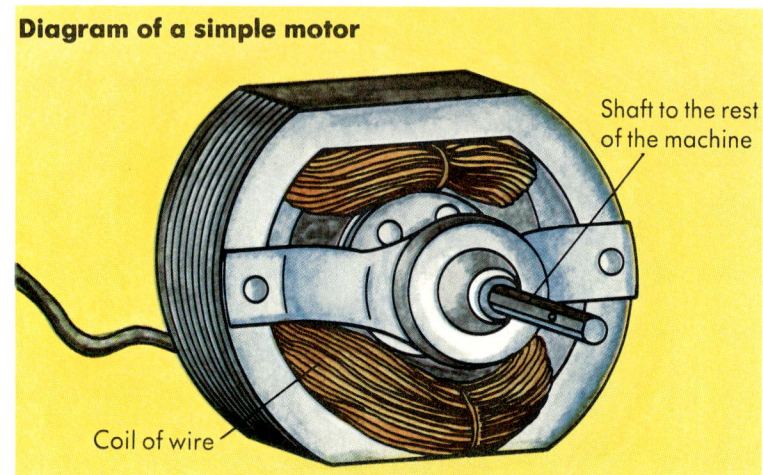

Diagram of a simple motor

Shaft to the rest of the machine

Coil of wire

The diagram above shows a simple motor in a toy racing car. A d.c. current passes through a coil, making the coil an electromagnet. The poles of the coil are attracted to the opposite poles of a movable magnet. The coil is connected to the current so that the current the coil receives reverses just before the opposite poles of the two magnets line up. Reversing the current reverses the poles of the coil. Then the poles can repel the poles of the other magnet. Again the movable magnet turns until its two poles find the opposite poles of the magnet in the coil.

The shaft is connected to the axle in the car. It turns the axle to move the car.

Review It

1. How do you show that electric current causes magnetism?
2. What happens when you change the direction of current in an electromagnet?
3. What does an electromagnet do in a motor?

Clock motor

Activity

Electromagnets

Purpose
To observe an electromagnet's behavior.

Materials
- two 1.5-volt dry cells
- insulated wire, with stripped ends
- switch
- large nail
- small metal objects

Procedure

Part A
1. Wind ten turns of wire around the large nail. Do not overlap them.
2. Connect one dry cell terminal to one switch terminal with a wire.
3. Connect one end of the coil's wires to the dry cell and the other end to the switch, as in a. NOTE: The current in this circuit is too weak to hurt you.
4. Close the switch and try to pick up some small metal objects with the large nail, as in b. Record how many you pick up.
5. Open the switch. Observe and record what happens to the objects.

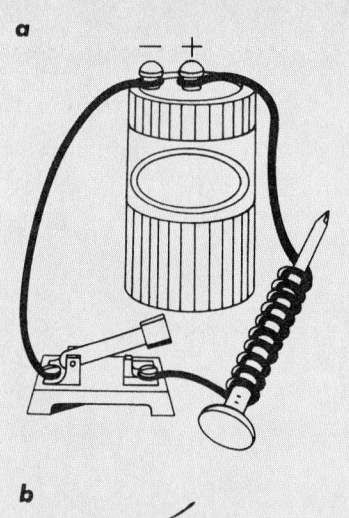

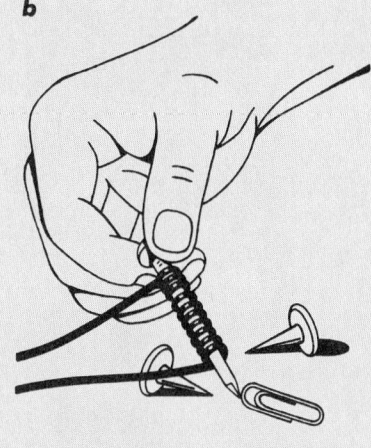

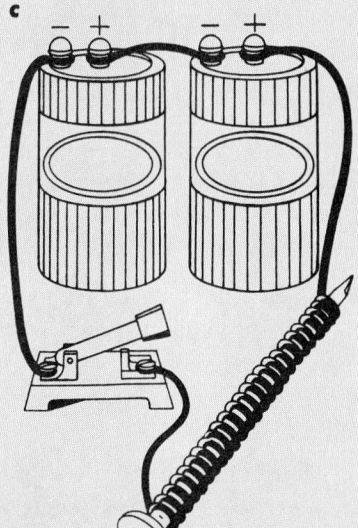

Part B
1. Repeat steps 1–5 with 25 turns of wire around the nail.
2. Write down how many objects you pick up and which is greater, the number in Part A or B.
3. Open the switch and disconnect the wires.

Part C
1. Connect both dry cells in series, by attaching a wire from one center terminal to the end terminal on the other cell.
2. Connect a second wire from the other terminal of a cell to the switch, as in c.
3. Connect one end of the electromagnet from Part B to the remaining cell terminal. Connect the other end to the switch. NOTE: The current in this circuit is too weak to hurt you.
4. Close the switch, and try to pick up nails or clips.
5. Record how many objects you pick up and which is greatest, the number in Part A, B, or C.
6. Open the switch, disconnect the wires, and unwind your electromagnet.

Analysis
1. What does the switch do to the electromagnet?
2. How did you strengthen the electromagnet?

22–4
Changing Magnetism into Electricity

In 1831, Joseph Henry and Michael Faraday found that moving a magnet through a wire coil caused a current to flow. When the Prime Minister asked Faraday what good this discovery was, Faraday answered, "Someday, Sir, you may tax it." He realized that it often takes years before we make practical use of basic discoveries. Now we use this discovery to produce electricity for our homes and factories. To understand how, ask yourself:

a. What proves that magnetism can produce an electric current?
b. How does a generator make electricity?

The instrument shown below measures the strength of electric current. You expect it to read "zero" when no current is present in its circuit. Yet in the first picture, the pointer moves when a magnet moves into the coil. Current must be flowing in the wire.

The magnet moves out of the coil in the second picture. The pointer moves again, but to the opposite side of the scale. Current is flowing in the wire, but in the opposite direction.

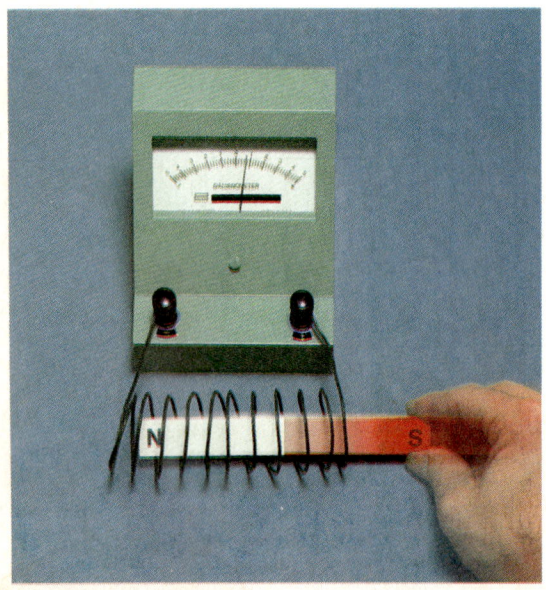
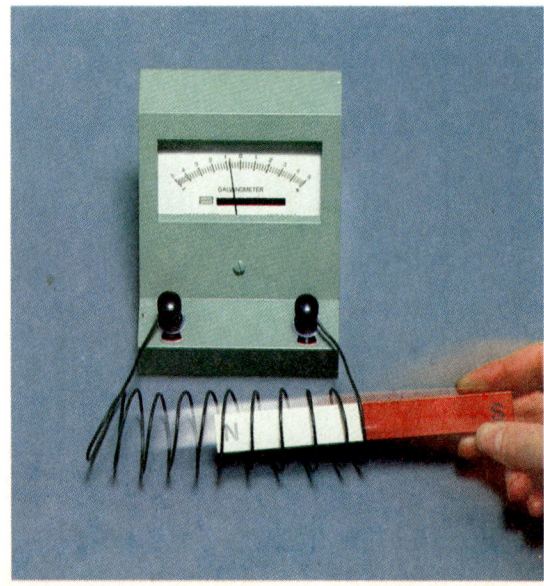

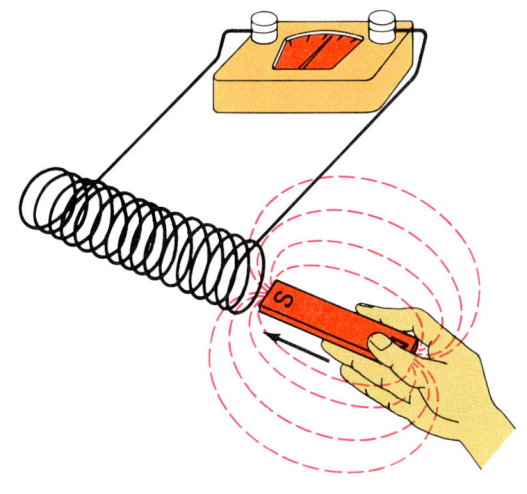
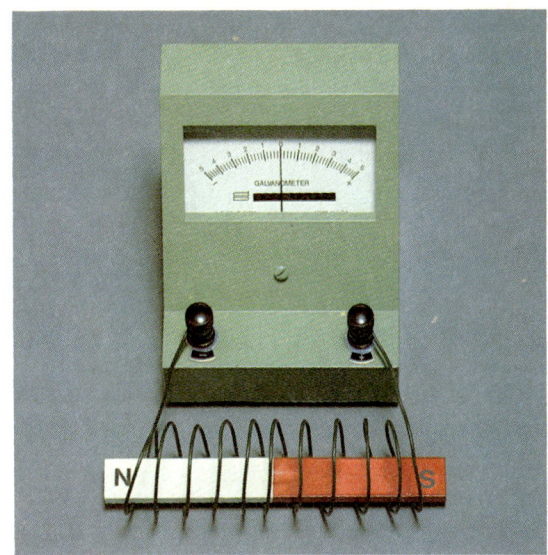

Current also flows in the wire if the coil moves instead of the magnet. Moving the magnet faster or increasing the number of coils will also increase the current. The process by which magnetism causes electric current is **electromagnetic induction.**

Electromagnetic induction occurs when a changing magnetic field and a coil are near each other. If the magnetic field that goes through the coil changes, a current flows in the coil. The field might be changing because either the magnet causing the field or the coil is moving. Or if the magnet is an electromagnet, its field might be changing because the amount of current flowing into it is changing. The last picture shows that a magnet whose field is not changing does not cause a current.

Since electricity can cause magnetism and magnetism can cause electricity, scientists link these two forces in a single theory. **Electromagnetism** is the study of the effects of electricity and magnetism. According to this theory, electricity and magnetism result from a single force.

The theory predicted that waves of electricity and magnetism would travel through space. These waves were discovered 100 years ago. Today we call them radio waves.

Have You Heard?

Scientists are trying to find a single theory that explains all the forces of nature. A theory now combines the force that causes electricity and magnetism with one of the forces inside an atom's nucleus.

Electric Generators Produce Electricity

An electric **generator** (jen′ə rā/tər) uses electromagnetic induction to produce electricity. Because generators change mechanical energy into electric energy, a generator and a motor perform opposite tasks. A motor uses current to produce motion, while a generator uses motion to produce current. A source of mechanical energy, such as a steam or fuel-burning engine, a water wheel, or a hand crank, must drive the generator. To do so, these mechanical devices move the magnet or coil. Most electricity is made this way, as the photographs show.

Diagram of a simple generator

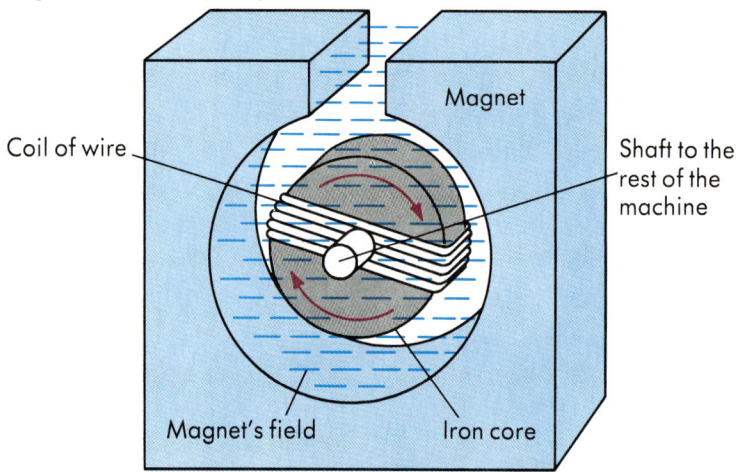

The diagram above shows how a bicycle generator works. As the back tire spins, it turns a wheel on the outside of the generator. The wheel turns a wire loop between the poles of a magnet. As the loop turns, it cuts through the lines of the magnet's field, which makes current flow in the loop. Wires carry the current to the lamp to light the bulb.

Review It

1. What is electromagnetic induction?
2. Why must an outside source of energy drive a generator?

Some examples of generators

Breakthrough

Supertrain

Imagine this scene. Late at night, the lights at the train crossing are flashing as you wait to pass. The train silently appears, sweeps by with a swish of air, and is gone before you can draw another breath. The usual roar of the engine, shaking of the ground, and clanking of the steel wheels are absent. People sleeping in nearby houses are unaware that a train with two hundred passengers has come and gone.

Is this science fiction? Can there be a train with no wheels? Does the train move by magic? No, the train moves silently using special kinds of magnets and electric currents. The magnets are small but powerful.

The magnets are super-cooled by refrigeration units on the train. These cold magnets are called superconducting magnets. They will stay powerful without a constant source of electricity.

An electric current turns the track into an electromagnet. The magnets in the train repel the magnets in the track. So the train rises a short distance off the rails. Then the train rides the moving electromagnetic wave as if it were an "electromagnetic surf." The faster the wave, the faster the train will go. At about one and one-third its highest speed, the train will rise

above the track about 10 centimeters. Such trains should reach speeds of 250 kilometers per hour. Imagine traveling quietly and comfortably from New York to Chicago in three hours. Our present kind of train makes this trip in about 15 hours.

Electromagnetic trains will ease environmental problems in several ways. In addition to making far less noise, they will not pollute the air with smoke. In big cities, fewer people might drive their cars if they could ride these trains. In this way, we can save oil and reduce pollution.

The chances are very good that you will be able to ride such a train. Experimental trains powered by superconducting magnets have already been tested successfully in a few countries.

For Discussion
1. What does an electromagnetic train use in place of wheels?
2. How are electromagnetic trains easier on the environment than our present trains?

Chapter Summary

- Like magnetic poles repel, and unlike poles attract. (22–1)
- Magnetism is the force between magnetic objects. (22–1)
- The magnetic field of a magnet is strongest at the poles. (22–1)
- Atoms in magnetic substances gather into domains. (22–2)
- An electric current produces a magnetic field. (22–3)
- Moving electrons in atoms cause magnetic fields. (22–3)
- An electromagnet is a magnet made of a current flowing through a wire coil. (22–3)
- A motor uses an electromagnet to perform work. (22–3)
- Electromagnetic induction is a changing magnetic field causing an electric current. (22–4)
- Electromagnetism is the study of the effects of electricity and magnetism. (22–4)
- Electromagnetic induction in a generator produces electricity. (22–4)

Interesting Reading

Kentzer, Michal. *Collins Young Scientist's Book of Power.* Silver Burdett, 1979. Discusses the roles of electricity and magnetism in changing man's style of life and the environment in general.

Math, Irwin. *Wires and Watts.* Scribner, 1981. Explains how to build your own models to demonstrate some principles of electricity and magnetism to yourself.

Renner, Al G. *How to Make and Use Electric Motors.* Putnam, 1974. Gives clear diagrams and instructions for making three kinds of battery-powered electromagnetic motors.

Questions/Problems

1. If you break a bar magnet in two, what magnetic properties do the halves show?
2. Compare the magnetic fields of a bar magnet and the earth.
3. How can you magnetize a steel screwdriver?
4. Why do paper clips hanging from a magnet pick up other paper clips? What would happen if you carefully removed the magnet?
5. How are electromagnets like permanent magnets? How are they different?
6. If a magnet is inside a wire coil, how can the magnet or coil cause a current in the wire?
7. Explain how electricity and magnetism affect each other.
8. Compare and contrast how electric motors and electric generators work and change one kind of energy into another.

Extra Research

1. Use reference books to find out what Michael Faraday, William Gilbert, Hans Christian Oersted, and André Ampère contributed to the study of magnetism.
2. Visit an electric power plant to find out how electricity is generated. Then make a report to your class.
3. Read in an encyclopedia about how a transformer uses electromagnetic induction.
4. Use an encyclopedia to learn how electromagnetism is used to record information on tapes in tape recorders.

Chapter Test

A. Vocabulary   Write the numbers 1–10 on a piece of paper.
Match the definition in Column I with the term it defines in Column II.

Column I

1. region around a magnet where its force acts
2. group of atoms in magnetic substances
3. device that converts electric energy into mechanical energy
4. magnet made of a coil with a current flowing in it
5. theory that explains both electricity and magnetism
6. device that converts mechanical energy into electric energy
7. force between magnetic objects
8. device used to find north
9. process by which magnetism causes electricity
10. place on a magnet where its magnetism is strongest

Column II

a. compass
b. domain
c. electromagnet
d. electromagnetic induction
e. electromagnetism
f. generator
g. magnetic field
h. magnetic pole
i. magnetism
j. motor

B. Multiple Choice   Write the numbers 1–10 on your paper.
Choose the letter that best completes the statement or answers the question.

1. All magnets   a) are surrounded by a magnetic field.   b) have two south poles.   c) are shaped like a bar.   d) always retain their magnetism.

2. A compass points to   a) magnetic north.   b) geographic north.   c) geographic east.   d) magnetic east.

3. The cause of the earth's magnetic field is   a) partly the earth's spinning.   b) a bar magnet.   c) not known.   d) the aurora.

4. All electric motors   a) use a generator.   b) use electric energy to perform work.   c) convert mechanical energy into electric energy.   d) none of the above.

5. All electric generators   a) do not need a source of mechanical energy.   b) convert electric energy to mechanical energy.   c) convert energy as a motor does.   d) convert mechanical energy to electric energy.

6. The domains of a magnetized substance   a) line up in all directions.   b) balance each other.   c) line up mainly in one direction.   d) can never be shifted.

7. A permanent magnet loses its magnetism   a) after a very long time.   b) when its domains get out of line.   c) when you heat or strike it.   d) a, b, and c.

8. An electric current in a wire coil produces   a) another current.   b) a magnetic field.   c) a generator.   d) a source of voltage.

9. To strengthen an electromagnet,   a) put a nonmagnetic substance in the coil.   b) increase the number of turns in the coil.   c) decrease the voltage.   d) a, b, and c.

10. Electromagnetic induction can occur if a   a) current flows in a coil.   b) circuit has a battery.   c) magnet moves in a coil.   d) motor runs a device.

Chapter 23
The Electronic Revolution

The students in the picture are playing a computer game. In some games, a computer makes its own moves and announces the moves aloud. Just a few years ago such games were not possible. Computers that could play even silent games were as large as a room.

Smaller computers became available only after tiny parts were developed that could do the work of large parts. These tiny parts now appear not only in computers but also in stereos, televisions, and many other devices. This chapter describes the invention of these parts. Then it explains how these parts work, why their small size is important, and how they are used.

Chapter Objectives

1. Explain how electronics improved communications equipment.
2. Discuss how changes in making electronic components have affected the devices that contain them.
3. Explain the uses of chips.

23-1 Electronic Devices

During the 1860s, Pony Express riders carried the mail from Missouri to California in just eight days. People cheered the horses and riders that had set this speed record. But in just 18 months the cheers stopped. New discoveries in electricity had put the Express out of business. As you read on, think about these questions:

a. How do electronic components work?
b. How did the vacuum tube affect electronic devices?

Making Electronic Devices

In the last century, people learned how to speed communication by using electrons in wires or radio waves in air. A message could be made into a changing current and then sent as radio waves at the speed of light. But over long distances, the wires' resistance reduced the current. Also, radio waves weakened as they spread out from the sender, just as light dims as it spreads out from its source.

Scientists also discovered that electrons can flow through a glass tube emptied of air. This device, called a **vacuum tube,** worked like a valve—it controlled the flow of electrons. The drawing shows how a vacuum tube looks. The positive charge on the plate is large enough to attract negative electrons. The positive charge pulls electrons from the thin wire across the gap.

Later, scientists improved the control of the electron beam through the gap. They put a grid in the gap, as the second picture shows. The beam passes through the grid. Giving the grid an extra negative charge repels the negative electrons and weakens the electron beam. Giving the grid a positive charge helps the plate attract the electrons, which strengthens the beam. Feeding a weak incoming current to the grid causes a strong change in the electron beam. A vacuum tube that can magnify changes in current is an **amplifier** (am′plə fī′ər).

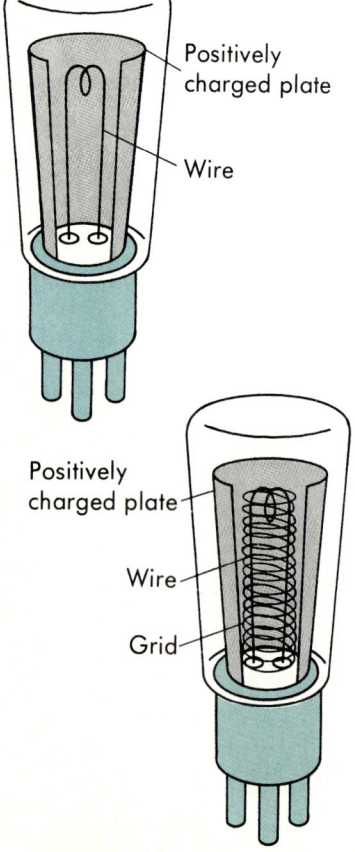

Two versions of the vacuum tube

The vacuum tube was a major advance in communications and **electronics,** which is the study of electrons and their use in radio, television, computers, and other such devices. Now messages could be sent over long distances because weak currents could be strengthened. Vacuum tubes made radios and other electronic devices practical.

Vacuum tubes and other electronic parts are known as **components** (kəm pō′nənts). Modern amplifiers are still important components in electronic devices. The diagram shows how amplifiers help produce television pictures. A television camera picks up the light reflected from an object. Video equipment changes the light into electric currents. A microphone changes sound into currents. Electronic amplifiers make these currents stronger. Finally, the currents are changed into radio waves.

The radio waves weaken as they spread out from the sender. The television antenna in your home picks up these weak waves, which are changed into a weak current. Amplifiers in the television strengthen the current. Other components change the current into pictures and sound.

An early computer

Have You Heard?

Around 1947, the first large electronic computer, ENIAC (*E*lectronic *N*umerical *I*ntegrator *a*nd *C*omputer) was built. It contained over 18,000 vacuum tubes. The air conditioning equipment needed to cool ENIAC was enough to cool the whole Empire State Building.

Vacuum Tubes Caused Problems

Even though vacuum tubes were a great advance in electronics, they had one main problem. They were too big. Some of the first tubes were as large as soda bottles. Some electronic machines used hundreds or thousands of these big tubes. Often several people were needed to find and replace just one bad tube. The picture shows part of an early computer. It contained rows of vacuum tubes and other large components wired together in huge metal boxes.

The size of early electronic devices limited their use. Because the tubes were so large, they used a lot of electricity. They also gave off large amounts of heat and wore out quickly. Researchers began looking for ways to solve these problems.

Review It

1. How does an amplifier affect current?
2. What problems did using vacuum tubes in electronic devices cause?

Activity

Sending Messages

Purpose
To send a message.

Materials
- paper cup
- plastic wrap
- aluminum foil
- tape or rubber band
- scissors
- flashlight
- paste or glue

Procedure
1. Cut out the bottom of the cup.
2. Attach the plastic to the outside of the cup with tape or a rubber band, as in *a*.
3. Darken the room.
4. Stand facing a wall at a distance of 2 m.
5. Ask your partner to stand to one side of you and shine the flashlight on the bottom of the cup as you say the alphabet into it. The situation looks like *b*.
6. Look for the pattern of each letter that forms on the wall.
7. Cut out bits of aluminum foil and paste them on the outside of the plastic, as shown in *c*.
8. Repeat steps 4–6.
9. Note whether the pattern was easier to observe with or without the aluminum bits.
10. Sketch the pattern for some letters of the alphabet. *NOTE: You may not be able to distinguish a pattern for every letter. Just sketch those patterns that are very different.*
11. Say some of the letters into the cup to form a simple word, such as "cat."
12. Send a message by sketching a pattern code for simple words.

Analysis
1. How did you change the message in this activity?
2. How could you increase the amount of change to make the patterns clearer?

a
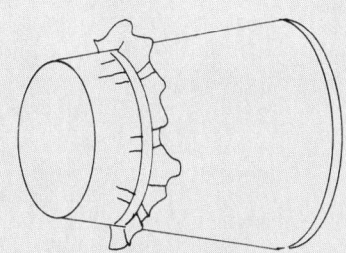

b

c
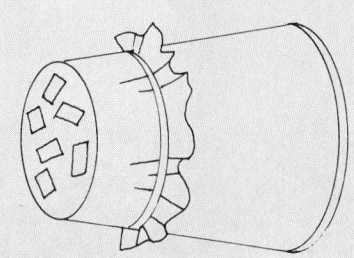

23–2
Making Electronic Components Smaller

The first radios were larger and heavier than many of today's televisions. Now a radio can be smaller than your hand. They do the same work as vacuum tubes did in the first radios, but they do it much better. Consider these questions as you read about such changes:

a. How did the transistor improve electronic circuits?
b. Why is the size of electronic components important?

Transistors Were the First Small Components

About 1950, scientists found a way of controlling electrons that move through certain solids—instead of a vacuum. These solids make **solid-state** electronic components.

The first solid-state component was a **transistor** (tran zis′tər), shown below. Like a vacuum tube, a transistor controls current. Transistors are made of materials that conduct electricity better than insulators but not as well as conductors. These materials are **semiconductors.** The elements silicon and germanium combined with other elements are two important semiconductors.

Transistors were a great improvement over vacuum tubes because they are so much smaller. They use less electricity, give off less heat, and last longer. Radios, record players, televisions, and toys were only a few of the many items that were "transistorized."

The first transistor

Modern transistors

One problem remained. Electronic devices made with transistors still needed wires to connect transistors and other components to complete the circuit. These connections took up much space and often came apart.

From Small to Smaller

During the 1970s, engineers took a giant step forward in electronics. On a small piece of silicon, they deposited materials not only for the transistors but also for the whole circuit! In this arrangement, called an **integrated** (in′tə grāt′id) **circuit,** all the connections and components are built into one piece. Now a space the size of a postage stamp holds thousands of components. Compare the sizes of the vacuum tube, transistor, and integrated circuit shown.

If you could see electric currents in an integrated circuit, they might remind you of streams of cars on a highway. Some cars travel in the main lanes. Others get on and off the highway. In an integrated circuit, electrons can move through many paths to do many different tasks.

The next step in the electronic revolution took place when engineers deposited thousands of integrated circuits on a tiny silicon **chip.** Because the chip is so small, circuits are much smaller. Electrons can travel the shorter distances much more quickly than they could before.

These improvements greatly changed the size, cost, availability, and performance of hundreds of products. They changed radios that were as big as televisions and computers that were as big as a room into objects that fit into your pocket.

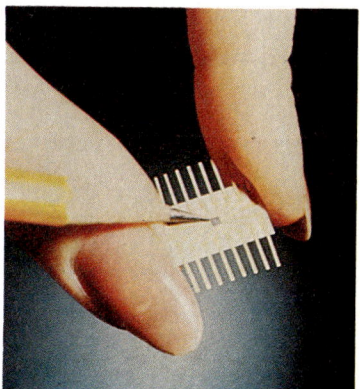

Comparing the sizes of a vacuum tube, a transistor, and an integrated circuit.

Review It

1. How is the transistor better than the vacuum tube?
2. How did making components smaller affect electronic devices?

Have You Heard?

Now a small computer is twenty times faster, has a larger memory, and performs thousands of times better than ENIAC. It occupies 30,000 times less space and costs 10,000 times less than ENIAC did. Finally, ENIAC used as much power as a locomotive. Today's small computer uses only as much power as a light bulb.

23–3
Using Chips

What kinds of devices use chips? Actually, just about any electronic device could work better using a chip. Compare the pictures of watches at the left. A few years ago, most watches contained gears and springs. Today, however, watches use chips to keep time more accurately. Remember these questions as you read about the uses of chips:

a. How are chips used in computers?
b. How can we use computers and microprocessors?

Inside a Computer

The electronic revolution has affected the computer probably more than any other device. As components became smaller, computers became smaller, faster, and able to do more difficult tasks.

A computer's main purpose is to manage large amounts of information. It is a machine designed to store and work with information. A computer works by recognizing two conditions in its circuits: "on" (high voltage) and "off" (low voltage). The computer must change all information into combinations of "on" and "off" circuits. The more circuits it has, the more information it can hold.

For example, the telephone company's computers store names, addresses, and telephone numbers. When you ask directory assistance for the Garcias' number, the operator enters the name into the computer as shown.

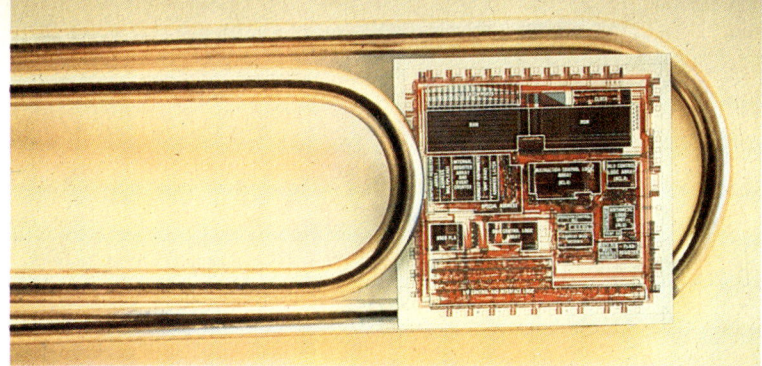

Instantly, the computer begins comparing the on-off combination for "G" with the combinations in its memory. When it finds "G," it repeats the process for the other letters. In less than a second, it finds the correct name and gives the operator the phone number.

Putting many integrated circuits on one chip was a great help in building computers. Engineers could design small computers with many abilities. They took an even bigger step when they designed a chip that contained all the computer's problem-solving circuits. This chip, called a **microprocessor** (mī′krō pros′es ər), controls the computer. The picture above shows a microprocessor.

Microprocessors are so small that they can be included in portable devices to measure a hospital patient's heartbeats. When the microprocessor detects a change in heart rhythm, it sounds an alarm.

Microprocessors have circuits to follow instructions. If you add memory circuits and circuits to bring information into and away from the microprocessors, you have a **microcomputer.** More and more people now use microcomputers at home or at work.

Computers do not "think." They can only follow exact, step-by-step instructions, called a **program.** The computer does only what you tell it to do. It will not become tired or forget. But if you give the computer an incorrect instruction, it will keep making the same mistake until you correct it. For example, perhaps your report card lists your history grade as a *D* when it should list a *B*. This mistake happened because a person typed the wrong letter into the computer. The computer will keep printing your *D* until someone corrects the error.

Challenge!
Look in an encyclopedia to find the difference between analog and digital computers.

Have You Heard?
People who work with computers have a saying about programs and computers: "Garbage in, garbage out" ("GIGO"). This saying means that the information and work we get *out* of computers is only as good as the information and work we put *in*to the computer.

427

Computers and Microprocessors at Work and Play

Today we use computers in many ways. Some computers are used by many people at the same time. These computers manage large amounts of information. For example, the National Weather Service computer receives information from all over the United States. As weather conditions change, stations revise the information. The computer compares information from all the stations. It then prints out reports for meteorologists to use to make forecasts.

Many scientists depend on computers in their research. Engineers use computers to plot graphs and maps and to make drawings. These pictures create or test designs of buildings, automobiles, and aircraft, as shown at the left. Computers that imitate flight situations help train pilots without risking lives or using fuel, as shown below.

Many people now use microcomputers at home. The microcomputers contain both teaching programs and games. You can use the same microcomputer and television to help with your homework or to play many different games.

Sometimes we need only a microprocessor chip to manage a single task. The chip in some microwave ovens is programmed to time and direct the oven to cook a roast. You program the chip to defrost the roast at a certain time, to cook it for another time interval, and to keep it warm until you are ready to eat.

Some cars, stoves, and clocks give spoken directions or warnings. The microprocessor runs the device and directs another device to produce the sounds. The students in the picture can use the "talking" microprocessor to study spelling, reading, and arithmetic.

The "brain" of a calculator is a microprocessor. The chip just follows instructions. It takes up only a small amount of space. The picture shows that the other components take up most of the space. Some calculators also have memory chips that store information for later use.

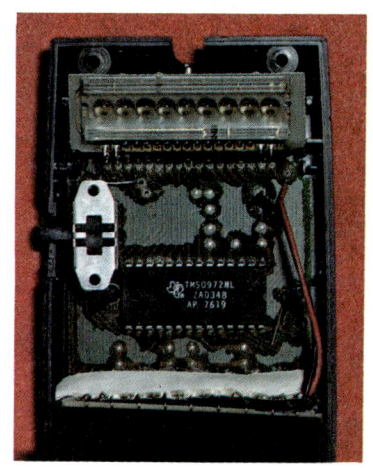

Many games now contain microprocessors to make them more exciting. A microprocessor in the game in the picture makes a different pattern of light and sounds each time you play it. You try to match its pattern. The microprocessor compares the buttons you press to see if you succeed. If you do, it makes a sound that means "you win."

Microprocessors and microcomputers are changing our lives. Each year more and more machines contain microprocessors.

Review It

1. How is a microprocessor different from a computer?
2. Name three uses for computers and three uses for microprocessors.

Activity

The Computing Process

Purpose
To show how a computer stores information and understands instructions.

Materials
- paper
- pencil
- scissors

Procedure

Part A
1. Draw a very simple picture of a house.
2. Write exact step-by-step instructions that explain how to draw the house.
3. Exchange instructions with your partner. Do not show each other your pictures.
4. Draw your partner's house by following the instructions word for word.
5. Correct your instructions and repeat step 4 until your partner draws the house as you did in your original picture. Record how the number and type of instructions changed.

Part B
1. Cut out 12 squares of paper, each 3 cm × 3 cm.
2. Label each square "ON" on one side and "OFF" on the other side.
3. Arrange three columns of four squares each on your desk, as in *a*.
4. Each column stands for one digit in a three-digit number. You can use different patterns of the off-on squares to represent each number, just as a computer does. Let the pattern for *zero* be "off-off-off-off." Let the pattern for *one* be "on-off-on-off." (The first "on" is at the top.)
5. Make up a list of patterns for the other digits (2–9). Each digit must have its own combination of offs and ons.
6. Use your code of patterns to represent the numbers 302, 498, and 756. The number 101 is shown in *b*.
7. Exchange codes with your partner and compare your systems.

Analysis
1. Compare your Part A instructions and Part B codes with how a computer follows a program.
2. What additional materials do you need to represent larger numbers in Part B?
3. What problems might occur if you did not have a standard code for all the digits?
4. How could you change Part B to represent negative numbers?

Breakthrough

The Bionic Revolution

Can you imagine using superhuman strength to lift a car? Today's science fiction story may soon become reality. The electronic revolution is making a big impact in medical research. Electronic devices are solving many human medical problems. Research in this area is known as *bionics*.

One of the earliest bionic devices was the heart pacemaker. In some people, the region of the heart that regulates the heartbeat is damaged. The damaged heart beats more slowly and may cause the person's death. Now a tiny electronic pacemaker can be placed inside the body. The electronic pacemaker takes over regulating the damaged heart.

Bionics may soon help diabetics too. A tiny device measures the amount of sugar in a person's blood. If the sugar level gets too high, the device automatically releases insulin, which lowers the amount of sugar to its proper level.

Exciting research is also producing artificial limbs, such as the arm shown. However, the limb might be clumsy and difficult to use. Bionics is changing this problem. A microcomputer is planted in a person's shoulder muscles. The microcomputer detects signals sent out by the shoulder muscles. The signals activate tiny motors in the artificial arm. The bionic limb then works like a natural arm.

Some ideas are still dreams. For example, some researchers hope to find ways to help blind people see and deaf people hear. In earlier research, tiny wires were placed in the vision center of the brain. A small electric charge was sent to the wires. As a result, blind people could see spots of light. Scientists are now looking for ways to convert these spots of light into images.

Similar methods may produce sounds for deaf people. The electronic revolution may begin a new medical revolution too.

For Discussion
1. What are some ways electronic devices help solve medical problems?
2. What are some future uses for bionics?

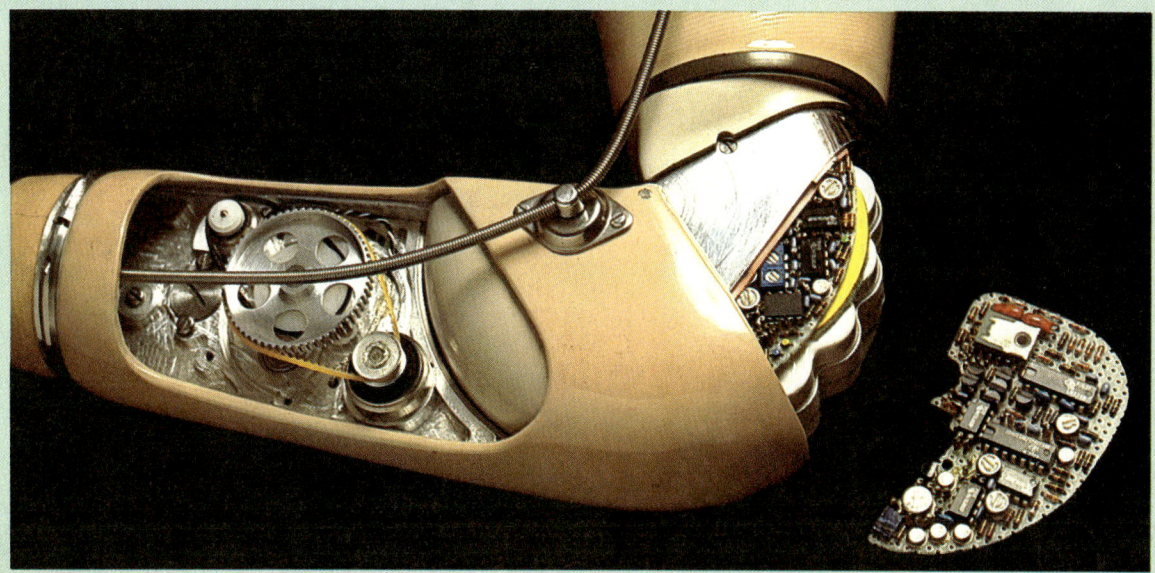

Chapter Summary

- A vacuum tube acts as an amplifier when it strengthens a current. (23–1)
- Electronics is the study of electrons in motion. (23–1)
- Early components were troublesome because of their size, the energy they required, and the heat they gave off. (23–1)
- The transistor is a solid-state component that controls electrons as the vacuum tube did. (23–2)
- Semiconductors conduct electricity better than insulators but not as well as conductors. (23–2)
- An integrated circuit contains thousands of components. (23–2)
- A chip contains thousands of integrated circuits. (23–2)
- A computer's main function is to manage large amounts of information. (23–3)
- A microprocessor is one chip that contains problem-solving circuits. (23–3)
- The computer and the microprocessor have many uses in business, industry, research, and the home. (23–3)

Interesting Reading

Corbett, Scott. *Home Computers: A Simple and Informative Guide.* Little, 1980. Discusses the history of computers, the ideas behind them, and their uses at home.

Englebart, Stanley L. *Miracle Chip: The Microelectronic Revolution.* Lothrop, 1979. Discusses the history and manufacture of electronic components and their uses.

O'Brien, Linda. *Computers.* Watts, 1978. Covers the history, construction, operation, and uses of computers.

Questions/Problems

1. How has electronics affected communication in the last one hundred years?
2. What effect did putting a grid in a vacuum tube have on the electron beam?
3. How does a television bring pictures and sound from a studio into your home?
4. What main change(s) in devices did the electronic revolution cause?
5. What is the difference between an integrated circuit and a chip?
6. Distinguish between those tasks that a computer can do and those that a microprocessor can do.
7. How have computers affected your life?
8. Think about all the things you and your parents do during one day. Then imagine that it is fifty years from now. How might computers have changed what you do? (You should consider things you cannot do now as well as what you already do.)

Extra Research

1. A television picture tube is a vacuum tube. Check an encyclopedia to learn how a picture appears on this tube. Diagram the process.
2. List the devices in your home that a microprocessor controls or could control.
3. Go to a local electronics store to learn more about microcomputers. Find out what kinds of tasks the microcomputer can do. Then report to your class.

Chapter Test

A. Vocabulary Write the numbers 1–10 on a piece of paper. Match the definition in Column I with the term it defines in Column II.

Column I

1. a chip that directs other components and chips
2. the first solid-state component
3. a kind of material used in making transistors and chips
4. an airless glass container that controls current
5. a piece of silicon containing layers of circuits
6. a part in an electronic device
7. component containing an entire circuit
8. step-by-step instructions for a computer
9. components in which an electric current moves through a solid
10. a vacuum tube that strengthens an electric current

Column II

a. amplifier
b. chip
c. component
d. integrated circuit
e. microprocessor
f. program
g. semiconductor
h. solid-state
i. transistor
j. vacuum tube

B. Multiple Choice Write the numbers 1–10 on your paper. Choose the letter that best completes the statement or answers the question.

1. Vacuum tubes   a) were invented in the 1970s.   b) are made of semiconductors.   c) were used in the first computers.   d) contain layers of circuits.

2. Transistors   a) are made of good conductors.   b) control current.   c) insulate circuits.   d) are larger than vacuum tubes.

3. Semiconductors are made of small bits of elements mixed with   a) gold or silver.   b) a vacuum.   c) silicon or germanium.   d) oxygen or neon.

4. Amplifiers are needed in communication equipment because   a) radio waves are so strong.   b) distances between senders and receivers are so short.   c) current weakens over long distances.   d) wires have little resistance.

5. Making electronic components smaller means that the devices that contain them a) work faster.   b) need fewer repairs.   c) use and give off less energy.   d) a, b, and c.

6. Currents can move through many different paths in a(n)   a) integrated circuit.   b) semiconductor.   c) vacuum tube.   d) transistor.

7. Reducing the negative charge on the grid in a vacuum tube   a) strengthens the current.   b) weakens the current.   c) stops the current.   d) does not affect the current.

8. A computer can   a) think for itself.   b) manage large amounts of information.   c) handle the needs of only one person at a time.   d) a, b, and c.

9. A computer can   a) correct errors in its program.   b) print correct information only.   c) correct errors in its information.   d) follow instructions in its program.

10. A microprocessor is used   a) to operate one system.   b) when many people need to work at one time.   c) to work with large amounts of information.   d) a, b, and c.

433

Careers

Computer programmer
Do you speak more than one language? Computer programmers do, but they do not use them to talk to people. Computer programmers understand and use the language of machines.

Programmers develop programs for computers. A programmer uses logic to understand a problem and create the steps for finding its solution. The programmer translates this program into computer language and gives the machine the instructions. If the program is exactly right, the computer can use it to solve problems or help keep records straight.

Most computer programmers have a college degree. They study computers in school and learn specific skills on the job.
Career Information:
American Federation of Information Processing Societies, 1815 N. Lynn St., Suite 800, Arlington, VA 22209

Electronics technician
How would you like to build a calculator, a radio, or a space shuttle? Electronics technicians can do all this and more.

These technicians specialize in many areas of electronics. They may work with engineers to build machinery models or to assemble computers and missiles. Some electronics technicians repair equipment for customers in electronics stores. Others make alarm systems to aid in crime prevention.

Most technicians choose their specific type of work while they are in technical school. They then spend a few years as apprentices in that area of electronics. After their official training, technicians study to keep up with new developments.
Career Information:
Electronics Industries Association, 2001 Eye St., NW, Washington, DC 20006

Broadcast technician
When you turn on the television, a picture and sound from somewhere else come to you. Those sights and sounds could not reach you without the work of a broadcast technician.

Broadcast technicians work in television studios. They control the microphones, tape recorders, lights, and sound effects that help create each program. The technicians also understand and operate the antennas that send messages to your television set.

Because each broadcast is the product of electrical equipment, technicians go to technical school to learn about electricity. The law requires that technicians also pass a test to get a first class radiotelephone operator's license.
Career Information:
National Association of Broadcasters, 1771 N St., NW, Washington, DC 20036

 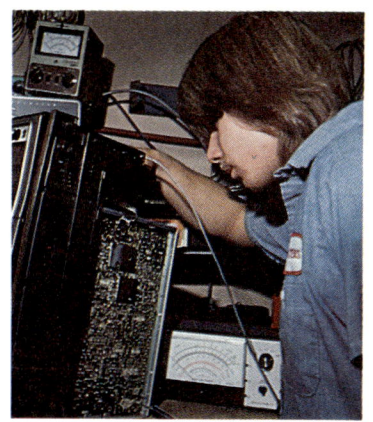 

Telephone lineperson

What if you could not use your phone? How would you talk to your friends? How would you get help in an emergency? The telephone lineperson's job is to make sure that you almost never have these problems.

Severe storms, hurricanes, and tornadoes often knock down telephone lines. The linepersons are always on call to repair damage and restore phone service. They climb telephone poles to replace wires, reconnect electric circuits, and check current flow. Linepersons also use special equipment to install telephone poles and wires in areas without service.

You need no experience to become a lineperson. The telephone companies train linepersons both in classrooms and on the job.

Career Information:
Communications Workers of America, Training Dept., 1925 K St., NW, 8th floor, Washington, DC 20006

Service technician

Behind every television screen and stereo speaker, hundreds of wires, chips, and modules are hidden. Only service technicians who understand these parts can fix appliances when they break down.

Often, these appliances stop working because a faulty circuit or antenna stops the currents or waves coming to them. The technician finds the damage and repairs it. Technicians might use electronic equipment to locate the faulty part.

Technicians need to know every part of a device and how it works. They receive special training after high school and learn for several years on the job.

Career Information:
Electronic Industries Association, Consumer Electronics Group, 2001 Eye St., NW, Washington, DC 20006

Electrical engineer

What do the inventors of television, radar, and computers have in common? These people may not have realized it then, but they did the work of electrical engineers.

Electrical engineers design many kinds of electrical and electronic systems. They develop hospital machines for medical tests and design new kinds of military missiles. Because of energy shortages, many electrical engineers work to create different types of power generators.

In order to design this complex equipment, electrical engineers must understand electricity. They learn circuitry, electronics, wiring, and other special skills during four years of college.

Career Information:
Institute of Electrical and Electronics Engineers, 345 E. 47th St., New York, NY 10017

UNIT SEVEN
FRONTIERS

What could the subject of this photograph be? The round objects might be burners on a stove. Maybe they are the holders you put under hot plates or pots to protect your table.

The photograph really shows rows of solar cells—devices that change sunlight into electricity. Using solar cells may be one way to produce energy. Studying Unit Seven will help you apply what you have learned in physical science to the challenges in your future.

Chapter 24 Energy Resources
We depend on large amounts of fuel to provide energy. We know that our fuels will not last forever. Our next step must be to save all the fuel we can until we find new sources of energy.

Chapter 25 Exploring the Universe
The universe contains many objects, including stars in different stages of development and systems of stars. Scientists study these objects to learn about the universe and to predict its future.

Chapter 24
Energy Resources

To make work easier, people use many tools and complex machines. Energy resources, such as coal, oil, and natural gas, make these conveniences possible. Many of the things that you like most about the world happen because of these three fuels. Although coal was discovered many centuries ago, we have known about oil and gas for only two hundred years. Within this short time, we have become dependent on these fuels. The photograph shows how great our dependence has become.

This chapter will explore the importance of these fuels in your life and the problems of using them. Next it will discuss the advantages and disadvantages of other energy resources. Finally, the chapter examines how to decide which energy resources are best for you.

Chapter Objectives

1. Explain how you use energy in your daily life.
2. Explain how electrical energy is produced.
3. Define fossil fuels and identify some problems in using them.
4. Describe the advantages and disadvantages of alternative energy sources.
5. Discuss how to use energy resources wisely.

24–1
Energy in Your Life

Energy is important to our lives. Energy heats our homes, runs our cars, and operates the machines we use. Often energy comes to us in the form of electricity. As you read, consider these questions about energy:

a. How do you use energy in your daily life?
b. How is electrical energy produced?

Energy Has Many Uses

Think of all the ways you use energy. The photographs illustrate some of these uses. Energy warms your home in winter or cools it in summer. Energy supplies the heat for warm showers, dishwashers, and washing machines. Stoves, ovens, refrigerators, and freezers use energy.

Office buildings and factories depend on energy too. Like your home, these buildings need heating, cooling, and lighting. Office machines, such as typewriters, calculators, and computers, use energy. Factories need energy to make new products for your use.

Without realizing it, we use large amounts of energy daily. Energy supplies the fuel needed by buses, trains, trucks, ships, and airplanes. We even use energy to grow and ship our food.

Energy in your life

Electricity is a familiar form of energy. Some of its uses are shown at the right. Every time you turn on a radio or flip a light switch, you use electrical energy. Appliances, such as irons, toasters, televisions, and hair dryers, use electrical energy.

If you multiply the amount of energy you use by the number of people in your neighborhood, you can imagine how much energy we use. Just think what would happen if we ever ran out of energy! A city without enough electrical energy might look like the picture below. The photograph shows New York City one night in 1965 when its supply of electrical energy was cut off.

Challenge!

Count the number of cars in the parking lot of your school. Suppose each car uses an average of 4 liters of gasoline every day for one year. How much gasoline will all these cars use in one year? Find out how many barrels of oil were needed to make this much gasoline.

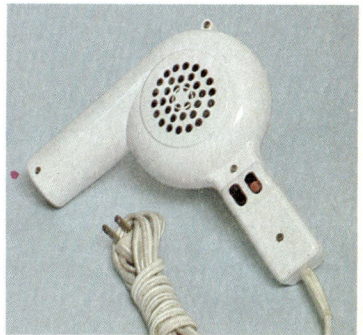

The 1965 New York City blackout

The electricity we use is not a source of energy. Electricity does not come from mines, as coal and uranium do, or from wells, as oil and natural gas do. We must use energy sources to make electricity.

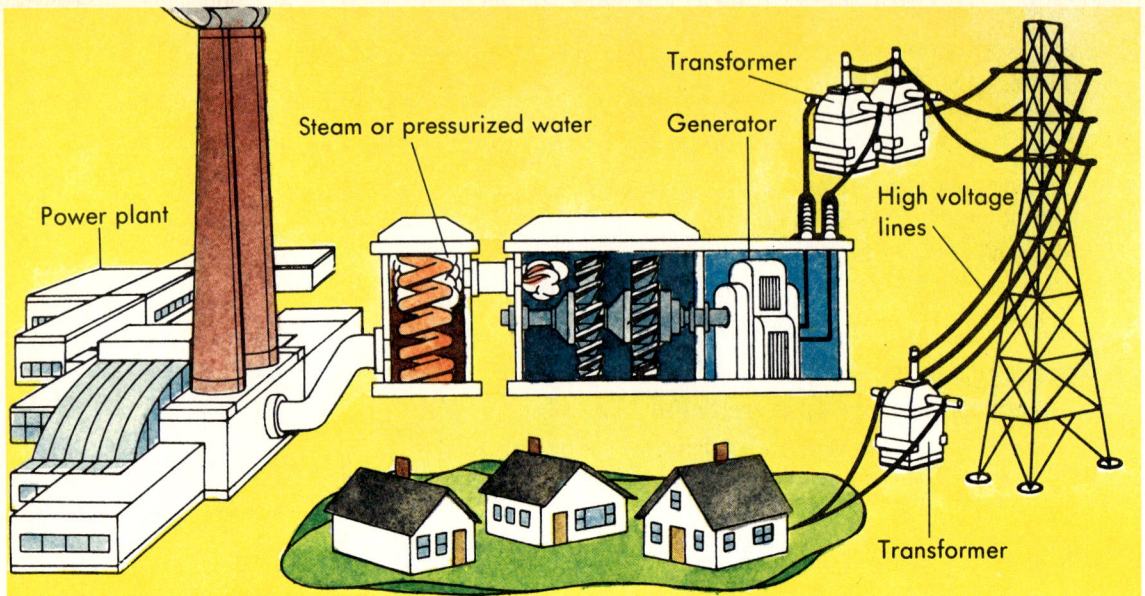

How electrical energy is produced

Making Electrical Energy

Coal, oil, uranium in a nuclear reactor, hot water from inside the earth, and falling water are energy sources that can be used to make electricity. The diagram shows how we produce electricity. Often energy from a source in the power plant heats water. The hot water or steam turns fanlike blades. The spinning blades turn a generator that produces electric current. Then a transformer boosts the voltage so wires can carry the electrical energy to houses, offices, and factories.

Electrical energy has many advantages over other forms of energy. At the flip of a switch, electricity is ready to work. Using electricity also has some problems. In the process of making electricity, energy is lost. Only about one-third of the energy stored in coal, oil, or uranium is changed into electricity. Two-thirds are lost as "waste energy" in the generating plant.

Review It

1. List eight ways you use energy.
2. Explain why electricity is not an energy source.

Activity

Investigating Energy Use

Purpose
To study how energy is used in your community.

Materials
- notebook
- pen or pencil

Procedure
1. Ask your parents and a neighbor if you can investigate their homes to make a report on energy use for your class.
2. Choose an area in your community to study with your partner.
3. Record the ways you see energy being used, saved, and wasted in the homes and area you have chosen. The diagram reminds you to check insulation, windows, and thermostats.
4. Find out and record what energy sources are used for heating, cooling, cooking, and cleaning.
5. Write down any problems that the energy use may cause, such as air or noise pollution.
6. Write down any ways your family or neighbors save energy.
7. Collect the information obtained by other students in the class. Make a chart that summarizes the energy sources, energy problems, and energy uses of your community.

Analysis
1. What are the three most common ways energy is used in your community?
2. According to your study, what three energy sources are used most often?
3. What are three problems that using energy causes in your community?
4. Suggest ways to correct the problems you observed and save energy.

b
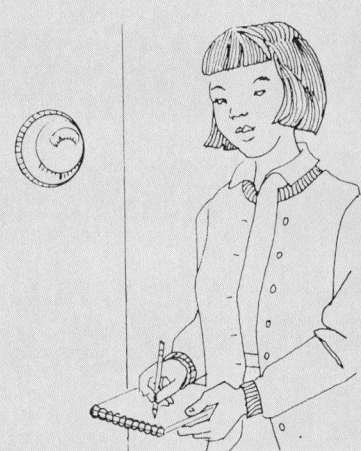

a

24–2 Fossil Fuels

What would we do if gasoline were too expensive to buy or had to be rationed? As you study about fuels such as gasoline, keep these questions in mind:

a. What is the importance of fossil fuels?
b. What are the problems of using fossil fuels?

Fossil Fuels Run the Nation

This drawing resembles the habitat of plants and animals that lived hundreds of millions of years ago. When these organisms were alive, they trapped energy from the sun as they made food. The organisms died and their remains decayed. As time passed, the plant and animal remains became coal, oil, and natural gas. These fuels are called **fossil** (fos′əl) **fuels** because they formed from the remains of fossil organisms. Burning fossil fuels releases the energy organisms obtained from the sun long ago.

Fossil fuels take millions of years to form. When we use these resources to heat our homes, to run cars, and to make fibers, plastics, and electricity, we are reducing our supply of fossil fuels. Fossil fuels starting to form now will not be ready to use for another hundred million years.

A coal-forming forest

Disadvantages of Using Fossil Fuels

Using fossil fuels presents serious problems. Because fossil fuels are not replaced once the supply is drained, they are known as **nonrenewable resources.**

Since the United States uses about one-fourth of the world's supply of oil, we worry about running out of fossil fuels. If we continue to use fossil fuels at our present rate, the world's oil and gas supplies may last only a few dozen years.

The United States has more than twenty-five percent of the world's coal supply, which should last for a few hundred years. But coal is dangerous and difficult to mine.

Other problems occur with the use of fossil fuels. When they burn, fossil fuels release harmful waste products—**pollutants** (pə lüt′nts)—into the air. Some pollutants are carbon monoxide, compounds of oxygen joined with nitrogen and sulfur, and small particles. The photographs show some effects of air pollutants on buildings and living things. Air pollutants can also cause health problems such as pneumonia, emphysema, and other lung illnesses. Burning fossil fuels also releases carbon dioxide, which may change the climate.

Obtaining fossil fuels creates problems. Strip mines tear up huge sections of land. Sometimes, ocean-drilled oil wells leak and pollute the water for kilometers in every direction. Changing oil into other fuels, such as gasoline, causes other pollutants.

Nonrenewable fossil fuels will not meet our future energy needs. We must use the remaining fossil fuels wisely and develop new energy sources.

Review It

1. How are fossil fuels important in your life?
2. List three problems of using fossil fuels.

24-3
Looking for More Energy Sources

As our fossil fuel supply decreases, we must look for other ways to produce energy. The search is on for new energy sources. As you read, answer these questions:

a. How can we use energy from the atomic nucleus?
b. How else can we use existing resources?
c. What is a renewable energy source?

Energy from the Atomic Nucleus

The nucleus of an atom contains a large amount of energy. One way of obtaining this energy is nuclear fission. During fission, the nucleus of a heavy atom, such as uranium or plutonium, is split apart. The splitting releases a large amount of energy. More than 10 percent of the electricity used in the United States comes from fission power plants such as the one shown.

Scientists have worked hard to solve many of the problems of using fission energy. However, some problems remain. Nuclear power plants are expensive to build, and dangerous radiation might accidentally escape from them. Disposing of the radioactive wastes that fission reactors produce is another serious problem. We need new ways to remove and store these wastes. The supply of uranium is still another problem. The uranium we use in the present kind of reactors will last only for a certain number of years.

In the future, we may be able to obtain energy from the nucleus with another method. During nuclear fusion, the nuclei of light atoms, such as hydrogen and helium, combine and release large amounts of energy. Fusion uses a cheap raw material—water—from which we remove deuterium. Also, fusion produces little radioactivity and fewer dangerous waste products than fission.

Loading nuclear fuel

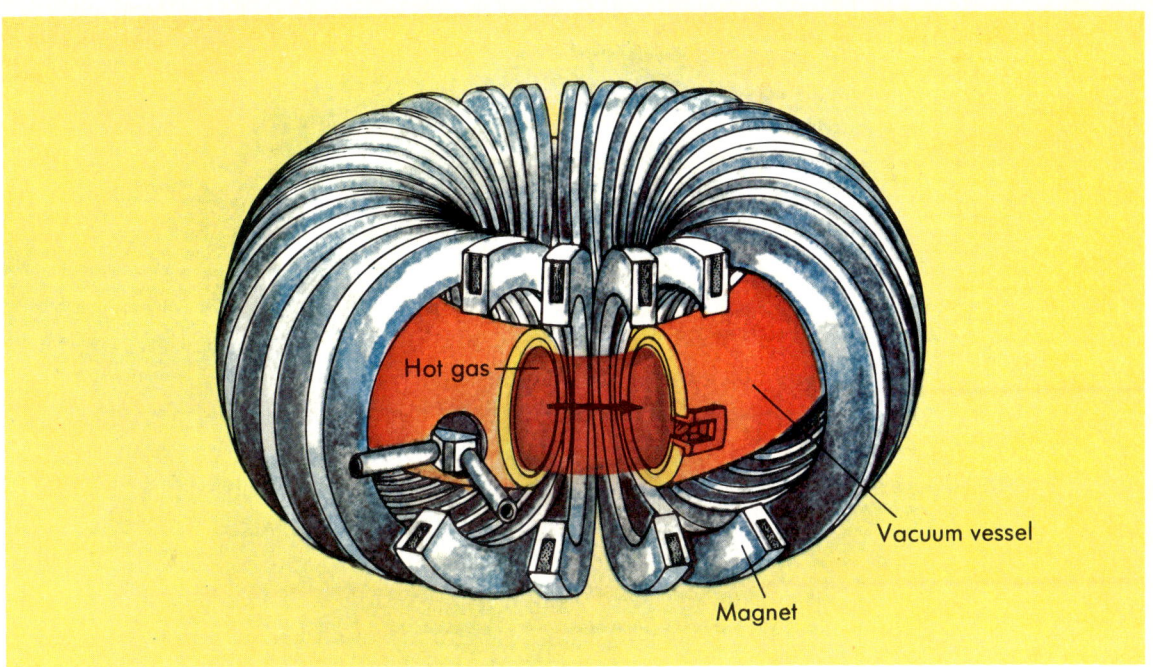

So far, people have not produced fusion energy practically. A gas heated as hot as fusion requires would cool too much if it touched any surface. So scientists are trying to contain the gas with magnetic fields, as shown above. Other scientists shoot intense laser and particle beams at hydrogen pellets to produce fusion. It may be many years before a commercial fusion plant is built.

Extending Existing Resources

Scientists search for ways to extend current resources. We can remove fuels from coal in a chemical process. First, coal is heated to a high temperature under high pressure. Next, steam passes over the coal. Hydrogen in the steam joins with carbon in the coal to make hydrocarbons. Finally the fuels are cleaned and separated. Methane, motor and jet fuel, and fuel oil result. Such fuels are called **synthetic fuels.**

Carbon dioxide is also a product of making and using synthetic fuels. Scientists worry about adding carbon dioxide to the atmosphere, since it affects climate.

Have You Heard?

Some nuclear power plants make new fuel as they operate. In a breeder reactor, the element plutonium is produced during the fission of uranium. The plutonium can then be used as a fuel for fission reactions. Using breeder reactors would extend our nuclear energy supply. But some people are not convinced that breeder reactors are safe.

Solar panels capture solar energy

Renewable Energy Sources

Some energy sources are **renewable:** either they do not run out or they can be reproduced in a few years. The following are all renewable.

The sun is one promising energy source. Every day the sun sends us five hundred times more energy than the total we use. Solar energy will be available long after other energy sources have been used up.

Solar energy can be used to heat houses in several ways. **Solar collectors** on roofs heat water that circulates through them. The hot water can be used for cooking, washing, and heating the house. **Solar cells** in panels on the roof shown at the upper left change sunlight into electricity.

Solar cells were used on Skylab, shown at the lower left. They are expensive and inefficient for a family to use. Also, they only work when the sun shines. If scientists can make solar cells that are less expensive and more efficient, solar cells could be an important future source of energy.

The sun also supplies energy through living materials, wind, and water. **Biomass** is any living material that can be burned as fuel. Wood is probably the best known and most widely used form of biomass. Wood-burning stoves heat many homes in the Northeast. Other forms of biomass are garbage, farm wastes, sugar cane, and other plants.

Biomass can serve as another energy source, but problems accompany its use. Using biomass produces pollutants similar to those that fossil fuels produce. Cutting down trees can speed the erosion of our land.

The earth provides us with hot water from geysers and hot springs. The heat comes from molten rock deep in the earth. This energy is called **geothermal** (jē′ō thėr′məl) **energy**. It can be used only at the few places where geysers and hot springs occur. At The Geysers, California, the geothermal plant has operated for over ten years. Because Iceland is a volcanic island, its people use geothermal energy to produce heat and electricity.

For centuries, people have used the energy of falling water to do work. The water turns a turbine that runs a generator. Producing electricity by water power is called **hydroelectric** (hī/drō i lek/trik) **power.** In Washington, Oregon, and California, hydroelectric plants such as the one above produce one-third of the electricity. Water power can be used only where there are rivers and dams.

The oceans are a possible source of energy too. In a few bays, very high tides produce hydroelectric power. The dam across the Rance River in France has produced energy in this way for over fifteen years. Even the motion of waves, up and down and back and forth, can produce energy. The temperature difference between layers of sea water may one day produce energy.

Windpower is another old energy source. The first windmills were built in Persia in the sixth century. The picture at the right shows a modern windmill. In regions where the wind blows regularly, windmills may be an important energy source for the future.

A modern windmill

Review It

1. List the advantages and disadvantages of using nuclear energy.
2. What are synthetic fuels?
3. Name four ways we can use the sun's energy.

Activity

Light Energy Collector

Purpose
To study how solar energy can be collected.

Materials
- thermometer
- white paper
- black construction paper
- desk lamp with incandescent bulb
- thin cardboard, 23 cm × 14 cm (light gray or brown)
- stiff, clear plastic, 20 cm × 12 cm
- watch or timer to measure seconds
- tape
- scissors

Procedure

Part A
1. Cut and fold the cardboard and tape one corner, as in *a*.
2. Cut some white paper to 12 cm × 3 cm. Tape the paper to one end of the plastic. Tape the plastic to the cardboard, as in *b*. You have a heat collector.
3. Stand a book on edge. Lean the collector against the book with the cardboard side to the book.

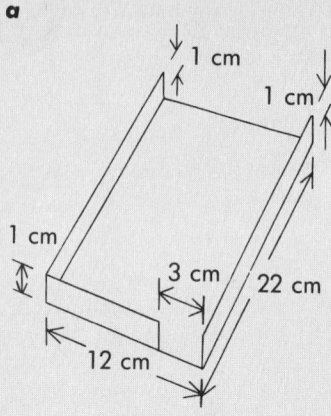

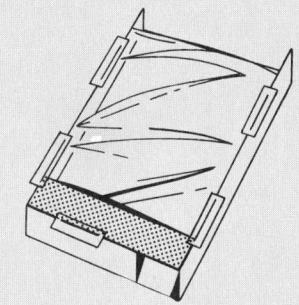

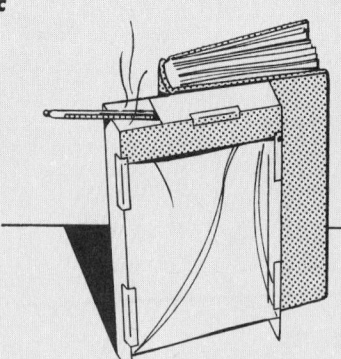

4. Place the lamp 20 cm in front of the plastic. Aim the lamp directly at the plastic. Turn on the lamp.
5. To measure the temperature of the air entering the collector, place the thermometer at the bottom of the collector's opening.
6. Wait 3 minutes. Record the temperature.
7. Cut some white paper to 20 cm × 11 cm. Slide the sheet into the collector.
8. Place the thermometer in the top of the collector, as in *c*.
9. Repeat step 6.
10. Repeat steps 7–9, but use the black paper.

Part B
1. Turn the collector until the light hits it at an angle.
2. Repeat Part A, step 6.

Analysis
1. Compare the results from the two colors of paper and the two angles. Explain any differences in temperature.
2. How would you redesign your collector to gather the most light energy?

Breakthrough

Alternative Energy Sources for Airplanes

Flying through the sky on a sunbeam sounds like a fairy tale. However, this story is about a new idea for flying airplanes using sunlight. The first solar-powered aircraft, the *Gossamer Penguin*, was flown in 1980.

The first futuristic airplane design appeared in 1961. In that year, a glider pilot named Derek Piggott flew the first pedal-driven airplane for a distance of 50 meters.

A later model of the bicycle-powered plane was the *Gossamer Condor*. In 1977, the *Gossamer Condor* flew just over 1.6 kilometers. Its speed was about 14 kilometers per hour. The plane was made of balsa wood, cardboard, thin metal tubes, piano wire, tape, and a thin, light plastic.

This plastic is the lightest aircraft skin ever developed. These thin, light qualities gave the plane one part of its name, *Gossamer*. The *Condor* part of the name refers to a very large bird that lives in the mountains of California and South America.

The *Gossamer Condor* weighed only 32 kilograms. Its lightness allowed flight using only *pedal-power*. Another plane, the *Gossamer Albatross,* made the longest flight by a pedal-powered plane. The plane was pedaled across the English Channel in 1979. The flight took nearly three hours at a speed of about 13 kilometers per hour.

The *Gossamer Penguin* used sunlight as a source of energy. Solar cells captured and converted sunlight into the power needed to fly the aircraft. The latest plane, the *Solar Challenger*, pictured above, crossed the 35 kilometers of the English Channel in July, 1981. The *Challenger's* wings and horizontal stabilizer are covered with 16,128 solar cells. The cells provided enough electricity to fly the plane at 70 kilometers per hour for 5 hours and 23 minutes.

For Discussion
1. What materials are used in building a *Gossamer*-type airplane?
2. What are some of the ways in which a solar-powered plane might be used in the future?

24—4
Making Decisions About Energy Sources

Each energy source has advantages and disadvantages. This section presents some ideas to help you decide how to use energy sources wisely. As you continue reading, think about these questions:

a. What are some ways to choose an energy source?
b. How can we better use our energy sources?

Choosing Among Energy Sources

Deciding which energy source to use is a new problem for Americans. For a century, we thought we had all the coal, oil, and natural gas we needed. Now fossil fuels are scarcer and more expensive. As fuel prices increase, we are using energy sources that once seemed too expensive. For example, making synthetic fuels is expensive. When we thought we had a lot of oil, few people thought about making synthetic fuels. Now we need to obtain energy from any possible source. The graph below predicts how we may be using our energy sources by 1990.

We can decide which energy source to use by considering location, environment, and the purpose for the energy. For example, because the Northwest has many rivers, hydroelectric power is a good energy source there. Wind might be a better energy source in Oklahoma, which has strong winds. The Southwest could use solar energy because the sun's rays are most directly overhead and the sky is clear during most of the year there.

Energy use in the United States

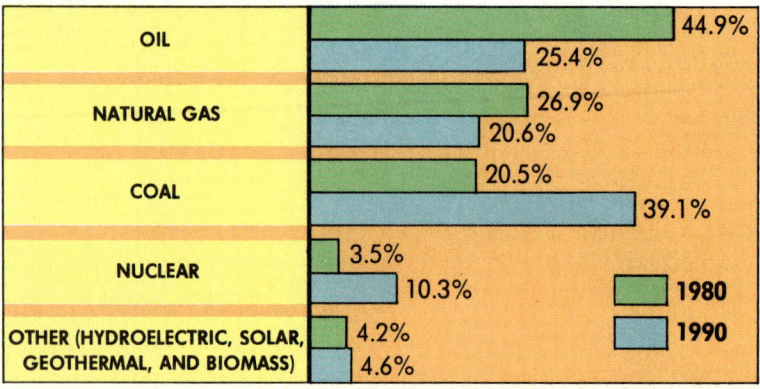

Source	1980	1990
OIL	44.9%	25.4%
NATURAL GAS	26.9%	20.6%
COAL	20.5%	39.1%
NUCLEAR	3.5%	10.3%
OTHER (HYDROELECTRIC, SOLAR, GEOTHERMAL, AND BIOMASS)	4.2%	4.6%

Damage caused by obtaining fossil fuels

In choosing an energy source, we must also consider its effect on the environment. For example, nuclear power plants produce large amounts of electricity. But their wastes must be stored and their hot water release must be controlled. If we use oil, spills from tankers could harm the environment, as shown above. Power plants run on coal also release harmful pollutants into the air. Some solar collectors and windmills require a lot of land space that might be used to grow food instead. Mining materials for solar cells can also be dangerous.

We must consider how we want to use the energy. For example, solar energy can provide heat and hot water for private homes. But it is usually too expensive to generate electricity with solar energy for a whole community.

We do not have to depend on just one source of energy. We can use a combination of energy sources to meet our energy needs. The combination of energy sources should provide the most energy with the fewest disadvantages.

Saving Energy

Conservation means saving or using less energy. For a long time, we have used energy carelessly. We have acted as though fossil fuels would last forever. Now we are beginning to conserve energy. For example, lowering the thermostat and turning out unneeded lights save energy. Window shades and curtains can keep the sun out on summer days and the heat in on winter nights. You can save a kilogram of coal by not using a one-hundred-watt bulb for twenty hours. The man in the picture is saving money and energy by insulating his home.

Businesses are beginning to save energy too. To warm offices, new buildings use the heat given off by lights and people. Computers control room temperatures by adjusting fans and vents to move warm and cool air where it is needed. Some factories use "waste" heat to warm the buildings.

If we use energy sources wisely, energy supplies will last longer. Then we will have more time to develop new energy sources for the future.

Challenge!

Private cars use a great deal of energy. What other ways of transportation could you suggest for your community?

Review It

1. What three factors should you consider in choosing an energy source?
2. How can conservation lessen our energy problem?

Chapter Summary

- Energy is used daily in so many ways that we often overlook it. (24–1)
- Electricity is a form of energy obtained from many energy sources. (24–1)
- Fossil fuels—coal, oil, and natural gas—are nonrenewable. (24–2)
- Fossil fuels release harmful pollutants into the air. (24–2)
- Nuclear energy, solar energy, hydroelectric power, geothermal energy, wind power, and biomass are alternative energy sources. Using them has advantages and disadvantages. (24–3)
- Synthetic fuels can be made from coal in a chemical process. (24–3)
- Solar energy, hydroelectric power, geothermal energy, wind power, and biomass are renewable resources. (24–3)
- Location, environment, and use must be considered when choosing an energy source. (24–4)
- Conservation promotes wise energy use and provides time to develop new energy sources. (24–4)

Interesting Reading

Coombs, Charles. *Coal in the Energy Crisis*. Morrow, 1980. Provides information about many effects of coal mining and use.

Pringle, Lawrence. *Nuclear Power*. Macmillan, 1979. Surveys the development of nuclear power and discusses the effects of using it.

Watson, Jane Werner. *Alternative Energy*. Watts, 1979. Future possibilities for nuclear, solar, wind, water, and other alternative sources of energy.

Weaver, Kenneth F., and others. "Energy. A Special Report in the Public Interest." *National Geographic*, 1981. Deals with the nation's present energy problems and possible energy future.

Questions/Problems

1. What is the difference between electricity and an energy source?
2. Suppose we had enough fossil fuels to last 10,000 years. Would that mean we had no energy problem today? Explain.
3. Solar energy is easy to get. Why do we use this energy source so little, compared to fossil fuels?
4. Suppose we began an active conservation program tomorrow. List some ways in which such a program would affect the habits of governments, businesses, industry, and individuals. Would a conservation program solve our "energy crisis"? Why or why not?
5. You live on a volcanic island in the South Pacific. The island is poor in fossil fuels. Describe your options for producing energy.

Extra Research

1. Make a list of all the ways your family uses energy. Find out which energy source is used for each one.
2. Visit the plant that makes electricity for your community. Find out what energy problems exist at the plant.
3. Collect newspaper articles about energy questions in your community. Find out which topics are mentioned most often in these articles.
4. Find out if solar energy or wind power is being used anywhere in your community. If so, visit the location and find out how the system works.
5. Have a classroom debate about the advantages and disadvantages of using a particular energy source. Be sure to search carefully for facts to back up your viewpoint.

Chapter Test

A. Vocabulary Write the numbers 1–10 on a piece of paper. Match the definition in Column I with the term it defines in Column II.

Column I

1. changes sunlight into electricity
2. unwanted gases and particles that harm the environment
3. energy sources produced by the remains of buried organisms
4. produced by running water
5. produced from heat within the earth
6. energy source that can be used up
7. made artificially from chemical changes of coal
8. living material that is used as fuel
9. saving energy or using less energy
10. energy source that can be reproduced in a few years

Column II

a. air pollutants
b. biomass
c. conservation
d. fossil fuel
e. geothermal energy
f. hydroelectric power
g. nonrenewable resource
h. renewable resource
i. solar cell
j. synthetic fuel

B. Multiple Choice Write the numbers 1–10 on your paper. Choose the letter that best completes the statement or answers the question.

1. Which of the following is *not* an energy source? a) geothermal energy b) coal c) electricity d) nuclear power

2. How much energy is lost when we make electricity from coal, oil, or uranium? a) 0% b) 67% c) 33% d) 100%

3. The only one of the following which is *not* a fossil fuel is a) coal. b) oil. c) natural gas. d) uranium.

4. Our oil and gas supplies will probably last us about a) a year or two. b) 25 years. c) 500 years. d) forever.

5. Air pollutants are likely to cause which of the following diseases in humans? a) lung b) heart c) diabetes d) ulcers

6. A good way to conserve energy is to a) turn thermostats up in winter. b) insulate your home. c) turn on more lights. d) raise shades on the sunny side of your house during the summer.

7. Choose the nonrenewable energy source. a) hydroelectric power b) uranium c) biomass d) geothermal energy

8. Which of the following is a source of nuclear energy? a) uranium b) oil c) hydrogen d) a and c

9. The most widely used energy source is a) nuclear. b) solar. c) fossil fuel. d) electricity.

10. Using biomass a) produces pollutants. b) uses a nonrenewable resource. c) uses energy from the sun directly. d) none of the above.

Chapter 25
Exploring the Universe

The picture shows a glowing cloud of gas and thousands of shining stars in the night sky. The stars and gas form the shape of North America. Many of the stars, which look like tiny points of light, are actually much larger than the sun. Although you cannot see all the activity, amazing events are happening among the stars. Some stars explode in great bursts. Others collapse. Huge groups of stars speed outward through space. Although these events may seem fantastic, they result from the same laws of physical science that cause events on earth.

This chapter explains how stars form and change. It also describes the huge groups of stars, gas, and dust in the universe. In addition, the chapter discusses theories about the development and the future of the universe.

Chapter Objectives
1. Explain how stars form and how they shine.
2. Explain how stars change and age.
3. Describe the shape and movement of galaxies.
4. Discuss theories about how the universe developed and how it may change in the distant future.

25-1
The Development of Stars

Many people think the space between stars is empty. Actually, space contains much gas and dust. In some spots, large amounts of the gas and dust collect and form stars. The photograph shows this kind of huge dust cloud. Consider these questions as you read about stars:

a. How do stars form?
b. What makes stars shine?
c. How do astronomers study the sun and other stars?

How Stars Form

The huge cloud of gas and dust in the picture is a **nebula** (neb′yə lə). The small dark spots within it are clumps of gas and dust that have started collapsing. Notice in the diagrams that a star forms from such a nebula. Gravity draws the gas and dust together. As matter in the clump concentrates, gravity between the particles increases. The gas becomes more and more compressed as this process continues. The material in the center of the clump becomes very hot.

The Eagle Nebula

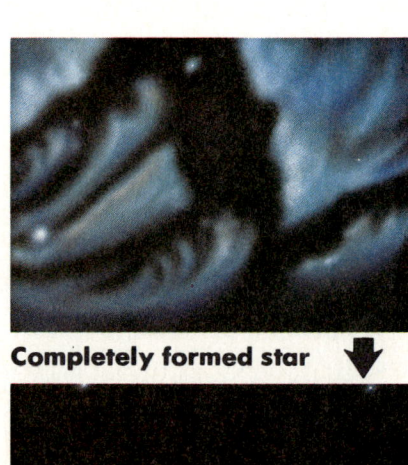

Completely formed star

The final event in the birth of a star happens as the pressure and density at its center become very great. The temperature rises above 1,000,000°C. Atoms in the gas break down into nuclei and electrons. The diagram shows how the nuclei and electrons whirl around at high speeds. Some of the nuclei pass so close together that they combine through nuclear fusion. This process causes a high pressure in the center of the star. When pressure pushing out balances gravity pulling the gas in, the star is born.

Nuclei combining during fusion in a star

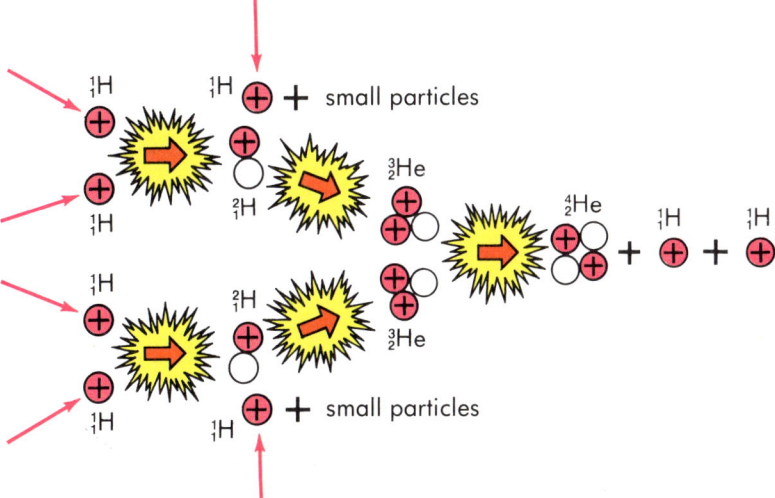

Nuclear Fusion Makes Stars Shine

In stars, hydrogen nuclei combine to form a helium nucleus. The helium nucleus contains slightly less matter than the hydrogen nuclei that formed it. Some of the extra matter changes into energy that flows outward to the star's surface. Stars shine because they give off some of this energy as light.

The greater a star's mass, the more quickly it uses up its hydrogen. The sun, an average-sized star, has been shining for about five billion years. Astronomers believe it will keep shining for another five billion years. Some stars have as much as fifty times the mass of the sun. They use up their hydrogen much more quickly than the sun. Other stars have only about one-tenth of the sun's mass. These stars should shine much longer than the sun.

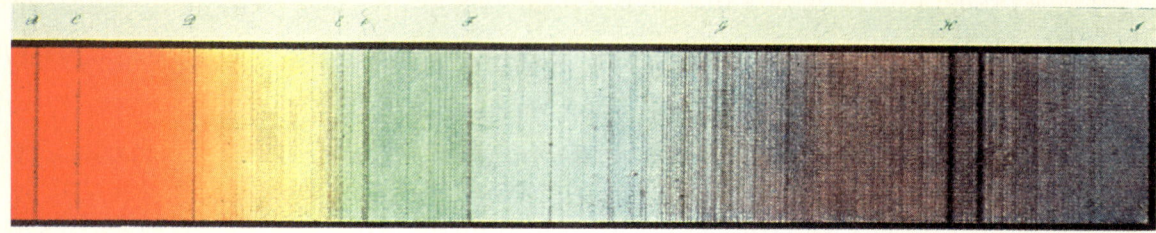

Studying the Sun

Each element in a star gives off its own set of colors. An element's set of colors is different at different temperatures. Astronomers study a star by breaking up its light into a spectrum of colors. Astronomers use photographs of a spectrum to identify the elements in a star and the star's temperature.

From studying the sun's spectrum, shown above, astronomers know that it is mainly hydrogen. The sun also contains helium and small amounts of all other elements. The spectrum shows at least seventy elements.

All stars have several layers. But only the sun is close enough for us to observe these layers directly. About once a year, the moon passes directly between the earth and the sun. This event, called a solar eclipse, allows us to see the outermost layers of the sun. As the picture shows, the sky becomes dark during an eclipse, even though it is day. A halo of sunlight appears around the moon. The halo is the outermost layer of the sun. This layer, called the **corona,** is normally fainter than the daylight sky. For this reason, the corona can only be seen clearly when an eclipse darkens the sky. During an eclipse, astronomers study the corona to determine its shape and temperature and how the gas in it moves.

Review It

1. What is the role of gravity in forming a star?
2. Where does the energy of a star come from?
3. What can we learn from a star's spectrum?

Activity

Using a Spectroscope

Purpose
To make a device for viewing a spectrum.

Materials
- 2-cm square of clear diffraction grating
- paper towel tube
- black construction paper
- clear tape
- sharp blade or knife
- ruler
- scissors
- fluorescent lamp
- ordinary lamp
- colored pencils

Procedure
1. Trace the open end of the tube on construction paper. Repeat.
2. Draw a circle 2 cm bigger in diameter around each of the tracings. Cut out the two bigger circles.
3. Make several cuts between the bigger circles and the smaller ones, as in a.
4. Use the knife to cut a slit about 2 cm × 1 mm in the center of one disk.
5. Place the disk over one end of the tube and tape the flaps to the tube, as shown in b.
6. Cut a 1-cm-square window in the other disk. Tape the diffraction grating over the opening, as in c.

a
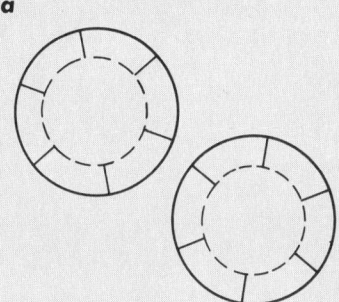

b
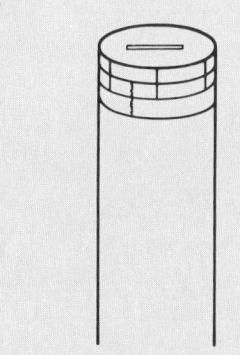

c
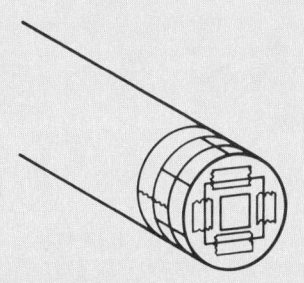

7. Darken the room.
8. Hold the disk with the diffraction grating over the open end of the tube. Look through the grating and line up the slit with a light source so that the light will strike the grating. *CAUTION: Do not look through your spectroscope directly toward the sun. You could seriously damage your eyes.*
9. Rotate the disk until you see a clear spectrum. Tape the disk in this position. You have made a spectroscope.
10. Look through your spectroscope with the slit pointed toward an ordinary light bulb. You should see at least one spectrum.
11. Sketch what you see with colored pencils.
12. Look through your spectroscope at a fluorescent lamp and sketch what you see.

Analysis
1. Contrast the spectrum of one bulb with that of the other.
2. How do you think a spectrum's colors help determine the elements that make up a star?

25–2
Aging Stars

Stars go through different stages of life, just as people do. But stars live much longer. Think about these questions as you read about how stars age:

a. How do white dwarfs and supernovas form?
b. What are neutron stars and black holes?

White Dwarfs and Supernovas

A constant battle between gravity pulling inward and pressure pushing outward happens during a star's lifetime. At some stages, these forces are balanced. At other times, one or the other becomes stronger. These changes determine what happens to the star as it ages.

As a star gradually burns up much of its hydrogen, fusion slows down. Eventually, all the hydrogen in its center is used. The diagrams show what happens in a star after this stage. When fusion stops, the pressure at the star's center decreases. Gravity then becomes overpowering. As a result, the star begins to collapse toward its center. This compression of matter, however, heats up hydrogen in the star's middle layers. Fusion of this hydrogen begins. The heat produced causes the outer layers to expand. The star brightens as it gets bigger. However, like any gas, the outer layers cool as they expand. As the temperature drops, the star gives off redder light. The star is a **red giant.**

The development of a red giant

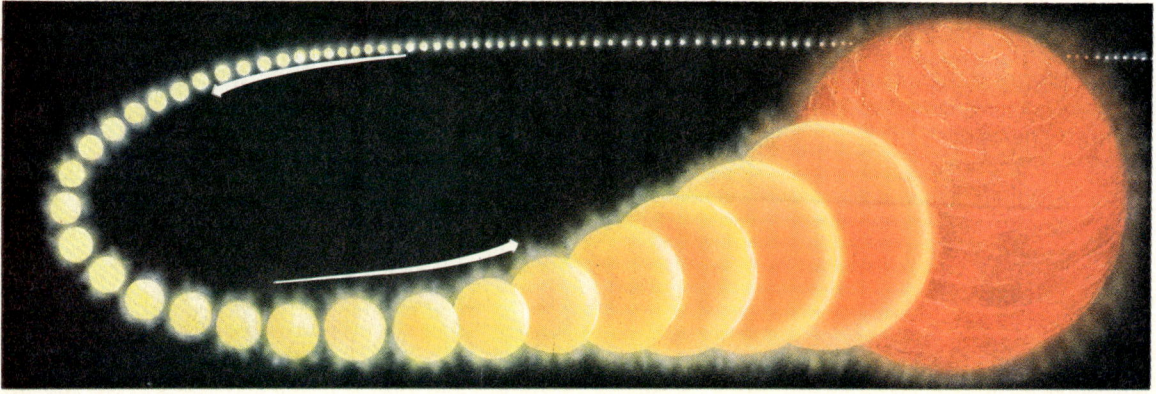

464

The Crab Nebula

While the outer layers expand and cool, gravity continues to compress the center of a red giant. The temperature and pressure in the center increase so much that the fusion of helium nuclei can begin to form heavier elements, such as carbon. What happens to the star next depends mainly on its mass.

Many stars with about as much mass as the sun throw off their outer layers after becoming red giants. The picture at the right shows the Ring Nebula, in which the central star has released its outer layers of gas. Gravity causes the remaining part of such a star to collapse further and become very dense. However, the gravity is not strong enough to pack the electrons closer than a certain distance. Gravity inward and the pressure of the electrons outward balance and hold the star together. A small, faint **white dwarf** results.

A white dwarf is a dead star. Fusion no longer occurs inside it. But it still has so much energy that cooling takes billions of years.

Stars that have more mass than the sun keep swelling after they become red giants. These stars are called **supergiants.** Fusion inside a supergiant produces heavier elements up to iron. The star then explodes violently as a **supernova** (sü′pər nō′və). The picture above shows the Crab Nebula, the remains of a supernova that the Chinese observed about 900 years ago. Elements heavier than iron are formed in supernovas.

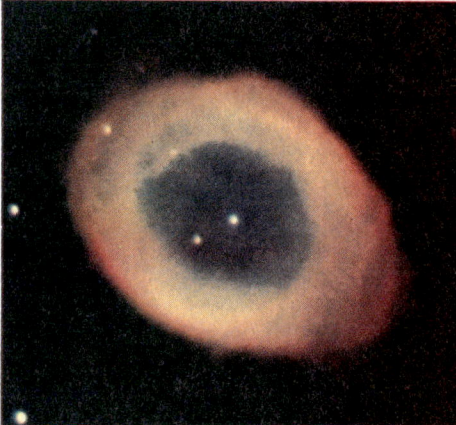
The Ring Nebula

Have You Heard?

The time a white dwarf needs to cool completely is longer than the estimated age of the universe! Therefore, astronomers believe all white dwarfs that formed in the universe are still shining.

465

Have You Heard?

The diameter of the sun is about 1,400,000 km. A neutron star with the same mass would have a diameter of only 20 km. At the surface of such a star, gravity would be 100 billion times as strong as the earth's!

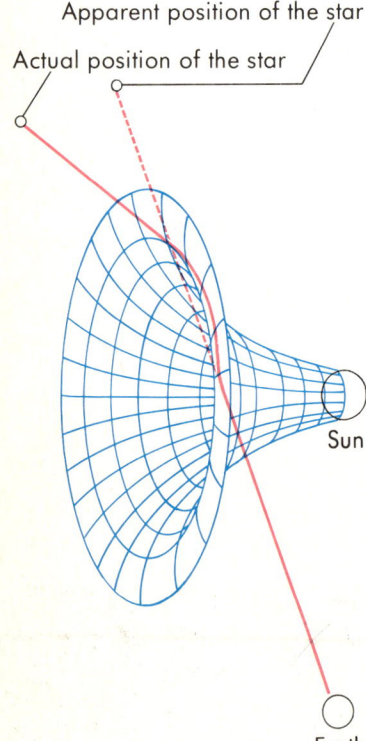

Apparent position of the star
Actual position of the star
Sun
Earth

Neutron Stars and Black Holes

Supernovas leave behind part of the star. If the remainder is less than three times the mass of the sun, gravity causes it to collapse. Gravity becomes so strong that it presses the electrons into the nuclei. Now the gas consists mainly of neutrons. When neutrons are tightly packed, they push away from one another, just as electrons do in a white dwarf. Eventually, the outward pressure of the neutrons balances the inward pull of gravity, and a **neutron star** results.

Some neutron stars give off radio waves in a narrow beam. As the star turns, the beam sweeps past the earth. Regular pulses of radio waves, like flashes of light from a lighthouse, reach the earth. Stars that give off such radio pulses are called **pulsars** (pul′särz).

A few neutron stars also give off pulses of visible light. The Crab Nebula contains such a star. Other neutron stars give off X rays.

After a supernova, gravity causes stars with the most mass to collapse completely. So much mass packed tightly into a small space causes extremely strong gravity. Astronomers think the pull of gravity from such an object would be strong enough to keep anything, even light, from escaping the dying star. Because no light could escape, we could not see the object. It is a **black hole.**

To understand black holes, astronomers study what happens when gravity is very strong. An object with a huge mass has such strong gravity that it bends space around it. The curved space bends any light that passes by. Astronomers have already observed this effect in starlight passing by the sun. The diagram at the left shows how the sun's gravity curves space and bends the light from a star behind it. Black holes have much more mass than the sun. Astronomers believe the space around black holes is so curved that it keeps light in the hole.

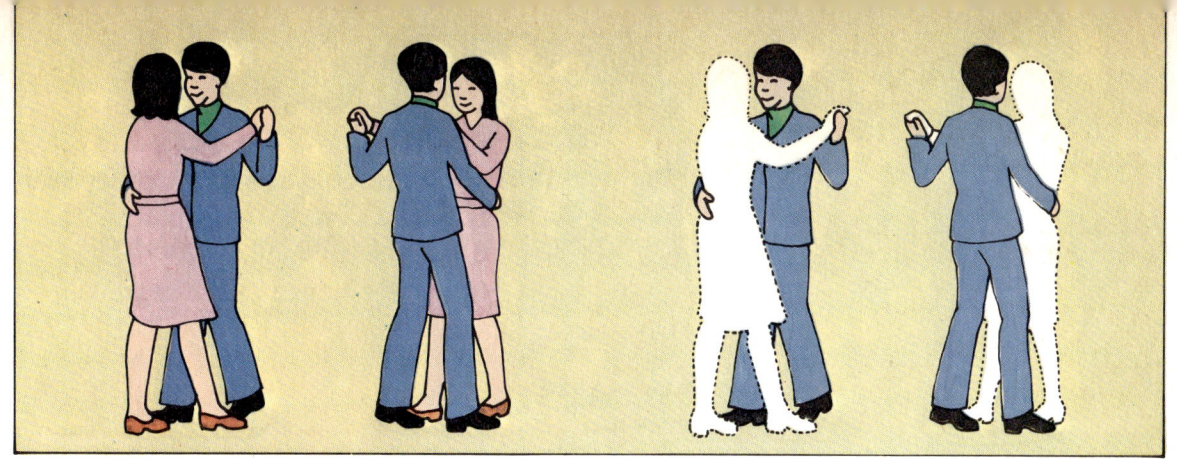

To find a black hole, astronomers look for events that show the effects of its gravity. Use the diagrams above to imagine two dancers swinging each other. If one dancer were invisible, you could still see the other one moving back and forth. In a similar way, astronomers look for stars that move back and forth as if they were being pulled by an invisible "dancer"—a black hole.

In the constellation Cygnus, astronomers have found a star that seems to be pulled by something invisible. Bursts of X rays come from this spot in the sky. Astronomers believe the gravitational force of a black hole holds the star close and pulls off gas from the star's outer layers. As the drawing below shows, the gas forms a disk around the black hole, heats up, and gives off X rays before falling into the black hole.

Review It

1. Explain how white dwarfs and supernovas form.
2. How do astronomers look for black holes?

25–3
A. Universe of Galaxies

The stars you see in the sky at night are only a few of the great number of stars in the universe. Most stars are too far away to be seen without a telescope. By using powerful telescopes, astronomers have found that all stars occur in giant systems. Answer these questions as you read about these giant star systems:

a. What are galaxies?
b. How is the Doppler effect used in studying galaxies?
c. What are quasars?

Galaxies Differ in Shape

A large system of stars held together by gravity is a **galaxy** (gal′ek sē). The universe has millions of galaxies. Many galaxies, such as our own Milky Way, contain gas and dust as well as stars.

Galaxies differ in shape. Some are spiral—they look like pinwheels with arms that unwind from a central area. The Milky Way is a spiral galaxy. It probably resembles the Great Galaxy in the constellation Andromeda, pictured at the left. The two smaller galaxies in this picture have oval shapes. The middle picture shows Centaurus A, another oval galaxy. Unusual dark lanes of dust are wrapped around it.

Not all galaxies have regular shapes. The fuzzy area in the last picture is an irregularly shaped galaxy called the Large Magellanic Cloud.

The Motion of Galaxies

Imagine that you are on a raisin in some raw bread dough that is bigger than your city. The dough is baking in a giant oven. As the dough bakes, it rises, carrying the raisins away in all directions. Even though the raisins stay the same size, the dough expands. From your raisin, you can see this expansion in all directions.

Another person on another raisin sees exactly what you see. But the dough is so big that neither of you can see its end or find its center. You can see only that all the other raisins are moving away from you.

In the universe, the galaxies are like the raisins, and the dough is like space. Astronomers observe the galaxies moving as they do, because space is expanding in all directions.

The evidence that the universe expands comes from studies of the Doppler effect in the galaxies' light. The diagram shows how scientists use an object's spectrum to measure this change in wavelength. If an object is moving away from us, fewer wave crests reach us, so the wavelengths seem longer. The light "shifts" toward the red end of the spectrum, the area of longer wavelengths. You can see this **red shift** in the spectrum below. The red shift increases if the galaxy is moving away from us at a greater speed.

A Doppler-shifted spectrum

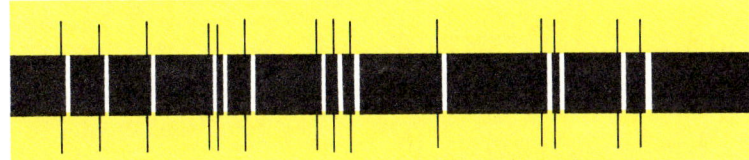

Laboratory spectrum for comparison

Actual spectrum

Laboratory spectrum for comparison

Scientists have found that all distant galaxies have red shifts. The most distant galaxies show the greatest red shifts. In every direction, the galaxies speed outward through space, and the most distant ones move the fastest.

Optical view of quasar 3C 273

X-ray view of quasar 3C 273

Challenge!

Light travels 9.6×10^{12} kilometers per year. Multiply this number by 15 billion (15×10^9) years to find out the distance to a quasar such as the one mentioned on this page.

The Most Distant Objects in the Universe

The stars near us (those in our galaxy) are too close to have large red shifts caused by the expanding universe. Thus, astronomers were surprised in 1963 to discover objects that looked like stars in our galaxy but had huge red shifts. These objects gave off large amounts of radio waves. They are quasistellar (starlike) radio sources—**quasars** (kwā′särz).

Quasars have the largest red shifts known. For this reason, they are probably the most distant objects in the universe. Some quasars are so far away that their light takes fifteen billion years to reach us.

Even though they are far away, we can easily detect many quasars with telescopes. Therefore, quasars must give off amazing amounts of energy. No one knows for certain where all the energy comes from. Many astronomers think the energy might come from a giant black hole in the middle of each quasar. According to this theory, as the black hole draws the surrounding matter toward it, the matter heats up. What we observe is the energy the matter gives off before the black hole swallows it.

Astronomers study quasars by measuring the light, radio waves, and X rays they give off. The pictures show two different views of the same quasar. Above is the quasar seen through a telescope. Below is the way it appears in an X-ray "photograph" taken by a satellite.

Review It

1. How do galaxies differ in shape?
2. Why do astronomers think the universe is expanding?
3. Why do astronomers think quasars are the most distant objects in the universe?

Activity

The Expanding Universe

Purpose
To demonstrate how the universe expands.

Materials
- balloon (not red)
- red marker pen
- black marker pen
- tape measure

Procedure
1. Copy the table shown in *a*.
2. Ask your partner to blow up a balloon just enough to make it taut and to hold it closed. Measure the balloon's diameter.
3. To represent the galaxies, draw red dots about 2 cm apart all over the balloon.
4. Circle one dot with the black marker.
5. Choose 6 other dots, some close and some far away from the circled dot. Number the dots 1 through 6, as shown in *b*.
6. Measure the distances from the circled dot to the 6 other dots. Record the distances in Column 2 of the table.
7. Ask your partner to blow into the balloon until the diameter is doubled. Hold the balloon's end closed.
8. Measure the distances from the circled dot to the 6 other dots again. Record these distances in Column 3 of the table.
9. Subtract the numbers in Column 2 from those in Column 3. Record your answers in Column 4.
10. Divide the numbers in Column 3 by the numbers in Column 2. Record your answers in Column 5.

Analysis
1. Blowing up the balloon represents the expansion of the universe. Explain how the changes in distances between galaxies depend on how far apart the galaxies were originally. (Compare the numbers in Columns 2 and 4 of your table.)
2. Does the factor by which the distances between galaxies changed also depend on how far apart the galaxies were originally? (Compare the numbers in Columns 2 and 5 of your table.) Explain your answer.

a. Data Table

Dot No.	Original Distances	Distance After Expansion	Change in Distances	Factor by which the Distances Changed
1				
2				
3				
4				
5				
6				

b

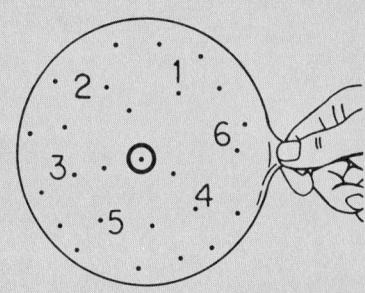

25–4
Theories About the Universe

Throughout the ages, people have wondered about how the universe formed, how old the universe is, and how the universe might end. Consider these questions as you read:

a. What is the big bang theory?
b. Why do astronomers think the universe has a large amount of invisible mass?
c. What are two scientific theories about the future of the universe?

The Big Bang Theory

The drawing shows one idea of how the universe began. Many astronomers think all the matter in the galaxies was once packed together in a dense mass. They believe it exploded about 15 billion years ago in a big bang. This explosion had no center. It occurred everywhere in space at the same time.

According to the **big bang theory,** this giant explosion sent matter and strong radiation in all directions. Most of the radiation quickly changed into matter. Galaxies eventually formed from this matter. As the galaxies developed, they continued to speed through space in all directions. The lightest elements, such as hydrogen and helium, were probably formed soon after the big bang.

The big bang

Much later

Still later

Cluster of galaxies in the constellation Hercules

Scientists predicted that if the big bang theory were correct, the universe should have weak radio waves coming from all directions. These radio waves would be the remainder of the radiation from the big bang. Around 1965, scientists discovered such waves. Their discovery supports the big bang theory.

Invisible Mass in Galaxies

Galaxies occur in large groups called **clusters.** The picture shows a cluster of galaxies. All the objects that are not round are galaxies.

Astronomers determine how much mass a cluster of galaxies has by studying the motion of each galaxy in the cluster. If each galaxy moved rapidly enough, it would escape from the cluster. Since the cluster has not broken apart, it must have strong gravity to hold its galaxies together. Because strong gravity results from a large mass, the cluster must have a large mass.

Astronomers are surprised by the amount of mass in clusters of galaxies, because only a small part of this mass is visible. Astronomers do not know why the rest of the mass would be invisible. They think some of it may be hidden in black holes or some may be hidden in the form of particles that are hard to detect.

The Future of the Universe

Scientists do not know what will happen in the universe billions of years from now, but they have some ideas. Many astronomers think the universe is **open:** it will continue to expand forever. Others believe the universe is **closed:** it will eventually stop expanding and start contracting.

How closely mass is packed determines whether the universe will ever contract. If matter is not spread out too much, it could produce enough gravity to stop the present expansion and pull the galaxies back together. All matter would one day be tightly packed again. Some astronomers think another big bang would then occur. They wonder if our universe is only the most recent in a series of big bangs.

If matter is spread out enough, the universe will keep expanding forever. Many astronomers are trying to learn how spread out matter is. Someday they may find an answer to their question.

At the left is a photograph of a distant cluster of stars. You might look at the photograph and wonder how and why this object exists. Or you might see lightning and wonder how and why it happens. In your lifetime, many of the questions of physical science will be answered. Also, many new questions will be asked. As long as people are curious about the universe, they will search for answers to these questions. Perhaps you will become one of the many people who contribute to our knowledge of how our world and our universe work.

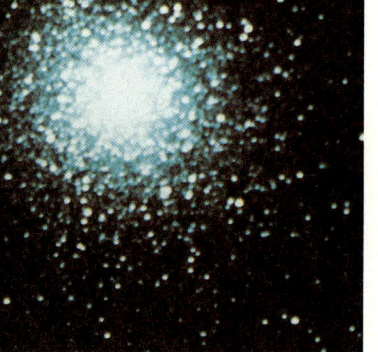
Cluster of stars in the constellation Hercules

Review It

1. What discovery supports the big bang theory?
2. How do astronomers account for invisible mass?
3. Contrast the theories of an open and closed universe.

Issues in Physical Science

Messages from Outer Space

"Long distance calling from 10,000,000,000 kilometers away!"

Many scientists think we will be receiving such a message some day. These scientists are convinced that other forms of intelligent life exist in our universe. With the many trillions of stars in the universe, some stars might have planets on which life could survive. Many scientists believe that intelligent beings from these planets might try to contact us.

Other scientists think the idea of other life in the universe is nonsense. They claim that extraterrestrial life, that is, life on other planets, is just science fiction. They could be right. However, many ideas that once seemed fantastic are now reality. No one in the sixteenth century really believed in Leonardo da Vinci's flying machines. Jules Verne's stories about trips to the moon certainly could not be taken seriously. Or could they?

To say "never" or "impossible" is dangerous in science. The lead story in this month's science fiction magazine may be next year's scientific breakthrough. So, the search for *extra*terrestrial *i*ntelligence (SETI) was begun. In 1960, some astronomers began conducting a number of special projects. The Australian radio telescope shown was one of those used to listen for messages from other intelligent beings. Earth messages have also been sent to other parts of the universe. These messages tell about us. Messages were also carried by the *Pioneer* and *Voyager* spacecraft.

Of course, many questions arose during the SETI project. For example, what language should we use to talk with other intelligent beings? In what parts of the sky should we search? What happens if we do contact other intelligent life? NASA (National Aeronautics and Space Administration) scientists are improving the high-powered, special radio receivers they use for SETI. Should we spend tax money on SETI? Or should we solve problems on earth first?

For Discussion
1. What efforts have been made to communicate with extraterrestrial beings?
2. Give some possible answers to the questions arising from SETI research.
3. Would you spend tax money on SETI research?

Chapter Summary

- Gravity pulls together gas and dust which gradually form stars.   (25–1)
- Nuclear fusion in a star gives off energy that makes the star shine.   (25–1)
- A star becomes a red giant when fusion moves from its center to the star's middle layers.   (25–2)
- A star's mass determines whether it becomes a white dwarf, a neutron star, or a black hole.   (25–2)
- A black hole can be detected by its effect on space and matter around it.   (25–3)
- Galaxies can have regular or irregular shapes.   (25–3)
- Distant galaxies have large red shifts, which indicate that they are moving away from us very quickly.   (25–3)
- Quasars are believed to be the most distant objects in the universe.   (25–3)
- According to the big bang theory, the universe formed about 15 billion years ago in a giant explosion and has been expanding ever since.   (25–4)
- Clusters of galaxies seem to have a huge amount of invisible mass.   (25–4)
- The universe's future depends on how much the matter in it is spread out.   (25–4)

Interesting Reading

Adler, Irving. *The Stars: Decoding Their Messages*. Revised ed. Crowell, 1980. Explains how astronomers get their information about stars, galaxies, and space in general.

Apfel, Necia. *It's All Relative*. Lothrop, Lee, and Shepard Books, 1981. Discusses Einstein's theory of relativity and its predictions about the future of the universe.

Jaber, William. *Exploring the Sun*. Messner, 1980. Describes the sun comprehensively. Includes a history of our understanding of the sun.

Questions/Problems

1. Explain how conditions of gravity and pressure change during a star's lifetime.
2. How would studying the sun help us learn about other stars?
3. How does mass affect the length of a star's lifetime?
4. Compare and contrast a neutron star and a black hole.
5. What might happen to a spaceship as it neared a black hole?
6. What would a blue shift in a galaxy's spectrum indicate?
7. What characteristics of quasars make them seem like stars? like galaxies?
8. Is it possible that the universe is not the only one that ever existed? Explain your answer.

Extra Research

1. Visit a planetarium in your area and attend its sky show. Report on what you learn.
2. On a clear night, observe the stars with a pair of binoculars for at least half an hour. Note the differences in color and brightness among the stars and how they move across the sky.
3. Obtain a map of the night sky for the current time of the year and find as many constellations and planets as you can. By carefully observing the night sky once each week, you can see the movement of some planets among the stars. Use reference books in the library to explain what you observed.

Chapter Test

A. Vocabulary Write the numbers 1–10 on a piece of paper. Match the definition in Column I with the term it defines in Column II.

Column I

1. a large group of stars held together by gravity
2. the universe, if it continues to expand forever
3. the idea that the universe began with a giant explosion
4. indicates that an object is moving away
5. a huge cloud of gas and dust in space
6. objects that have the largest known red shifts
7. an exploding supergiant
8. a neutron star that gives off regular pulses of radio waves
9. an extremely dense object from which no light could escape
10. halo surrounding a star

Column II

a. big bang theory
b. black hole
c. corona
d. galaxy
e. nebula
f. open
g. pulsar
h. quasars
i. red shift
j. supernova

B. Multiple Choice Write the numbers 1–10 on your paper. Choose the letter that best completes the statement or answers the question.

1. A star is born when   a) gravity balances gas pressure.   b) it throws off its outer layers.   c) electron pressure balances neutron pressure.   d) matter expands.

2. The fusion of hydrogen in a star   a) occurs at low temperatures.   b) produces helium.   c) is rare.   d) lasts only a few years.

3. Stars with much more mass than the sun   a) live longer than the sun.   b) burn up hydrogen slowly.   c) will explode.   d) become white dwarfs.

4. Fusion does *not* occur in a   a) normal star.   b) red giant.   c) white dwarf.   d) supergiant.

5. Quasars   a) give off little energy.   b) give off light waves, radio waves, and X rays.   c) look like galaxies.   d) a, b, and c.

6. Galaxies   a) have oval shapes only.   b) have spiral shapes only.   c) do not move.   d) usually move away from each other.

7. Clusters of galaxies   a) are small objects.   b) have large masses.   c) have weak gravity.   d) break apart.

8. Scientists believe the universe   a) is five million years old.   b) was once a dense mass.   c) began in an explosion.   d) b and c.

9. The discovery of weak radio waves coming from all directions   a) supports the big bang theory.   b) proves the universe is closed.   c) proves the universe is open.   d) none of the above.

10. Scientists believe the universe is expanding because   a) distant galaxies have large red shifts.   b) most galaxies are speeding outward through space.   c) distant galaxies move the fastest.   d) a, b, and c.

Careers

Oil tanker crew member
Try to picture a ship the size of the Sears Tower—the tallest building in the world. Supertankers, which are floating oil carriers, are about that large. Without tanker crews, these big ships and their energy cargo would never get to their destinations.

Tanker crew members go with their ship from country to country as it picks up or delivers oil. The crew operates the equipment that pumps huge amounts of oil in and out of the tanker. At sea, the crew lives on board and tends, cleans, and repairs the ship.

Tanker crew members start in the lowest position or rank. As they gain experience, they take exams to get licenses and move up in responsibility.
Career Information:
Office of Maritime Manpower, Maritime Administration, U.S. Department of Commerce, Washington, DC 20230

Petroleum engineer
When drillers strike oil, it does not come gushing out of the ground. In fact, very little oil comes up on its own. Petroleum engineers design ways to bring the sluggish oil up from below.

Petroleum engineers work to increase oil recovery. They plan equipment that forces petroleum out of the earth. The engineers might inject water or chemicals into a well to move the oil, or they might use suction to vacuum up the oil. Engineers test their plans in the oil field. If the machines work, more fuel is available for everyone.

Petroleum engineers spend four years in college, where they learn about geology, oil rigs, chemistry, and machines.
Career Information:
Society of Petroleum Engineers of AIME, 6200 N. Central Expressway, Dallas, TX 75206

Nuclear plant technician
We build nuclear plants to make electricity. But we must use them very carefully. A nuclear plant technician oversees these powerhouses and keeps them running smoothly.

Hundreds of instruments keep track of the energy nuclear power plants make. Technicians check the dials, gauges, and controls in a plant for signs of trouble. They spot problems by comparing their instrument readings with past records. If they find something wrong, technicians repair the damage at once.

Nuclear power plant equipment is complex. Some technicians go to college to learn how power is produced at these energy sources. Others train on the job.
Career Information:
Publications Office, Atomic Industrial Forum, Inc., 7101 Wisconsin Ave., Bethesda, MD 20814

Astronomy technician

What does the inside of a moon crater look like? You could ask astronomy technicians. They help astronomers study objects in space.

Astronomy technicians work in observatories or universities. Some design and build equipment for the telescopes or run computers. Some technicians work with astronomers during long hours of night observation. Others help astronomers with calculations using the data recorded at the telescope and computers.

Astronomy technicians have varied backgrounds. Some are electronics technicians or machinists with two-year degrees from vocational and technical schools. Others are electronics or optical engineers with college degrees in physics, engineering, or astronomy. Technicians gain experience on the job.

Career Information:
American Astronomical Society, Education Officer, Sharp Laboratory, University of Delaware, Newark, DE 19711

Environmentalist

The world's population is increasing. The earth's size is not. For all of us to survive, we must learn to use our resources and our environment wisely. An environmentalist studies and does research to help us do these things.

Environmentalists record data on how people affect their surroundings. They might study air or water pollution, deforestation, energy use, or other problems. From their data, environmentalists set standards and plan ways to improve how people use their natural resources and habitat.

The study of the environment is a branch of science all its own. Students learn the principles of ecology and conservation in college.

Career Information:
Environmental Protection Agency, Public Information Office, 401 M St., SW, Washington DC 20460

Uranium mine inspector

The discovery of radioactivity gained importance in the 1940s when scientists developed nuclear power. Today, nuclear power plants run on uranium ore, which is removed from the earth by miners.

Working with uranium can be very dangerous. Uranium mine inspectors visit mines to insure that conditions in the mines comply with safety laws. If the inspectors find any hazards, they discuss the problems with the managers of the mines. Inspectors also prepare reports and make sure that the hazardous conditions are corrected.

Mine inspectors work for the government. They have either a four- or five-year college degree in mining engineering or about five years experience in mining.

Career Information:
Mine Safety and Health Administration, Office of Information, 4015 Wilson Blvd., Arlington, VA 22203

Acknowledgments

All photographs not credited are the property of Scott, Foresman and Company.
XVI, Courtesy Office National d'Etudes et de Recherches Aérospatiales; **2,** Joe Munroe/Photo Researchers; **4,** Joyce Stoner, The Henry Francis du Pont Winterthur Museum; **5,** Andy Levin/Black Star; **6,** Dan McCoy/Rainbow; **7,** VANSCAN™ Thermogram by Daedalus Enterprises, Inc. Courtesy National Geographic Magazine; **8, 9,** Dan McCoy/Rainbow; **10,** NASA; **17,** Courtesy Goodyear Rubber Co.; **22,** M. M. Bruce/Tom Stack & Associates; **33,** Jay M. Pasachoff; **31,** Roger Tory Peterson/Photo Researchers; **38,** Hank Morgan/Rainbow; **40,** Duguay/Focus on Sports, Inc.; **41,** NASA; **44,** Jerry Wachter/F.O.S. Inc.; **45,** Kevin Fitzgerald/F.O.S. Inc.; **46,** Used by permission. COURTESY OF TWENTIETH CENTURY-FOX. Copyright © 1976 TWENTIETH CENTURY-FOX FILM CORP. ALL RIGHTS RESERVED. **48,** (l) Dr. Harold Edgerton; **49,** (l) NASA, (r) Joe Di Maggio/Peter Arnold; **56,** Dennis Yeandle/Black Star; **61,** J. Kelly/F.O.S. Inc.; **63,** (c) L. V. Smith/Prototype Pix/Design Photographers International; **72,** Shostal Associates; **74,** Dr. Harold Edgerton; **75,** Dave Waterman/Woodfin Camp; **79, 81** NASA; **88,** (l) Cletes Reaves/Freelance Photographers Guild/Alpha, (c) Taurus Photos, (r) George Hunter/FPG/Alpha; **89,** (l) FPG/Alpha, (r) Bonnie Freer/Peter Arnold; **90,** Dan McCoy/Black Star; **92,** Lawrence B. Dodge/Uniphoto; **110,** Craig Varden/Taurus; **113,** Courtesy Ford Motor Company; **114,** (tl) Russell Kelly/F.O.S. Inc., (tr) E. R. Degginger, (bl) John G. Ross/Photo Researchers, (bc) John Webber/Tom Stack & Associates, (br) Don James/Photo Researchers; **116,** Don Marley/F.O.S. Inc. **120,** Andrew Bernstein/F.O.S. Inc.; **121,** Tom Myers/Tom Stack & Associates; **125,** Philip J. Bailey/Taurus Photos; **128,** Dan McCoy/Black Star; **132,** Cameramann International; **133,** VANSCAM™ Thermogram by Daedalus Enterprises, Inc. Courtesy National Geographic Magazine; **141,** Wil Blanche/DPI; **143,** Dan McCoy/Rainbow; **148,** (l) W.J. Choroszewski/FPG/Alpha, (r) Cameramann International; **149,** (l) Tom Tracy/FPG/Alpha, (c) Earl Roberge/Photo Researchers;

150, Photograph by M. Ohtsuki, courtesy of A. V. Crewe, University of Chicago; **152,** Karl Kummels/Shostal; **154,** Kitt Peak National Observatory; **155,** Kenneth Fink/Photo Researchers; **157,** Copyright by California Institute of Technology and the Carnegie Institution of Washington. Reproduced by permission of the Hale Observatories; **160,** (l) E. R. Degginger; (r) John H. Gerard; **161,** (bl) E. R. Degginger, (bc) John H. Gerard; **162,** G. D. Plage/Bruce Coleman; **165,** E. R. Degginger; **170,** J. Alex Langley/DPI; **172,** (l) John H. Gerard, (r) E. R. Degginger; **173,** (tl) John H. Gerard, (tc) Runk/Schoenberger/Grant Heilman, (tr) John H. Gerald/Field Museum, Chicago, (cl, cc) E. R. Degginger, (cr) Runk/Schoenberger/ Grant Heilman, (bl) E. R. Degginger, (br) John H. Gerard/Illinois State museum; **174,** (t) Fred Ward/Black Star, (b) Don Brewster/Bruce Coleman; **175,** (l,c,r) E. R. Degginger; **180,** E. R. Degginger; **181,** E. R. Degginger; **183,** University of California, Berkeley; **186,** Erich Lessing/Magnum; **190,** Courtesy Sargent-Welch Scientific Company; **191,** Fritz Goro; **200,** (l) John H. Gerard/Field Museum, Chicago, (c,r) John H. Gerard; **201,** (all) Life Science Library, *Matter* © 1963 Time Inc., Time-Life Books, Publisher; **203,** Stichting Johan Maurits Van Nassau Mauritshuis, The Hague; **206,** Jay M. Pasachoff; **209,** Jon Brenneis/FPG; **210,** (t) Dan McCoy/Rainbow, (b) Courtesy Brookhaven National Laboratory; **211,** Dan McCoy/Rainbow; **214,** Lester Sloan/Woodfin Camp; **218,** NASA; **221,** California Institute of Technology; **223,** Stanford University; **226,** (l) Cameramann International, (c) FPG/Alpha, (r) Harvey Shaman/FPG/Alpha; **227,** (c) Jacqueline Durand, (r) Cameramann International; **228,** Herb Comess; **243,** Courtesy Trilling Resources Ltd.; **246,** Lee Boltin; **253,** E. R. Degginger; **257,** Lowell Georgia/Photo Researchers; **259,** NASA; **262,** Joe Munroe/Photo Researchers; **265,** Llewellyn/Uniphoto; **266,** Linda Moore/Rainbow; **269,** Dean Brown/Nancy Palmer Agency; **270,** Courtesy A.C. Spark Plug Division, General Motors; **272,** Roland Michaud/Woodfin Camp; **279,** Ivan Massar/Black Star; **282,** Photo courtesy of the Higgins Armory Museum, Worcester, Massachusetts; **292,** Courtesy Micro Essential Laboratory, Inc.; **293,** Ted Spiegel/Black Star; **300,** (l) George Hunter/FPG/Alpha, (c) St. Vincent's Hospital/Peter Arnold, (r) Martin M. Rotker/Taurus; **301,** (l) Craig Hartley/FPG/Alpha, (r) Courtesy of International Molding Machine Company; **302,** Dan McCoy/Rainbow; **304,** Jay M. Pasachoff; **307,** Jon Ker/Shostal; **314,** Robert Weinreb/Bruce Coleman; **316,** Jeff Apoian/Photo Researchers; **318,** NASA; **319,** Rainbow; **322,** H. Trachman/Shostal; **325,** Photo Trends; **328,** Runk/Schoenberger/Grant Heilman; **329,** Jay M. Pasachoff; **333,** Dan McCoy/Rainbow; **334,** Fritz Goro/LIFE Magazine © Time Inc.; **335,** Fritz Goro; **336,** (l) Michael Pettypool/Uniphoto, Anthony Howarth/Woodfin Camp; **337** (both) J. Pavlovsky/Sygma; **340,** Judith Aronson/Peter Arnold; **347,** Rainbow; **354,** N. Smythe/Photo Researchers; **357,** (l) Yerkes Observatory, University of Chicago, (r) Pasachoff Educational Trusts and I.M. Kopylov; **358,** Runk/Schoenberger/Grant Heilman; **362,** Dennis Hallinan/FPG; **366,** (t) Martin M. Rotker/Taurus, (b) John D. Dunning/Photo Researchers; **378,** (l) Yoram Kohana/Peter Arnold, (c) Windsor Duel/FPG/Alpha, (r) FPG/Alpha; **379,** (l) FPG/Alpha, (c) Cameramann International; **380,** Courtesy Western Electric; **382,** Dave Baumhefner/NOAA; **386,** Jay M. Pasachoff; **394,** Brian Parker/Tom Stack & Associates; **397,** Bill Pierce/Sygma;

400, E. R. Degginger; **404,** NOAA; **406,** Courtesy Bell Laboratories; **409,** L. L. Rhodes/Taurus; **414,** Shostal; **415,** Courtesy The Budd Company; **422,** University of Pennsylvania; **424, (l)** Courtesy Bell Laboratories, **(r)** Cameramann International; **425, (t)** Courtesy Bell Laboratories, **(b)** Courtesy Western Electric; **426 (t, c)** Cameramann International, **(b)** Courtesy A.T. & T.; **427,** Courtesy Bell Laboratories; **428, (l)** James Sugar/Black Star, **(r)** Singer Aerospace and Marine Systems, Link Flight Simulation Division; **429,** Cameramann International; **431,** Dan McCoy/Rainbow; **434, (c)** Judith Aronson/Peter Arnold **(r)** FPG/Alpha; **435, (l)** Gerard Fritz/FPG/Alpha, **(c)** Sybil Shelton/Peter Arnold, **(r)** Tom Carroll/FPG/Alpha; **436,** Dan McCoy/Rainbow; **438,** Bart Bartholomew/Uniphoto; **440,** Michael J. Pettypool/Uniphoto; **441,** DPI; **444,** Courtesy Field Museum of Natural History, Chicago; **445, (l)** Shostal, **(c,b)** Ted Spiegel/Black Star; **446,** Camera Press, London/Photo Trends; **448, (t)** N. Groffman/FPG, **(b)** NASA; **449, (t)** Courtesy Power Authority of the State of New York, Tom Sobolik/Black Star; **451,** Jean Guichard/Sygma; **453,** Dave Baird/Tom Stack & Associates; **455,** William Hubell/Woodfin Camp; **458, 460,** California Institute of Technology; **462, (t)** Deutsches Museum, Jay M. Pasachoff, **(b)** Tersch Enterprises; **465, (l,r)** California Institute of Technology; **468, (l)** California Institute of Technology, **(c)** © Association of Universities for Research in Astronomy, Inc., The Kitt Peak National Observatory, **(r)** Kitt Peak National Observatory–Cerro Tololo Inter-American Observatory; **470, (t)** California Institute of Technology, **(b)** Courtesy Harvey Tananbaum, Harvard/Smithsonian Center for Astrophysics; **473,** California Institute of Technology; **474,** Ben Mayer; **475,** Jay M. Pasachoff; **478, (l)** David Bartruff/FPG/Alpha, **(c)** FPG/Alpha, **(r)** Peter Beck/FPG/Alpha; **479, (l)** Lee Foster/FPG/Alpha, **(c)** Cameramann International, **(r)** Courtesy American Petroleum Institute.

Art and photograph direction by: Ligature Publishing Services, Inc.

GLOSSARY

a hat	i it	oi oil	ch child		a in about
ā age	ī ice	ou out	ng long		e in taken
ä far	o hot	u cup	sh she	ə =	i in pencil
e let	ō open	u̇ put	th thin		o in lemon
ē equal	ô order	ü rule	ᴛʜ then		u in circus
ėr term			zh measure		

acceleration (ak sel′ə rā′shən): rate of change of velocity, usually expressed in meters/second/second

acceleration due to gravity: rate at which the velocity of a freely falling object changes as it falls under the influence of gravity alone; about 10 meters/second/second

acid: substance that can give up hydrogen ions to form hydronium ions in water

action force: any force applied to an object; a term used to distinguish this applied force from the reaction force of Newton's third law

air resistance (ri zis′təns): frictional force from air molecules hitting an object as it moves through the air

alternating current (a.c.): electric current that reverses its direction many times each second

ampere (am′pir): unit used to measure an amount of current

amplifier (am′plə fī′ər): electronic device that increases the strength of changes in electrical signals it receives

amplitude (am′plə tüd): height of a wave from its mid position to its crest or trough

astronomy: branch of science that deals with observing and interpreting the motion and structure of and radiation from heavenly objects

atom: smallest bit of an element that has all the characteristics of that element

atomic mass: mass of an average atom of an element, taking into account the percentages of its isotopes that exist naturally

atomic number: number of protons in an atom's nucleus

aurora: light given off as charged particles move in the earth's atmosphere and magnetic field; also known as the northern or southern lights, because they are most often seen near the poles

average acceleration: total change in an object's velocity during the time the object took to make that change

average speed: speed at which an object would have to move steadily to travel a distance in a given time; found by dividing the total distance by that time

axis (ak′sis): imaginary straight line around which an object rotates

balance: instrument used to measure the amount of mass of a substance by comparing its mass with that of a known mass

balanced forces: forces that cancel each other out because they are equal in strength and opposite in direction

base: substance that accepts hydrogen ions and forms hydroxide ions in water

big bang theory: well-established theory that the universe started expanding from a very hot, dense condition about 15 billion years ago

biomass: possible fuel that comes from living organisms

black hole: region of space in which so much mass is concentrated that nothing, not even light, can escape; the result of the death of massive stars

boiling point: temperature at which a substance changes from a liquid to a gas under normal atmospheric pressure

calorimeter (kal′ə rim′ə tər): closed container that is used to measure the amount of heat given off or taken in by a material

catalyst (kat′əl ist): substance that changes the rate at which a reaction takes place without itself being affected

Celsius temperature scale: system of measuring temperature in which water boils at 100° and freezes at 0°

center of gravity: place on an object around which torques are balanced and from which the object could be suspended and remain motionless; the single point in an object at which gravity can be assumed to be acting

centripetal (sen tri′pə tel) **force:** force pulling an object inward toward the center of a curving path

chain reaction: reaction that, once started, keeps itself going as one individual reaction causes adjacent ones to take place

chemical bond: attractive force that binds atoms together

chemical change: change in a substance or substances that regroups the atoms to make a new substance with different chemical properties

chemical energy: form of potential energy that depends on the arrangement of atoms, molecules, or ions in a substance

chemical equation: set of symbols identifying the substances involved before and after a chemical reaction

chemical formula (fôr′myə lə): arrangement of symbols that indicates the kind of atoms and their numbers in a compound

chemical property: characteristic of a substance that describes how the substance reacts with other substances

chemical reaction: one step in or part of a chemical change between substances

chemically stable: condition of an atom when its outer electron level is filled

chemistry: branch of science that deals with the properties, composition, and structure of matter, changes in matter, and any accompanying changes in energy

chip: tiny piece of silicon containing layers of impurities to make the equivalent of thousands of circuits

circuit (sėr′kit): closed path around which a current flows

circuit breaker: safety device that switches off when too much current flows in a circuit

closed universe: universe that is dense enough to stop expanding eventually and begin to contract

cluster: large group of galaxies

colloid (kol′oid): suspension in which the particles of one substance remain mixed in another

compass: object used to tell direction; a magnetic compass is a free-swinging magnet allowed to align itself with the earth's magnetic poles

component (kəm pō′nənt): any electronic part
compound (kom′pound): substance that results when two or more elements join together chemically in definite amounts
compound machine: combination of two or more simple machines
compression (kəm presh′ən): place in a compressional wave where the density of the material's particles is relatively high
compressional wave: way of transmitting energy through slight motions of matter back and forth along the direction the energy is traveling
computer: electronic device that receives, processes, and presents data with great speed and accuracy; usually consists of input-output devices, storage, arithmetic, and logic units, and a central processor unit
concave lens: lens whose middle part is thinner than its edges so that it spreads out light rays
concave mirror: mirror whose middle part curves away from an object so that it reflects light rays together from that object
concentrated (kon′sən trā′tid) *concentrated solution:* solution with a relatively large amount of solute dissolved in it
condensation (kon′den sā′shən): process by which a gas changes to a liquid even though the gas has not reached its boiling point
condensation reaction: chemical reaction between hydrocarbons or their derivatives leading to the formation of a large molecule and a simple one such as water
conduction: transfer of energy by direct contact between particles of a substance
conductor (kən duk′tər) *of heat:* material through which heat passes easily
of electricity: material through which electric current passes easily
conservation: saving energy, using less energy, or using energy wisely
constant: unchanging; remaining the same
contraction: decrease in the size of an object or material
convection: transfer of energy by the movement of rather large masses of a liquid or gas
convection current: liquid or gas moving during the process of convection
convex lens: lens whose middle part is thicker than its edges so that it bends light rays together
convex mirror: mirror whose center curves toward an object so that it spreads out light rays from the object
cornea (kor′nē ə): the eye's transparent covering that helps focus light rays
corona: outermost layer of the sun which becomes visible during a total solar eclipse
covalent (kō vā′lənt) **bond:** chemical bond that results when two atoms share a pair of electrons
cracking: process of breaking down large, single-bonded hydrocarbons into smaller, double-bonded ones
crest: top of a transverse wave
crystal (kris′tl): structural unit in most solids that has an orderly, repeating arrangement
current (kėr′ ənt): flow, such as a flow of water or electric charge, the latter expressed in amperes

data: recorded observations or facts about a substance or an event
deceleration (dē sel′ə rā′shən): rate at which an object's velocity is decreasing; a kind of acceleration that indicates a slowing down
decibel (des′ə bel): one-tenth of a bel, a unit used to express the intensity of sound; an increase in intensity of ten decibels means that the intensity of the sound is multiplied by 10
decomposition (dē′kom pə zish′ən) **reaction:** chemical reaction in which one substance breaks down into simpler substances
density: amount of mass in a standard volume; found by dividing mass by volume and usually expressed in grams/centimeter^3 or grams/milliliter
diatomic (dī′ə tom′ik) **molecule:** molecule made up of two atoms that are covalently bonded
dilute (də lüt′) *dilute solution:* solution with only a small amount of solute dissolved in it
direct current (d.c.): electric current that flows in a single direction
displacement: measure of the distance and direction that an object moves
distance: length of a path over which an object travels, usually expressed in meters or kilometers
domain: small region in a magnet in which the magnetism of most of the atoms lines up
Doppler (dop′lər) **effect:** apparent change in the frequency or wavelength of a wave caused by motion of the observer or the source of the waves relative to each other
dry cell: electrical cell in which current is contained in a gelatin or paste so as not to spill
ductile (duk′təl): able to be drawn into a wire, as some metals are

efficiency: percentage found by comparing the amount of useful work obtained from a machine to the amount of work put into the machine
effort arm: distance on a lever from the point at which a force is exerted to the fulcrum
electrical energy: kind of energy carried by moving charges
electromagnet: temporary magnet formed by passing electric current through a coiled wire
electromagnetic (i lek′trō mag net′ik) **induction:** process by which a changing magnetic field causes an electric current to flow in a wire loop
electromagnetic spectrum: entire range of electromagnetic waves arranged in order of their frequencies or wavelengths
electromagnetic wave: energy traveling through space as a result of changing electricity and magnetism
electromagnetism: theory that shows how electricity and magnetism result from a single force
electron (i lek′tron): negatively charged particle, often found in an atom, that usually moves around the nucleus
electronics: study of moving electrons and their use in devices such as radios, television, and computers
element: any of the simplest substances of matter that cannot be decomposed by chemical means

ellipse (i lips′): kind of curve resembling a squashed circle; planets travel in ellipses

emulsifier (i mul′sə fī′er): substance that keeps the particles of one liquid mixed in another liquid

emulsion (i mul′shən): colloid in which one liquid is mixed in another

endothermic (en′dō thėr′mik) **reaction**: chemical reaction that takes in energy

end point: point at which enough base has been added to an acid to produce a neutral solution

energy: ability to do work, usually expressed in joules

energy level: path of an electron around an atom's nucleus that corresponds to a certain amount of energy

equilibrium (ē kwə lib′rē əm): condition of an object when no net force acts on it

evaporation (i vap′ə rā′shən): process by which fast-moving molecules escape from a liquid, changing the liquid into a gas, even though the liquid has not reached its boiling point

excited *excited state:* condition of an atom when at least one of its electrons is in an energy level above its usual, lowest possible energy level

exothermic (ek′sō thėr′mik) **reaction**: chemical reaction that gives off energy

expansion: increase in the size of an object or material

farsighted: condition in which the eye does not properly focus objects that are relatively nearby

fission (fish′ən): splitting of heavy nuclei into lighter ones

focal point: single location to which parallel rays of light are brought together

force: any action that accelerates an object, expressed in newtons

forced vibration: vibration of an object not at its natural frequency but at the frequency of another object that is causing the vibration

fossil (fos′əl) **fuel**: substance formed over millions of years from the remains of dead organisms that can be burned to release energy; coal, oil, and natural gas

fraction: any of the components that can be separated out of crude oil on the basis of its boiling point

fractional distillation (dis′tl ā′shən): process of separating components of a mixture that have different boiling points by vaporizing each one at its boiling point

frame of reference (ref′ər əns): set of points of view that are not moving in relation to an object or observer

freezing point: temperature at which a substance changes from a liquid to a solid

frequency (frē′kwən sē): number of waves passing a specific point within a unit period of time, expressed in hertz

friction: force between surfaces that resists the motion of one object or surface past another

fulcrum (ful′krəm): point at which a lever is supported

fuse: safety device in which a metal strip melts to open a circuit when too much current flows through it

fusion (fyü′zhən): joining of light nuclei to make a heavier nucleus

galaxy (gal′ek sē): large, distinct group of stars, gas, and dust held together by gravity

gamma waves: electromagnetic waves having frequencies of about 10^{18} hertz and higher, the highest frequencies (and energies) of all the electromagnetic waves

gas: physical state of matter that does not have a definite shape or volume

generator (jen′ə rā′tər): device that uses electromagnetic induction to change mechanical energy into electricity

geothermal (jē′ō thėr′məl) **energy**: energy obtained from the earth's interior heat

graduated cylinder: cylinder marked with regular divisions (graduations) that is used to measure the volume of a liquid

gram, *g:* metric unit of mass

gravitation: attraction between one object and another due to gravity

gravity: force that exists between any two objects because the objects have mass; one of the fundamental forces of nature

group: elements located in a column of the periodic table and having similar chemical properties

half-life: time it takes for half the nuclei in a quantity of a radioactive substance to decay

heat: transfer of energy from one substance to another so that the energy appears as kinetic energy of the atoms in the substance, expressed in joules

heat source: anything that can give off heat because its temperature is higher than that of its surroundings

hertz (hėrts): one cycle per second, a unit used to measure a wave's frequency

hologram (hol′ə gram): image made by using the interference properties of radiation, often made with a laser; holograms show more information than a photograph does

hydrocarbon (hī′drō kär′bən): any of a group of compounds that contain only carbon and hydrogen

hydroelectric (hī′drō i lek′trik) **power**: power from moving water turning a turbine that turns a generator to produce electricity

hydronium (hī drō′nē əm) **ion**: H_3O^+, an ion containing the equivalent of an ordinary water molecule and an extra proton; forms when an acid is added to water

hydroxide ion: OH^-, an ion containing oxygen and hydrogen; forms when a base is added to water

hypothesis (hī poth′ə sis; plural, hī poth′ə sēz′): reasonable guess about how or why an event happens; hypotheses form the basis for further testing

inclined plane: simple machine consisting of a slanting flat surface

indicator: dye that changes color at a certain pH value

inertia (in ėr′shə): tendency of an object to maintain its motion or its condition of rest unless something disturbs it

infrared rays: electromagnetic waves having frequencies of about 10^{12}–10^{14} hertz.

insulator (in′sə lā′tər) *of heat:* material through which heat does not pass easily
of electricity: material through which electric current does not pass easily

integrated (in′tə grāt′id) **circuit**: electronic chip that contains the equivalent of thousands of electronic components and circuits

intensity (in ten/sə tē): strength of light falling on a certain area

interference (in/tər fir/əns): situation resulting from overlapping waves in which the waves subtract and cancel each other or add and reinforce each other

internal energy: total energy of all the particles that make up an object

ion (ī/on): atom or molecule that has lost or gained one or more electrons and so has a positive or negative charge

ionic (ī on/ik) **bond:** chemical bond that results when one or more electrons is transferred from one atom to another

isotopes (ī/sə tōps): forms of an element with the same number or protons but different numbers of neutrons

joule (joul or jül), *J:* one newton-meter, a unit of measurement used to express an amount of energy or work

Kepler's first law: statement that each planet orbits the sun in an ellipse, with the sun at one focus

Kepler's second law: statement that the speed of a planet in its orbit varies so that the line joining the planet and the sun sweeps out equal areas in equal times; the effect is that planets move fastest when they are closest to the sun and slowest when they are farthest away

Kepler's third law: statement that the square of the period of revolution of a planet is proportional to the cube of its average distance from the sun; the law tells by how much a planet with a larger orbit has a longer period than a planet with a smaller orbit

kinetic (ki net/ik) **energy:** energy of motion, which depends on an object's mass and the square of its velocity ($\frac{1}{2}mv^2$), expressed in joules

kinetic theory of matter: theory that all matter is made up of small particles that are always in motion

laser (lā/zər): device that amplifies light by causing atoms to give up light of a certain wavelength; it results in light of a single wavelength with all the waves lined up

latent (lāt/nt) **heat:** energy needed to change a substance from one physical state to another

law of conservation of energy: statement that energy cannot be created or destroyed; it can change from one form into another, but the total amount of energy stays the same; valid only in cases where matter and energy are not transformed into each other

law of conservation of mass: statement that matter can change its form but cannot be created or destroyed; valid only in cases where matter and energy are not transformed into each other

law of conservation of mass-energy: statement that the total amount of energy and matter remains the same, though matter and energy can change into each other

law of conservation of momentum: statement that if no outside forces act on a system, the total momentum of all the particles in the system remains the same

lever (lev/ər): simple machine made of a rigid bar making contact with a single pivot point called a fulcrum

light: electromagnetic waves having a frequency of about 10^{14} hertz; also known as the visible spectrum because it acts on the retina of the eye to cause sight

liquid: physical state of matter that has a definite volume but no definite shape

lubricant (lü/brə kənt): substance that reduces the friction that occurs when one surface moves past another surface

luminous (lü/mə nəs): giving off light

luster (lus/tər): property that describes the amount of shine of a surface

magnetic field: region around a magnet in which an object feels a magnetic force

magnetism: one of the forces of nature that exists near a magnet

malleable (mal/ē ə bəl): able to be hammered, rolled, or shaped without breaking, as some metals are

mass: amount of matter that an object has, usually expressed in grams or kilograms

mass number: total number of protons and neutrons in the nucleus of an atom

matter: in large quantities, anything that has mass and takes up space; energy in a different form

mechanical advantage (M.A.): number of times by which a machine increases the force applied to it

mechanical energy: energy an object has because of its motion and the forces acting on it; mechanical energy occurs as both kinetic and potential energy

melting point: temperature at which a substance changes from a solid to a liquid

meniscus (mə nis/kəs): curved surface of a liquid that results from attracting forces between the molecules in the liquid

metal: class of elements that are usually solid at room temperature, exhibit luster, and are good conductors of heat and electricity

metalloid (met/l oid): class of elements that have some properties of metals and some properties of nonmetals

meter, *m:* unit for measuring length in the metric system

metric system: system of measurement, originally set up in the 18th century, based on a few standard units and prefixes that reflect the multiplication of these units by powers of ten

microcomputer: small-sized computer made up of a microprocessor, memory circuits, and circuits that bring information to and from the microprocessor

microgravity (mī/krō grav/ə tē): condition that occurs when objects are accelerating equally so that gravity seems to disappear

microprocessor (mī/krō pros/es ər): chip that contains all a computer's problem-solving circuits

mixture (miks/chər): combination of two or more substances in variable amounts in such a way that they are not chemically combined

molecule (mol/ə kyül): group of atoms held together by covalent bonds; the smallest example of a covalently bonded substance

momentum (mō men/təm): measure of the strength of an object's motion, depending on both an object's mass and its velocity

motion: *See* relative motion

natural frequency: an object's characteristic frequency of vibration that depends on an object's size, shape, and material

nearsighted: condition in which the eye does not properly focus objects that are relatively far away

nebula (neb′yə lə): cloud of gas and dust in space

net force: resulting force on an object after all the forces acting on it and their directions are added up

neutral *neutral solution:* having a pH of 7, because the number of hydronium ions equals the number of hydroxide ions

neutralization (nü′trə lə zā′shən) **reaction:** chemical reaction in which an acid and a base destroy each other, forming a salt and water

neutron (nü′tron): particle having no electric charge and found either by itself or in a nucleus

neutron star: star in which the inward force of gravity is balanced by the outward pressure of tightly packed neutrons

newton, N: metric unit of measurement that tells the strength of a force; 1 N = 1 kg·m/s/s

newton-meter, $N \cdot m$: unit of measurement of work

Newton's first law of motion: statement that an object will stay at rest or continue to move in the same direction and at the same speed unless a force acts on it

Newton's law of gravity: statement that the force of gravity between two objects increases with the masses of the objects and decreases with the square of the distance between them

Newton's second law of motion: statement that an object accelerates because a force acts on it; F = ma

Newton's third law of motion: statement that for every action force, there is a reaction force of equal strength but opposite in direction

noise: unpleasant or unwanted sound

nonmetal: class of elements whose properties are opposite to the properties of metals

nonrenewable resource: any energy resource that is not being replaced in a reasonable amount of time as it is being used up

north pole *of a magnet:* end that always points north when the magnet is allowed to turn freely

nuclear energy: energy that is released when mass is transformed into energy as atoms split apart (undergo fission) or join together (undergo fusion)

nucleus (nü′klē əs; plural, nü′klē ī): positively charged core of an atom that contains neutrons and protons

ohm (ōm): unit used to express the amount of electrical resistance

Ohm's law: statement that the current in a circuit is determined by the voltage and the resistance in the circuit; I = V/R

open universe: universe that would keep expanding forever

optical fiber (op′tə kəl fī′bər): narrow, transparent tube through which light travels, even around curves, because of internal reflection

optical pyrometer (op′tə kəl pī rom′ə tər): instrument that measures the temperature of a glowing object by the color of light the object gives off

orbit (or′bit): periodic path of one object around another resulting from a force between them

parallel circuit circuit connecting several objects, in which the removal of one object does not break the circuit to the other objects

period *of an orbit:* length of time it takes to complete one revolution around a central object
in the periodic table: row of elements

periodic (pir′ē od′ik) **table of elements:** arrangement of all the known elements in a table in order of their atomic numbers

pH scale: range of numbers from 0-14 that indicates the relative acidic or basic character of a solution; a pH lower than 7 indicates an acid, a pH of 7 indicates a neutral solution, and a pH greater than 7 indicates a base

physical change: change in a substance that does not alter its identity or nature

physical property: property of a substance that remains as long as the substance does not undergo a chemical change

pitch *of a screw:* number of threads that are in a unit length of a screw
of sound: how high or low a sound seems, resulting mainly from the sound's frequency

photon (fō′ton): bundle of energy given off or taken in by an atom or nucleus as it changes its energy state

physical state: condition in which matter exists, such as solid, liquid, or gas

physics: branch of science that deals with the fundamental principles and laws that govern the behavior of matter and energy in the universe

plane mirror: smooth, flat surface that reflects light

polarized (pō′lə rīzd′) **wave:** set of waves that are all vibrating in the same plane

pole *of a magnet:* location where the magnetism is strongest; the poles of a magnet are often at its ends

pollutant (pə lüt′nt): any substance that is harmful to the environment

polymerization (pol′ə mər ə zā′shən) **reaction:** chemical reaction in which chains of double-bonded hydrocarbon molecules join together to form one giant single-bonded molecule

potential (pə ten′shəl) **energy:** energy an object has because of its position (mgh when due to gravity), expressed in joules

power: rate at which work is done or energy is used; found by dividing work (or energy) by time and expressed in watts

pressure (presh′ər): force per unit of area

prism (priz′əm): wedge-shaped transparent object that separates light into its different wavelengths (colors)

product: newly formed substance that results from a chemical reaction

program: step-by-step instructions that tell a computer what to do

projectile (prə jek′təl): any freely falling object that was thrown, hurled, hit, or shot forward

proton (prō′ton): positively charged particle found free or in a nucleus

pulley: simple machine consisting of a rope that passes over a grooved wheel
pulsar (pul′sär): dying or dead star from which we receive regular pulses of radio waves; pulsars are thought to be rotating neutron stars

quark (kwôrk): basic unit that makes up nuclear particles
quasar (kwā′sär): object in the sky whose image is close to that of a star but whose spectrum reveals that it is traveling away very fast; quasars are thought to be exceptionally bright cores of distant galaxies

radiant (rā′dē ənt) **energy:** energy that is carried through space in the form of electromagnetic waves
radiation (rā′dē ā′shən): either radiant energy or subatomic particles given off during radioactivity or fission
radio waves: electromagnetic waves having frequencies less than about 10^{11} hertz, the lowest frequencies (and energies) of all the electromagnetic waves
radioactivity (rā′dē ō ak tiv′ə tē): spontaneous decay of a nucleus into a lighter nucleus, very tiny particles, and radiation
rarefaction (rer′ə fak′shən): any of the places in a compressional wave in which the density of the material's particles is relatively low
reactant (rē ak′tənt): substance that interacts chemically with another substance during a chemical reaction
reaction force: force, called for by Newton's third law, that is equal in strength but opposite in direction to an action force
real image: image formed by a lens or mirror for which the rays of light actually come together and can be focused on a screen
red giant: first stage of a star as it begins to die; a rather large star that glows slightly reddish, since it is relatively cool
red shift: apparent increase in the wavelength (the Doppler effect) of light, resulting from the object or the observer moving away from the other
reference point (ref′ər əns): fixed point with respect to which relative motion can be determined
reflection (ri flek′shən): bouncing of a wave off the surface it hits
refraction (ri frak′shən): change in direction of a wave as it moves from one material to another and the wave's speed changes
relative motion: changing position with respect to the position of another object
renewable resource: energy resource that is being replaced as it is used in a reasonable amount of time
replacement reaction: chemical reaction in which one free element replaces an element that is part of a compound
resistance (ri zis′təns): measure of how much the flow of electric current is opposed, expressed in ohms
resistance arm: distance between the resistance on a lever (the object to be moved) and the fulcrum
resonance (rez′ə nəns): phenomenon that occurs when the vibrations of one object reinforce the vibrations of another object because they have the same natural frequency

retina (ret′n a): in the eyeball, the back, inside lining that is sensitive to light
revolution: motion of an object along a closed orbit around another object
rolling friction: weak, backwards force that arises as a round object rolls over a surface
rotation: spinning of an object about its axis

salt: substance that is formed when a positive ion from a base bonds to a negative ion from an acid as a result of neutralization
saturated (sach′ə rā′tid) *saturated solution:* containing as much solute as can be dissolved at a specific temperature and pressure
science: study of natural events through observation and experimentation resulting in an organized body of knowledge
scientific law: accepted explanation that should apply over and over again throughout the universe
scientific method: logical method often used by scientists to solve problems; involves collecting data, making and testing hypotheses, and drawing conclusions
screw: inclined plane in a spiral form
second, *s:* metric unit to measure time
semiconductor: material that conducts electricity better than an insulator but not as well as a conductor, often made of silicon and germanium combined with other elements, and used in electronic circuits
series circuit: circuit connecting several objects one after the other so that the current follows in a single path
shadow: area that is not lit or is only partially lit because an object is blocking light from reaching it
short circuit: situation that occurs when the current in a circuit is drawn into another path so that not enough current goes into the original circuit
SI: initials for the French words meaning International System of Measurements, the version of the metric system now internationally agreed upon
simple machine: any tool that is made up of only one or two parts
sliding friction: backwards force that exists between the surfaces of objects that are sliding over each other
solar cell: object that gives off electricity when it is hit by sunlight
solar collector: object that traps the sun's energy and heats up a fluid
solar energy: renewable energy source that involves using sunlight to heat a fluid or object or converting sunlight into electricity
solid: physical state of matter that has a definite shape and a definite volume
solid-state: electronic components that are made of solid semiconductors instead of vacuum tubes
solute (sol′yüt): substance that is dissolved by a solvent
solution (sə lü′shən): uniform mixture of particles too small to be seen in which one or more substances is dissolved in another
solvent (sol′vənt): substance that dissolves other substances
sound compressional wave that vibrates the particles of matter; form of mechanical energy that a sound wave has; that which we hear

south pole *of a magnet:* end that always points south when the magnet is allowed to turn freely

specific heat: measure of a material's ability to take in or give off heat; the amount of heat necessary to raise the temperature of 1 gram of a material by 1°C

spectrum (spek′trəm; plural, spek′trə): characteristic pattern of energies, often visible, given off by an element's atoms as they release energy

speed: rate of motion; found by dividing the distance an object moves by the time the object takes to go that distance and usually expressed in meters/second

spontaneous (spon tā′nē əs) **reaction:** chemical reaction that happens without any outside help

star: a heavenly object made up of a large, luminous mass of hot gas held together by its own gravity

static (stat′ik) **friction:** force in the direction that opposes any motion between the surfaces of objects that are touching but not moving past each other

strong force: one of the basic forces of nature; the force that holds the particles in a nucleus together

substitution reaction: chemical reaction in which hydrogen atoms of single-bonded hydrocarbons are replaced with other elements

supergiant star more massive than the sun in a late stage in its life in which it becomes even larger and brighter than a red giant

supernova (su′pər nō və): explosion of a massive, dying supergiant

suspension (sə spen′shən): mixture in which the particles of one substance are scattered in another but are not dissolved; the particles of one substance may separate from the other

sympathetic vibration: vibration of an object at its natural frequency that occurs when another object vibrates at the same frequency

synthesis (sin′thə sis) **reaction:** chemical reaction in which two or more elements, compounds, or both join to make a new compound

synthetic fuel: oil or gas made from coal

temperature: number that is a measure of the average kinetic energy of all the particles in an object or material, expressed in degrees Celsius

terminal (tėr′mən nəl) **velocity:** final velocity that a falling object reaches when air resistance balances the force of gravity

theory (thē′ər ē): well-tested explanation of observations that can make predictions about the outcomes of other tests

thermogram (thėr′mə gram): image of an object made by measuring the infrared rays it gives off

thermostat: instrument that maintains a constant temperature in some object by turning a heat source on and off in response to changing temperatures in the surroundings

thread: spiral path around a screw

torque (tôrk): how strongly an object is turning, which depends both on the strength of a force on the object and the force's distance from the object's axis

transformer (tran′sfôr′mər): device that raises or lowers voltage

transistor (tran zis′tər): solid-state electronic component made of a semiconductor

transverse (tranz vėrs′) **wave:** way of transmitting energy in which the crests and troughs of the wave are perpendicular to the direction in which the wave is traveling

trough (trôf): bottom of a transverse wave

ultrasonic (ul′trə son′ik): sound frequency above 20,000 hertz, which the human ear cannot hear

ultraviolet rays: electromagnetic waves having frequencies of about $10^{15}-10^{16}$ hertz

universal indicator: mixture of indicators that turns a different color at different pH values

vacuum (vak′yü əm): space that contains no matter at all

vacuum tube: early device used to control the flow of electrons; consisting of a glass tube emptied of air containing a positively charged plate that pulls electrons away from a negatively charged wire

vaporization (vā′pər ə zā′shən): process by which a substance at its boiling point changes from a liquid to a gas

velocity (və los′ə tē): quantity giving both the speed and the direction that an object is moving

vibration (vī brā′shən): rapid back-and-forth motion, often resulting in a sound wave

virtual (ver′chü əl) **image:** image at which the light rays appear to meet but actually do not; therefore, a virtual image does not appear on a screen placed at the image's position

viscosity (vi skos′ə tē): property of a liquid that causes it to resist flowing

visible spectrum: part of the electromagnetic spectrum that people can see

volt: unit used to express an amount of voltage

voltage (vōl′tij): push needed to move an electron from one place to another, expressed in volts

wavelength: distance between two similar points in a wave's cycle

wedge: simple machine that consists of an inclined plane with one or two sloping sides

weight: force that gravity exerts on a mass, expressed in newtons

wheel and axle: simple machine that consists of a wheel and a shaft that turns with the wheel and supports its center

white dwarf: rather faint, dead star about the size of the earth but containing about as much mass as the sun; the end state of stars containing as much mass as the sun or less

work: product of the force exerted on an object and the distance the force moves the object in the direction of the force, expressed in newton-meters or joules

X rays: electromagnetic waves having frequencies of about $10^{16}-10^{18}$ hertz

INDEX

absolute zero, 133, 145
acceleration, 32–34
  and gravity, 74–76
  and Newton's second law, 44–46
accelerators, 211
acetic acid, 284, 287
acids, *illus.* 284
  and bases, 284–285
  defined, 287
  and indicators, 290–291
  and neutralization, 294–296
  and pH scale, 292
  properties of, 284, 286–287
  See also Bases
acoustics, 369
action, 48–49
activity series, 274
aerospace engineer, 148
air resistance, 60
alchemy, 203
alpha particle, 213
alternating current (a.c.), 394–395
aluminum, 174, 175, 388
ammonia, 271, 288
amperes, 391
amplifier, 420
amplitude, 307, 365
Andromeda galaxy, 468, illus. 468
anechoic room, 369
Archimedes, 345
architect, 88
Aristotle, 10, 11, 190
assayer, 226
astronomy, 5
astronomy technician, 479
atom
  bonding, 248–250, 254–255
  and chemical equations, 256–257
  and compounds, 232–234, 256–257
  defined, 155, 172
  domains, 406, 407
  and electrical charge, 384–385
  energy levels of, 189, 190–191, 325, 333
  excited, 190, 325, 333
  and ions, 252–253
  and isotopes, 193, 194
  and light, 325, 326, 333
  models of, 188–189, 191 *illus.* 186, 188, 189, 191
  and molecules, 155, 254–255
  and nuclear fission, 216–218, 446–447
  and nuclear fusion, 220–221, 446–447, 461, 464–465
  nucleus of, 189, 208–210, 446
  and radioactivity, 212–214
  in star formation, 461
  See also Matter; Molecules; Nucleus
atomic energy, 188–191, 216–218, 220–221, 446–447
atomic mass, 193–194, 197
atomic number, 192–193, 197
auroras, 404, *illus.* 400, 404
average acceleration, 34
average speed, 29
axis, 66–67
balance, 14
balanced forces, 58–60
bar magnet, 402, 404
bases, *illus.* 285
  and acids, 284–285
  defined, 288
  and indicators, 290–291
  and neutralization, 294–296
  and pH scale, 292
  properties of, 285, 286, 288
  See also Acids
Becquerel, Henri, 212
beta particle, 213
bicycle, 101
big bang theory, 472–473, *illus.* 472
biochemical technician, 300
biomass, 448
bionics, 431
black hole, 466–467, 470, *illus.* 467
block and tackle, 98
Bohr, Niels, 189, 191
boiling point, 165, 177
  fractional distillation, 277
bonding
  in atoms, 248–250
  in carbon compounds, 276–279
  chemical, 248–250
  covalent, 254–255
  ionic, 252–253
boron nitride, 429
broadcast technician, 434
bromine, 273
bubble chamber, 210, *illus.* 210
  particle tracks in, *illus.* 206, 210, 223
calcium, 172, 190
calculator, 429, *illus.* 429
Calorie, 134
calorimeter, 134
camera, 356, *illus.* 356
  ultrasound, 366
carbon
  in compounds, 276–279, *illus.* 278
  covalent bonding of, 255
  diamond, 247, 250, *illus.* 172, 246, 250
  in dry cell, 388
  as element, 172
  graphite, 250, *illus.* 172, 173, 250
carbon dioxide, 178, 249, 270, *illus.* 249
carbon monoxide, 270
carpenter, 88
catalyst, 267
catalytic converter, 270, *illus.* 270
cataracts, 353
Celsius temperature, 13, 132
Centaurus A, 468, *illus.* 468
center of gravity, 61, *illus.* 68
centripetal force, 53
Chadwick, James, 190
chain reaction, 217, *illus.* 217
change
  chemical, 181
  physical, 178
charge, electric, 384–386
chemical
  bond, 248–250
  changes, 181, *illus.* 181
  energy, 120, 137
  equations, 256–257
  properties, 180, 200
  technician, 226
chemical reactions
  in acids and bases, 287–288
  of carbon compounds, 276–277
  defined, 256–257
  in dry cell, 389
  of hydrocarbons, 278–279
  and neutralization, 294–296
  recognizing, 264–267
  replacement, 272–274
  synthesis and decomposition, 270–271
chemist, 226
chemistry, 5
chips, 425, *illus.* 427
chlorine, 177, 252–253, 254–255
chlorophyll, 267
circuit
  breakers, 395, *illus.* 395
  defined, 391, *illus.* 391
  integrated, 425, 427

parallel, 393, *illus.* 393
series, 392, *illus.* 392
short, 395
circular motion, 52–53
coal, 442, 444–445, 447, 453
colloid, 241, 243
color
and chemical reactions, 265
of indicators, 290–292, 296
as physical property, 177, 178
of rainbow, 316
in sky, 331
in spectrum, 328–329, *illus.* 329
in thermogram, 133
combustion, 137
compass, 403, 404
components, electronic, 421, 424–425
compound machines, 106–107, *illus.* 106
compounds, *illus.* 234
and atomic bonding, 249–250
carbon, 276–279
and chemical equations, 256–257
covalent, 254–255
of elements, 232–234
ionic, 253
See also Mixtures
compressional wave, 311, *illus.* 311
computer, 6, 336, 421, 425
microcomputer, 427
microprocessors, 427, 428–429
computer programmer, 434
concave lens, 350, 353, *illus.* 350
concave mirror, 344–345, 357, *illus.* 344–345
concentrated solution, 239
condensation, 165
condensation reaction, 279
conduction, 138
conductor
of electricity, 388–389
of heat, 138
conservation, 455, *illus.* 454
conservation of energy, 116–117, 125
conservation of mass, 257
conservation of mass-energy, 223
conservation of momentum, 50
constant speed, 29
contraction, 142–143
convection, 139
heating, *illus.* 139
convection current, 139

convex lens, 348–349, 352, 353, 357, 358, *illus.* 348–349
convex mirror, 346–347, *illus.* 346
Copernicus, Nicolaus, 84–85
copper, 172, 173, 174, 274, *illus.* 173, 200
in chemical reactions, 265
compound, 233
as conductor, 388
smelting, 273
cornea, 352–353, 462
corona, 462, *illus.* 462
cosmetologist, 301
covalent bond, 254–255
Crab Nebula, 465, 466, *illus.* 465
cracking, 278–279
crest, of wave, 307, 310, 313, 314, 332
crude oil. *See* oil
cryogenics, 145
crystals, 160–161, 177, *illus.* 160–161
bonding in, 250, 253
current
alternating and direct, 394–395, 410
and circuit, 391
convection, 139
electrical, 388–391
and magnetism, 408–410, 412–414
and resistance, voltage, 390–391
curved mirrors, 344–347
Cygnus, 467
Dalton, John, 188
data, 9
deceleration, 34
decibels, 365
decomposition reaction, 271
Democritus, 155
density, 15, 177
deuterium, 193, 194, 221
diamond, 160, 234
bonding in, 250, 255
characteristics of, 247
diatomic molecule, 254–255
dilute solution, 239
direct current (d.c.), 394–395, 410
displacement, 26–27, 94–95
dissolving substances, 178, 238–239
distance, 26–27
and displacement, 26–27
and gravity, 84–85

and planet's speed, 82–83
and rotation, 66–67
and time, speed, 28–29
domain, 406–407, *illus.* 406
Doppler effect, 366–367, 469, *illus.* 367, 469
Doppler, Christian, 367
dry cell, 389, *illus.* 389
ductile, 174
$E=mc^2$, 223
Earth, 84
atmosphere of, 158
magnetic field of, 404
efficiency, 106–107
effort arm, 96
Einstein, Albert, 43, 85, 154, 217, 222–223
electrical energy, 120, 136, 394–395
producing, 414, 441–442
electrical engineer, 435
electricity
and circuits, 391, 392–393
and current, 388–391, 394–395
and electrical charge, 384–386, 388–389
and electronic devices, 420–422
and magnetism, 408–410, 412–414
producing, 442, *illus.* 442
uses of, 394–395
electromagnet, 409
electromagnetic induction, 413, 414
electromagnetic spectrum, 316–317, *illus.* 316–317
electromagnetic train, 415
electromagnetic waves, 316–318
electromagnetism, 413
electron
bonding, 248–249, 252–253, 254–255
and current, 388–389
and electrical charge, 188, 189, 384–386, 388
and electronics, 420–421
energy levels of, 189, 190–191
and ions, 252–253
in nucleus, 208–209
and quarks, 209, 210
See also Neutron; Nucleus; Proton
electron cloud model, 191, *illus.* 191
electronic devices, 420–422

electronics
  and computers, 426–428
  and electromagnetism, 410, 413
  and integrated circuits, 425
  and transistors, 424–425
electronics
  assembler, 149
  technician, 434
elementary particle physicist, 227
elements
  and compounds, 232–234
  defined, 172–173
  metals and nonmetals, 174–175
  names of, 183
  and nuclear fusion, 220–221, 464–465
  and periodic table, 196–197, 200–201, 249, *illus.* 198–199
  and replacement reactions, 272–274
  in stars, 221, 465
  in sun, 462, 464–465
ellipse, 82–83, *illus.* 82
emulsifier, 241
emulsion, 241
endothermic reactions, 266
end point, of neutralization, 296
energy
  atomic, 188–191, 218, 446–447
  in chemical reactions, 266
  choosing resources of, 452–455
  conservation, 455
  conservation of, 116–117, 125
  defined, 4
  and Einstein's equation, 222–223
  electrical, 120, 136, 394–395, 414, 441–442
  and electromagnetic waves, 316–318, 410
  from fission, 216–218, 446–447
  forms of, 120–121
  and fossil fuels, 444–445
  and friction, 118
  from fusion, 220–221, 446–447, 461, 464–465
  and heat, 134, 136–137
  kinetic, 114–115
  and mass, 222–223
  and matter, 154
  measuring, 113
  and nonrenewable resources, 445
  potential, 116–117, 118
  and power, 122–123
  radiant, 120, 140–141
  renewable sources of, 448–449
  sources of, 157, 441–442, 452–453, 455
  transfer of, 122, 138–141
  uses of, 440–442
  of waves, 306–308
  and work, 112–113
  See also Work
energy levels, 189, 190–191, 248–249, 325, 333
ENIAC (Electronic Numerical Integrator and Computer), 422, 425, *illus.* 422
environmentalist, 479
equation
  acceleration, 33, 34
  chemical, 256–257, 270–271, 272–273, 274, 278, 287, 294–295
  efficiency, 107
  energy, 115, 117
  force, 45
  gravity, 78, 85
  mechanical advantage, 103
  in Ohm's law, 390–391
  power, 122, 123
  for speed, velocity, 28–29
  for wave speed, 308
  work, 94–95, 97
equilibrium, 59
evaporation, 165, 178
excited atoms, 190, 325, 333
exothermic reactions, 266
expansion, 142–143
  of universe, 469
eyes, *illus.* 352
  and lasers, 333, 337
  and lenses, 352–354, *illus.* 353
Faraday, Michael, 412
farsightedness, 353
fireblanket, 243
fireworks, 269
fission, 216–218, 446–447, *illus.* 216
fluorine, 249, 254
focal point, 344
force
  centripetal, 53
  defined, 41
  electric, 384–385
  of friction, 40, 62–64
  of magnetism, 403
  and Newton's second law, 44–45
  and rotating objects, 66–68
  strong, 208–209
  and work, 94–95
forced vibration, 374
formula, 233
fossil fuels, 444–445, 452
fractional distillation, 277, *illus.* 277
fractions, 277, *illus.* 277
frame of reference, 25
freezing point, 164
frequency, 307, 308
Fresnel lens, 350
friction
  air resistance, 60
  defined, 40, 41
  effects of, 63–64
  and efficiency, 107
  and energy, 118
  and heat, 63, 137
  kinds of, 62–63
fulcrum, 96
fuse, 395, *illus.* 395
fusion, 220–221, 446–447, 461, 464–465, *illus.* 220, 461
galaxies, 468–469, *illus.* 468
  clusters, 473, *illus.* 473
Galileo, 10, 84, 357
gamma rays, 213, 223, 316, 317
gases
  characteristics of, 158–159
  defined, 155, 156
  in periodic table, 201
generator, 414, *illus.* 414
geographic poles, 404
geothermal energy, 448
glaucoma, 353
gold, 172, 173, 177, 180, 388, *illus.* 173, 200
Goodyear, Charles, 17
graduated cylinder, 14
gram, 13
gravity
  and acceleration, 74–75
  and atmosphere, 158
  center of, 68
  and galactic clusters, 473
  law of, 84–85, *illus.* 85
  living under different conditions of, 81
  and planet's orbit, 82–83, 84–85
  in star, 461, 464–467
  and torque, 66–67, 68
  and universe, 474
  weight and mass, 78–79
groundwater, 293
group, of elements, 200–201
half-life, 212–213
hardness, 177

491

heat
　defined, 131
　and energy transfer, 138–141
　and expansion, contraction, 141–142
　and friction, 63
　latent, 166
　measuring, 133, 134
　producing, 136–137
　source, 136
　and temperature, 130–131
　See also Temperature
heating and air conditioning system installer, 149
Heisenberg, Werner, 191
helium, 145, 174, 220, 221, 461, 465
Henry, Joseph, 412
hertz, 307
hologram, 335, *illus.* 335
horsepower, 123
hydrocarbons, 276–279, 447
hydrochloric acid, 286–287
hydroelectric power, 449, *illus.* 449
hydrogen
　and ammonia, 270
　and covalent bonding, 254–255, *illus.* 254
　as element, 172, 174, *illus.* 192, 248
　in fusion, 461–462, 464
　isotopes of, 193–194, *illus.* 194
hydrogen peroxide, 271
hydronium ion, 287, 292
hydroxide ion, 288
hypothesis, 9, 10
inclined plane, 102–104, *illus.* 103
indicators, 290–291, *illus.* 290, 291, 296
inertia, 40–41, 44–45
infrared rays, 133, 316, 318
instruments, musical, 372–374
insulator
　electrical, 388–389
　heat, 138
integrated circuit, 425, *illus.* 425
intensity, 324–325
interference, 314, *illus.* 314
internal combustion engine, *illus.* 137
internal energy, 120
International System (SI), 12
iodine, 175, *illus.* 175

ions, 252–253
　and acids, 287, 291, 292
　and bases, 288, 291, 292
　and neutralization, 294–295
iron, 172, 174, 177, 192, 274
　as element, 170, *illus.* 173
　in fusion, 221
　as magnet, 402, 403
　and rust, 181, 232–233, 266
　smelting, 272
　in stars, 465
isotopes, 193, 194
　half-life of, 212–213
　radioactive, 214, 220
Josephson junction, 397
joule, 113, 134
Joule, James, 113
Jupiter, 84, 403
Kekulé, Friedrich, 255
Kepler, Johannes, 82–83, 84
Kepler's three laws, 82–83
kilowatt (kW), 123
kinetic energy, 114–115, 125
kinetic theory of matter, 130–131, 138, 142
laboratory safety, 18–19
Large Magellanic Cloud, 468, *illus.* 468
lasers
　nature of, 332–333, *illus.* 333
　uses of, 334–337
laser technician, 379
latent heat, 166
law
　of conservation of energy, 116–118
　of conservation of mass, 223
　of conservation of momentum, 50
　of gravity, 84–85, *illus.* 85
　Kepler's, 82–83
　Newton's, 40–41, 44–46, 48–50
　Ohm's, 390–391
　scientific, 10
Lawrence, E.O., 211
lead, *illus.* 173
lenses
　concave, 350
　convex, 348–349
　and the eye, 352–354
　uses of, 356–358
lens maker, 379
lever, 96–99, *illus.* 97
light
　and curved mirrors, 344–347

　and Doppler effect, 469
　and eyes, 352–354
　and lasers, 332–337
　and lenses, 348–350, 356–358
　nature of, 324–326
　and mass-energy, 222–223
　particles, 325–326
　and plane mirrors, 342–343
　speed of, 217, 222–223
　and visible spectrum, 328–329, 331
　as wave, 306, 314, 316, 318, 326
light year, 325
lightning, 386
limestone, 271
liquid
　defined, 155, 156
　viscosity of, 162
liter, 13
litmus, 284, 285, 290–291
loudness, 365, *illus.* 365
lubricant, 63
luminous object, 325
luster, 174, 175
machines
　compound, 106–107
　simple, 96–99
machinist, 89
magnesium, 274
magnet, 402–403, 406–407, 415
magnetic
　field, 403, 406–407, 408–410, 412–414, *illus.* 403, 404
　poles, 402–403
magnetism
　and electricity, 408–410, 412–414
　properties of, 177, 402–404
　substances of, 406–407
malleable, 175
Mars, 78, 331, 403
mass
　atomic, 193–194
　and black hole, 466–467
　defined, 13, 78
　and energy, 222–223
　in galactic clusters, 473
　and gravity, 74, 78–79, 84–85
　and kinetic energy, 115
　and Newton's second law, 44–46
　in stars, 460–461
　in universe, 474
　and weight, 78–79
mass number, 194
matter
　and atoms, 188–194

changes in state of, 164–166
chemical properties, changes, 180–181
defined, 4, 154
and electrical charge, 384–386
and elements, 172–175
kinetic theory of, 130–131
and nuclear fission, 216–218, 446–447
and nuclear fusion, 220–221, 446–447, 461, 464–465
and particles, 130–131, 156
physical properties, changes, 176–178
and radioactivity, 212–214
three states of, 155–156
See also Atom; Molecules
measurement, 12–15
mechanic, 148
mechanical advantage, 97
mechanical energy, 120, 137, 410, 414
mechanical engineer, 148
megawatt (MW), 123
melting point, 164, 177
in bonding, 250, 253
Mendeleev, Dmitri, 196, 197
meniscus, 14, *illus.* 14
mercury, 173, 174, *illus.* 173
Mercury, 403
metalloids, 175
metallurgist, 300
metals
active, 274
as conductors, 138, 174, 388–389
as elements, 174–175, 200
groups of, 200, 201
meter, 12
methane, 276, *illus.* 276
metric system, 12–15, *illus.* 12–13
microcomputer, 427
microgravity, 78, 81
microprocessor, 427, 428–429
microscope, 358, *illus.* 357
microwaves, 316, 318, 319, 428
Milky Way, 468
milligram, 13
milliliter, 14
millwright, 149
mirage, 355
mirrors
concave, 344–345
convex, 346–347
plane, 342–343
uses of, 356–358

mixtures
defined, 236–237
solutions, 238–239
suspensions, 240–241
See also Compounds
molder, 301
molecules, 155
and chemical reactions, 267
and covalent bonding, 254–255
and polymerization, 279
and speed of sound, 370–371
See also Atom; Matter
momentum, 44–46
conservation of, 50
Moseley, Henry, 197
motion
and acceleration, 32–34
circular, 52–53
defined, 24–25
distance, displacement, 26–27
and energy, 114–115
of falling objects, 74–76
of galaxies, 469, 473
Newton's first law of, 40–41
Newton's second law of, 44–46
Newton's third law of, 48–50
of particles, 130–131, 140–141, 156, 158–159, 164
of planets, 82–83, 84–85
speed, velocity, 28–29, 32–33
and work, 94–95
motors, 410, *illus.* 410
music, 372–374
natural frequency, 373
natural gas, 276, 444–445
nearsightedness, 353
nebula, 460–461, *illus.* 221, 459, 460, 465
net force, 59
neutralization, 294–296
neutron
and atomic mass, 194
defined, 190
and electrical charge, 384–385
in fission, 216–217
and isotopes, 193, 194, 212–213
and nucleus, 208–210
and quarks, 209, 210
star, 464–465, 466
See also Electron; Nucleus; Proton
newton, 45
Newton, Isaac, 40–41, 44, 84, 85, 357
newton-meter (N·m), 95

Newton's laws of motion, 40–41, 44–46, 48–50, 52–53
nitrogen, 145, 172, 175, 249, 254, 259
and ammonia, 270
noise pollution, 371
nonmetals, 174–175, 201
as insulators, 388–389
nonrenewable resources, 445
north pole, 402–403
nuclear
energy, 120–121
fission, 216–218, 446–447
fusion, 220–221, 446–447, 461, 464–465
power plant, 218, 442, 453
power plant technician, 418
waste, 446–447, 453
nucleus
of atom, 208–210
in fission, 216–217, 446–447
in fusion, 220–221, 446–447, 461, 464–465
and neutron, 190–191
and radioactivity, 212–214
See also Electron; Neutron; Proton
nylon, 279, *illus.* 279
odor, 176
ohm, 389
Ohm, Georg, 390
Ohm's law, 390–391
oil, 442, 444–445
crude, 277
oil tanker crew member, 478
optical fibers, 334, *illus.* 334
optical illusion, 355
optical pyrometer, 132–133
orbit, 82–83
oxygen
as element, 172, 192, *illus.* 192
and copper, 233
and covalent bonding, 254–255
and iron, 181, 232–233, 266
as rocket fuel, 145, 259
parallel circuits, 393
particles
in atomic nucleus, 208–210
and auroras, 404
in gases, 156, 158–159
of light, 325–326
in liquids, 156, 162
of matter, 130–131, 156
in quarks, 209–210
in radiation, 213–214

493

in solids, 156, 160–161
in solution, 238–239
in suspension, 240–241
percussion instruments, 372–374
period, of elements, 200
period of revolution, 83
periodic table of elements, 196–197, 200–201, 249, *illus.* 197, 198–199, 249
Mendeleev's, 196
permanent magnets, 407
petroleum. *See* Oil
petroleum engineer, 478
pharmacist, 300
pH paper, 292, *illus.* 292
pH scale, 292, 296, *illus.* 292
phenolphthalein, 296
phosphorus, 175, 178, 250, *illus.* 175
photographer, 378
photon, 324, 325, 326
and lasers, 333
physical change, 178
physical properties, 176–177, *illus.* 174, 176–177
and chemical bonding, 250
physical science, 4, 5, 6, 7
physical therapist, 89
physics, 5, 6
pilot, 88
pitch
of screw, 104
in sound, 366, 372–373
plane mirrors, 342–343, *illus.* 342
planets, 82–83
plasma, 157
plutonium, 446–447
polarization, 313, *illus.* 313
poles, 402–403, 410
pollution
from fossil fuels, 444–445, 448, 453, *illus.* 445, 453
from microwaves, 319
noise, 371
of water table, 293
polyester, 279
polymerization, 279
polyvinyl acetate, 279
polyvinyl chloride, 279
potential energy, 116–118, 125
power, 122–123, 125
preservatives, 267
pressure, of gases, 158–159
primary colors, 329
prism, 328, *illus.* 328

products, of reaction, 264–265
program, 427
projectile, 76, *illus.* 76
propane, 276, 278, *illus.* 276
propylene, 278
proton
and acids, bases, 286–288, 291
and atomic number, 192–193, 194
bonding, 248–250
and electrical charge, 188, 384–385
in fission, 216, 217
in fusion, 220–221
in nucleus, 208–210
See also Electron; Neutron; Nucleus
pulley, 98, *illus.* 98
pulsars, 466
quarks, 209–210, *illus.* 209
quartz, 160, 250, *illus.* 160
quasars, 470, *illus.* 470
radiant energy, 120, 140–141
radiation
defined, 213–214
of energy, 140–141
uses of, 214
radiation therapist, 227
radio, 420, 421
radio waves, 306, 316, 318, 347, 413, 420, 421
radioactivity, 212–214
and waste, 446–447, 453
rarefaction, 311
Rayleigh scattering, 331
rayon, 279
reactants, 264–265
reactions
chemical, 264–267
decomposition, 271
of hydrocarbons, 276–279
in motion, 48–49
replacement, 272–274
synthesis, 270–271
real image, 345
red giant, 464–465, *illus.* 464
red shift, 469, 470
reference point, 24–25
reflection
of light, 344–346
of sound, 365
of waves, 313, *illus.* 313
refraction
of light, 348–349, 355
of waves, 312, *illus.* 312

relative motion, 24
relativity theory
general, 85
special, 43
renewable resources, 448–449
replacement reaction, 272–274
research, 5, 6, 7
research tester, 379
resistance
air, 60
arm, 96, 97, 98, 99
and current, 389, 390–391
resonance, 374
retina, 337, 352, 353, 354
Ring Nebula, 465, *illus.* 465
rocket propellants, 145, 259
rolling friction, 63
rotation
and axis, 66–67
and center of gravity, 68
rubber, 17
rusting, 181, 232–233, 266
Rutherford, Ernest, 189
Sabine, Wallace Clement, 369
safety, in laboratory, 18–19
salt
as compound, 234
and ionic bonding, 252–253
and neutralization, 294–296
saturated solution, 239
Schroedinger, Erwin, 191
science, defined, 5
scientific law, 10
scientific method, 8–10
scrap handler, 227
screws, 104, *illus.* 104
semiconductors, 424–425
serendipity, 17
series circuit, 392
service technician, 435
SETI (Search for Extraterrestrial Intelligence), 475
shadow, 324
short circuit, 395
SI units, 12, 14
silicon chip, 425
silver, 173, 177, 270, 388, *illus.* 173, 200
simple machines, 96–99
inclined plane, 102–104
sliding friction, 62–63
sodium, 252–253
sodium chloride, 252–253, 256, *illus.* 252, 253
sodium hydroxide, 288

solar cells, 448, *illus.* 141, 448
solar collectors, 448
solar eclipse, 462
solar energy, 448
  and airplanes, 451
solar system, *illus.* 41, 83
solid
  characteristics of, 160–161
  defined, 155, 156
solid-state components, 424–425
solute, 239
solutions, 238–239
solvent, 239
sonar, 365, *illus.* 365
sonic boom, 371
sound
  characteristics of, 370–371
  Doppler effect, 366–367
  and music, 372–374
  as wave, 364–367
sound barrier, 371, *illus.* 371
south pole, 402–403
space, and gravity, 85
sparks, 386
specific heat, 134
spectrum
  of atom, 190, *illus.* 190
  and color, 328–329, 462
  and elements, 462
  electromagnetic, 316–318
  of sun, 462
speed
  of animals, 31
  of chemical reactions, 267
  defined, 28
  of light, 216, 222–223
  and mass, 222–223
  of people, 31
  of planet, 82–83
  of sound, 370–371
  and velocity, 28–29
  of waves, 307–308, 312
  See also Velocity
spontaneous combustion, 137
spontaneous reactions, 264–265
stars
  development of, 460–462
  elements in, 221, 465
  and galaxies, 468–470
  gravity in, 461, 464–467
  neutron, black holes, 466–467, 470
  and nuclear fusion, 220–221, 461
  white dwarfs, supernovas, 464–465

static friction, 62, 63
string instruments, 372–374
strong force, 208–209
studio engineer, 378
substitution reactions, 279
sugar, *illus.* 161
  bonding in, 255
  as compound, 234
  melting point of, 250
  in solution, 238
sulfur, 161, 175, 193, *illus.* 161, 175, 193
sulfuric acid, 287, 295
sun
  energy of, 121, 448, 453
  and nuclear fusion, 220–221, 461, 464
  and planets' orbits, 82–83
  and radiation, 141
  spectrum of, 462
  as a star, 461–462
superconductivity, 397
supergiants, 465
supernova, 464–465
surveyor, 88
suspensions, 240–241
sympathetic vibrations, 374
synthesis reaction, 270
synthetic fuels, 447, 452
taste, 176
telephone lineperson, 435
telescope, 347, 357, 470, *illus.* 357
television, 329, 421, *illus.* 421
temperature
  and changes in matter, 164–166
  and chemical reactions, 267
  and convection, 139
  defined, 131
  and energy production, 449
  of gases, 159
  and heat, 130–131
  measuring, 132–133
  in metric system, 13
  in stars, 461
  See also Heat
terminal velocity, 75
theories of atoms, 188–191
theories of the universe, 472–474
theory, 10
thermogram, 133, *illus.* 7, 128, 133
thermostat, 143, *illus.* 145
Thomson, J. J., 188, 189
thread, 104

time
  in metric system, 13
  and power, 122–123
  and speed, 28–29
tin, *illus.* 173
torque, 66–67, 68
transfer
  of energy, 138–141
  of work, 122–123
transformer, 394–395, *illus.* 394
transistors, 424–425, *illus.* 424, 425
transverse wave, 310, 313, *illus.* 310
tritium, 193, 194
trough, of wave, 307, 310, 313, 314, 332
ultrasonic sound, 366
ultraviolet rays, 316, 317
universal indicator, 292
universe
  closed, 474
  expansion of, 469
  open, 474
  theories of, 472–474
uranium, 212–214, 446
  and electrical energy, 442
  and nuclear fission, 216–218
  and radioactivity, 212–214
uranium mine inspector, 479
vacuum, 306
vacuum tube, 420–422, *illus.* 420, 425
vaporization, 165
Veil nebula, *illus.* 221
velocity
  and acceleration, 32–34, 44–46, 74–76
  defined, 29
  and kinetic energy, 115
  and momentum, 46
  and speed, 28–29
  terminal, 75
  See also Speed
Venus, 403
vibrations, 364, 372–374
video, 421
virtual image, 342–343
viscosity, 162
visible spectrum, 328–329, *illus.* 328
vision, and lenses, 352–353
volt, 389, 390–391
voltage, 389, 390–391, 394–395

volume, 14–15
  of gases, 159
vulcanization, 17
water
  changes in state of, 164–165
  and chemical changes, 181
  as compound, 233
  as energy source, 442, 448–449
  expansion, contraction of, 142–143
  molecules, 254–255, 257, *illus.* 255
  and neutralization, 294–295
  in nuclear power plant, 218
  as solvent, 239
water table, 293
watt, 123
waves
  behavior of, 312–314
  and Doppler effect, 366–367
  electromagnetic, 316–318
  light, 306, 314, 316, 318, 326
  motion, 310–311
  properties of, 306–308, *illus.* 307
  sound, 364–367, 370–371
wavelength, 307, 308
  compressional, 311
  of colors, 328
  of lasers, 332–333
weather satellites, 318
wedge, 103, *illus.* 103
weight, 78
welder, 301
wheel and axle, 99, *illus.* 99
white dwarf, 464–465
wind
  as energy source, 112–113, 449, 452
wind instruments, 372–374, *illus.* 373
windmill, 449, *illus.* 449
wood
  in chemical reaction, 266
  as energy source, 448
work
  calculating, 94–95
  and compound machine, 106–107
  defined, 94
  and electromagnetism, 410
  and energy, 112–113
  and inclined planes, 102–104
  and power, 122–123
  and simple machines, 96–99
X rays, 316, 317
  from black holes, 466–467
  from neutron stars, 466
  from quasars, 470
X-ray technician, 378
zinc, 172, 174, 388

Motley Crue sucks my right tit

FREMONT UNIFIED SCHOOL DISTRICT
Fremont, California